KEYNOTE guide to topics in your course

ABOUT THE AUTHOR

Donald S. Russell received his B.S. degree from Northwest Missouri State College and the degrees of M.S., M.Ed., and Ed.D. from the University of Southern California. He has had thirty-two years of teaching experience, the last fifteen of which have been with the Department of Mathematics at Ventura College in California. In 1966 he was nominated to and included in the current edition of *Who's Who In American Education*. Dr. Russell is the author of the standard college texts, *Elementary Algebra* and *Intermediate Algebra*.

CLIFFS KEYNOTE REVIEWS

Algebra Problems

by

DONALD S. RUSSELL

Ventura College

SECOND EDITION

CLIFF'S NOTES, INC. • LINCOLN, NEBRASKA 68501

ISBN 0-8220-1704-0

L.C. Catalogue Card Number: 67–16624

Printed in the United States of America

CONTENTS

TO THE STUDENT

Originally this book was written to serve several groups of students of mathematics. The revised KEYNOTE edition has been expanded so that the book will be of even greater service to those same groups. Students of intermediate algebra can use it as an additional, but permanent, source for problem solutions of the sort which appear on the blackboard and are promptly erased. Students in advanced mathematics and science courses will find the book useful for review since such courses depend heavily on the fundamental concepts presented here. Individuals preparing for competitive examinations can use the book to review quickly, yet thoroughly, the fundamental principles with which they are expected to be familiar. Anyone learning algebra on a self-study basis will find this book extremely helpful.

This KEYNOTE edition contains additional chapters, as well as extensions and reorganization within the original chapters. The book should facilitate the readers' acquisition of a fuller understanding of such topics as the algebra of sets, inequalities, functions, permutations and combinations, determinants, synthetic division, the various forms of the linear equation, the determination of the roots of higher degree equations, and many others. In addition, some 400 supplementary problems, with their answers, have been included so that the reader may test his mastery of the various concepts presented.

The solutions of more than 1,000 problems, covering all phases of intermediate algebra, have been shown in detail. Each chapter contains the development of a minimum amount of theory to assist the student in following from one step to the next in the solution of the problems. This book is not intended to replace the standard textbook in which extensive development of theory is presented. But, unfortunately, some traditional textbooks assume that every student has the background to follow the solutions to sample problems, and they frequently omit steps. Too often, the student has forgotten some of his basic elementary algebra and fails to see why certain operations have been performed. To help students overcome that difficulty, this KEYNOTE provides many explanations of even the simplest steps in the solution of problems.

Careful thought has gone into planning the sequence of topics. For this reason, the reader should be certain that he understands each concept he studies before proceeding to a new topic.

The author gratefully acknowledges his indebtedness to Miss Maureen Sanders of the staff of Barnes & Noble for her assistance and helpful suggestions in the preparation of the manuscript for publication.

Chapter 1

Fundamental Concepts and Operations

FUNDAMENTAL CONCEPTS

Numeral. A numeral is a symbol or group of symbols used to signify or name a number. Thus, for example, $\frac{1}{3} \times 9, 1 + 2, 10 - 7$, etc. are names for three. The symbols used in arithmetic, 0, 1, 2, 3, 4, 5, 6, 7, 8, 9, are taken from the Arabic numeration system; these numerals can be combined to represent any specific number. Preceded by a negative sign they have a negative value. In algebra, the letters of the alphabet are used to represent unspecified numbers; they are called **literal numerals.**

Sets. A set is a clearly defined collection or list of objects or things. The things or objects that make up a set are called the **members** or **elements** of the given set. A set is usually represented by a capital letter such as X or Y, and the elements of a set by lower case letters such as a, b, c. The elements of a specific set may be numbers or things, and may be denoted by numerals, words, or symbols.

Sets may be designated in three different ways. First, we may describe the set in sentence form such as: X is the set of state capitols; or Y is the set of students in a given class. Secondly, we may use an equation to designate a set. An **equation** or **equality** is a statement that two symbols or groups of symbols stand for the same thing. The part to the left of the equal sign is known as the **left member** of the equation, and the part following the equal sign is known as the **right member** of the equation. If a capital letter representing a set is used as the left member of the equation, and the names of the elements, separated by commas and enclosed within braces, are used as the right member of the equation, the result will state that the letter symbol names the same set as the listed elements. For example, if A is the set of the first five letters of the alphabet, the set could be represented as

$$A = \{a, b, c, d, e\}.$$

Thirdly, we may describe the set A using the **set-builder** form of symbolism. An example of this form is

$A = \{x \mid x \text{ is one of the first five letters of the alphabet}\}$.

When reading a statement in set-builder form, the vertical line is read "such that." Thus the statement can be read: "The set A equals the set of all x such that x is one of the first five letters of the alphabet."

If we want to show that an element is a member of a certain set, for example, that b is an element of set A, we write $b \in A$ (read "b is an element of A"). If b was not an element of A, this would be shown as: $b \notin A$ (read "b is not an element of A").

A set may have no elements. Consider the following sets: the set of all women who have served as president of the United States; the set of all past United States presidents who were nominated by the Socialist Party. Neither of these sets has any elements. In such cases, we say it is the **empty** or **null** set, and denote it by the symbol ϕ.

Natural Numbers. The natural numbers are also referred to as the **counting numbers** or **positive integers.** They are found to the right of zero on the number line shown in Fig. 1. Described in set terminology, the natural numbers $= \{1, 2, 3, 4, \ldots\}$. The three dots following the numeral, 4, indicate that additional numbers follow those named.

The set of natural numbers has an unlimited number of members. A set with an unlimited number of members is called an **infinite set.** Conversely, a **finite set** is one in which membership is limited. An example of a finite set is a set X where X is the set of even positive integers less than ten. This set could be written

$$X = \{2, 4, 6, 8\}.$$

Negative Numbers. For every positive number there corresponds a negative number. Negative numbers are found to the left of zero on the number line. A negative number is always preceded by a negative sign. The set of negative numbers $= \{\ldots -4, -3, -2, -1\}$.

Zero. The symbol 0 was included in our number system as a place holder. Thus, since there are no groups of tens in the number 703, the zero is used to fill the tens place in the numeral.

Zero is located between the negative and the positive numbers on the number line.

Zero has the following properties.

1. Zero added to any number is that number. For example, $3 + 0 = 3$ and $5d + 0 = 5d$. Therefore, zero is called the **identity element of addition.**
2. Zero times any number is equal to zero; thus, $3 \cdot 0 = 0$.

3. Zero divided by any number (except 0) is zero; for example, $\frac{0}{7} = 0$ and $\frac{0}{2a} = 0$ (if $a \neq 0$).
4. Division by zero is undefined and is *never permitted.* Thus, $\frac{9}{0}$ and $\frac{7c^2}{0}$ are both undefined.

Integers. The set of integers is the set of natural numbers, their negatives, and zero.

Rational Numbers. A rational number is any number that can be expressed as the ratio of two integers provided the second integer is not zero. Thus, the numbers $\frac{2}{1}$, $\frac{15}{2}$, $\frac{5}{3}$, and $-\frac{12}{7}$ are rational numbers. And in general, any number of the form $\frac{a}{b}$, where a and b are integers and $b \neq 0$, is a rational number. You will notice that, of the examples, only the first fraction can be expressed as a whole number. The others can only be expressed as fractions or decimals. Expressed as a decimal, $\frac{15}{2}$ equals 7.5. This is called a **terminating decimal;** it has an exact value which can be expressed by a certain number of digits. The numbers 1.24, 0.563, and 0.00000734 are also terminating decimals. If a decimal is followed by three dots to indicate that additional digits follow those shown, it is called a **non-terminating decimal.** There are two kinds of non-terminating decimals: repeating and non-repeating. **Repeating decimals** are those that contain a pattern of one or more repeated digits. The fractions $\frac{5}{3}$ and $\frac{2}{7}$ can be expressed as non-terminating repeating decimals. The fraction $\frac{5}{3}$ is equivalent to 1.66 . . . , and $\frac{2}{7}$ is equivalent to 0.285714285714. . . . An easy way to express a repeating decimal is to put a dot or a line over the digit or digits that are repeated. For example, $1.\dot{6}$ and $0.\dot{2}\dot{8}\dot{5}\dot{7}\dot{1}\dot{4}$, or $1.\overline{6}$ and $0.\overline{285714}$ are repeating decimals. Any terminating or repeating decimal is a rational number.

Irrational Numbers. Irrational numbers are those numbers that cannot be expressed as the ratio of two integers. Examples are: π, $\sqrt{2}$, $\sqrt[3]{5}$. The irrational numbers may also be described as the set of **non-terminating non-repeating decimals.** Sometimes irrational numbers are called **surds.** An irrational square root of a rational number is sometimes called a **quadratic surd.**

The Real Number System. The real number system is the set of integers, positive and negative rational numbers, and positive and negative irrational numbers. All numbers in the real number system can be located on the number line shown in Fig. 1. Conversely, every point on the number line represents a number in the real number system. On the number line, positive numbers are represented to the right of zero with the values growing progressively larger from left to right. Negative numbers are represented to the left of zero with the values becoming progressively smaller from right to left. The arrows at each end of the number line indicate that the line can be extended indefinitely both to the right and to the left of zero. Between any two real numbers there are an infinite number of real numbers. Thus, for example, between 1 and 0 we can locate $\frac{1}{2}$; between $\frac{1}{2}$ and 0 we can locate $\frac{1}{4}$; between $\frac{1}{4}$ and 0 we can locate $\frac{1}{8}$; and so on. This process can be continued indefinitely; and, therefore, there are an infinite number of real numbers between 0 and 1.

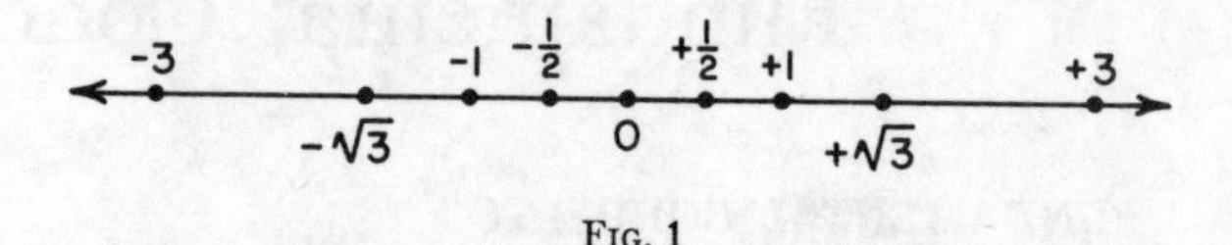

FIG. 1

There are times when we will be concerned only with the numerical value of a number. The numerical or **absolute value** of a number refers only to the magnitude of the number, and not to its direction from zero on the number line. Thus, for example, the absolute value of 5 is 5 and the absolute value of -5 is 5. The absolute value of any number is symbolized by placing the number between two vertical bars. For example,

$$| 5 | = 5.$$
$$| -5 | = 5.$$

The absolute value of 0 is 0.

Note that for each positive integer on the number line there is a negative integer located at an equal distance from and on the opposite side of zero. In other words, for each member in the set of positive integers there is a corresponding member in the set of negative integers. The converse is also true, for each member in the set of negative integers there is a member in the set of positive integers. When every element in one set can be paired with a single element in a second set and every element in the second set can be paired with a single element in the first set, we say that there is a **one-to-one correspondence** between the two sets. Thus, there is a one-to-one correspondence between the set of positive integers and the set of negative integers. If we consider a set of 15 students and a classroom with a set of 15 desks, the set of students and the set of desks are also in one-to-one correspondence.

We defined the integers as the set of natural numbers, their negatives, and zero. Then we defined the real numbers as the set of integers, rational numbers, and irrational numbers. Thus, the set of integers is contained in the set of real numbers; or, in other words, every element in the set of integers is an element in the set of real numbers. If every element of one set, say X, is also an element of another set, say Y, we say that X is a **subset** of Y. Thus, if $X = \{a, b\}$ and $Y = \{a, b, c\}$, then X is a subset of Y. This is denoted $X \subset Y$. The empty set is a subset of every set, and every set is a subset of itself.

Any given set from which we may be seeking specific elements or subsets is known as the universal set or the universe. The universal set is designated by the capital letter U. In some cases, every element in the universal set with which we are working may be a member of the set for which we are looking. In such cases, we represent the elements of the latter by the letter U. Note that the symbols for the empty set and the universal set are not enclosed within braces.

Consider again the example mentioned above.

$$X = \{a, b\}.$$
$$Y = \{a, b, c\}.$$

In this example, let Y be the universe. A number of subsets could be formed from this universal set. Namely, $\{a\}$, $\{b\}$, $\{c\}$, $\{a, b\}$, $\{a, c\}$, $\{c, b\}$, ϕ, and U. Note that the fourth subset, $\{a, b\}$, is the set X. Thus we see again that X is a subset of Y.

Any set may be designated as a universe. If the real numbers comprise a universal set, some of its subsets are the set of integers, the set of rational numbers, and the set of irrational numbers. The set of integers could also be designated as the universe. Two of its subsets would be the even positive numbers and the odd negative numbers.

Two sets W and Z are equal if and only if each element in W is an element in Z and each element in Z is an element in W. The order in which the elements are listed is immaterial. Thus, if $W = \{a, b, c, d\}$ and $Z = \{b, d, c, a\}$, then $W = Z$.

If $X = \{a, b\}$ and $Y = \{a, b, c\}$, then X does not equal Y. This is called an **inequality**. An inequality is a statement that two quantities are not equal. An oblique line drawn through the equal sign is read "does not equal". Thus, $3 \neq 4$, and $a \neq b$, if a is either larger or smaller than b. If a is larger than b, we denote this by the symbolism $a > b$ (read "a is greater than b"); and if a is smaller than b, we denote this by the symbolism $a < b$ (read "a is less than b").

> If A is the set of all even negative integers greater than -10, then
> $A = \{-8, -6, -4, -2\}$.
>
> If $B = \{\text{Tuesday, Thursday}\}$, then B is the set of the days of the week whose names begin with the letter T.

List the sets in problems 1–5.

1. A is the set of all even numbers greater than 15 and less than 25.
   $A = \{16, 18, 20, 22, 24\}$.
2. M is the set of all perfect squares less than 50.
   $M = \{1, 4, 9, 16, 25, 36, 49\}$.
3. D is the set of all odd natural numbers.
   $D = \{1, 3, 5, \ldots\}$.
4. X is the set of all natural numbers greater than 7 and less than 8.
   $X = \phi$.
5. Y is the set of the days of the week whose names begin with the letter S.
   $Y = \{\text{Saturday, Sunday}\}$.

Describe the sets in problems 6–10.

6. $A = \{2, 4, 6, 8\}$.
   A is the set of all even positive integers less than 10.
7. $Z = \{0, 1, 2, 3, 4, 5, 6, 7, 8, 9\}$.
   Z is the set of all single-digit positive integers.
8. $X = \{\ldots -2, -1, 0, 1, 2, \ldots\}$.
   X is the set of all integers.
9. $X = \{\text{red, blue, yellow}\}$.
   X is the set of primary colors.
10. $B = \{1, 4, 9\}$.
    B is the set of the one-digit positive integers that are perfect squares.

State whether the set in each of problems 11–15 describes a finite or an infinite set.

11. the set of all past vice presidents of the United States
    This is a finite set.
12. the set of all even natural numbers
    This is an infinite set.
13. $A = \{3, 6, 9, \ldots\}$.
    This is an infinite set.
14. the set of all couples who were married in the state of Nevada in 1966
    This is a finite set.
15. the set of all fractions greater than 1 and less than 2
    This is an infinite set.

In problems 16–20 list all of the subsets of each given set.

16. $X = \{1, 2, 3\}$.
    The subsets are: U, $\{1, 2\}$, $\{1, 3\}$, $\{2, 3\}$, $\{1\}$, $\{2\}$, $\{3\}$, and ϕ.
17. $Y = \{\text{Tom, Dick, Harry}\}$.
    The subsets are: U, {Tom}, {Dick}, {Harry}, {Tom, Dick}, {Tom, Harry}, {Dick, Harry}, and ϕ.
18. $B = \{1, 2, 3, \ldots\}$.
    This set has an infinite number of subsets, and it is impossible to list an infinite number of subsets.
19. $C = \phi$.
    The only subset of set C is the empty set ϕ.
20. $Z = \{x \mid x \text{ is a point on the number line}\}$.
    This set also has an infinite number of subsets, and it is impossible to list them.

Factors. A factor is one of two or more numbers which when multiplied together form a given product. Thus, 3 and 2 are factors of 6; and 6, x, and y are all factors of the expression $6xy$. In any expression consisting of two or more factors, any one or more of those factors can be considered the **coefficient** of the product of the remaining factor or factors. Thus, in the expression $7x^2yz$, 7 is the coefficient of x^2yz; $7z$, the coefficient of x^2y, and y, the coefficient of $7x^2z$. Unless otherwise stated, we shall assume that when we refer to the coefficient of an expression, we mean the numerical rather than the literal factor of that expression.

A **prime number** is any whole number greater than 1 that has no integral factors except itself and 1. Examples are: 2, 5, 13, and 29. A **composite number** is any whole number that is not prime. Examples are: 4, 18, 72, and 336.

Terms. An algebraic term is that part of the expression consisting of one or more factors that is separated from the rest of the expression by a plus (+) or minus (−) sign. In the expression $b^3 - 2ab + 5$, the terms are b^3, $-2ab$, and 5. A **multinomial** is a general name for an algebraic expression of two or more terms. Examples are: $c - 2d$ and $3m^2 + 4m - 2n + 4$. More specific names for algebraic expressions depend on the number of terms they involve. A **monomial** is an algebraic expression of one term. Examples are: $3x$, $5x^2y$, and $-mnp$. A **binomial** is an algebraic expression of two terms. Examples are: $x + y$, $2m^2 - 3p$, and $3 - 4z$. A **trinomial** is an algebraic expression of three terms. Examples are: $x + y - 3z$, $2a + 3b^2c + 1$, and $12 - 3m - 5p$. An algebraic expression of one or more terms in which the exponents of all literal factors are integral and positive is called a **polynomial.** Examples are: $7x^2 - 3y$, $\frac{2m^2}{3} - m + 3$, and $3a^3 - 2a^2 + 3a - 1$.

Basic Axioms of Algebra. An **axiom** is defined as a basic assumption. In working with the fundamental operations of algebra, we will assume the following.

1. **The Axiom of Comparison.** If a and b are any real numbers then either $a = b$, $a > b$, or $a < b$. This is sometimes referred to as the law or axiom of **trichotomy.** To determine which of $a = b$, $a > b$, or $a < b$ is true, we can use the number line. If a is to the right of b on the number line, then $a > b$; if a is to the left of b on the number line, then $a < b$; and if a and b occupy the same position on the number line, then $a = b$.
2. **The Axioms of Equality.**
   a. **The Reflexive Relationship.** If a is any real number, then $a = a$.
   b. **The Symmetric Relationship.** If a and b are any real numbers and if $a = b$, then $b = a$.
   c. **The Transitive Relationship.** If a, b, and c are any real numbers and $a = b$ and $b = c$, then $a = c$. This axiom states that quantities equal to the same or equal quantities are equal to each other.
3. **The Axioms of Addition.**
   a. **The Axiom of Closure.** The addition of real numbers may be shown on the number line. For example, if a is any positive real number we can locate that number to the right of zero (see Fig. 2). If b is a positive real number and is to be added to a, we start at a and move b units to the right to the point M. The sum of a and b then is the number M.

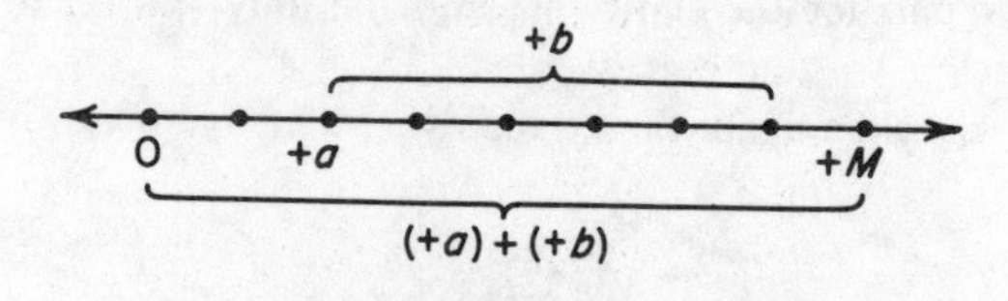

FIG. 2

In a previous discussion, we said that the number line could be extended indefinitely to the right of zero. This means that no matter how large a and b are we can always find their sum M on the number line. The closure axiom of addition states that if we add any two real numbers their sum will also be a real number; or, in other words, the set of real numbers is closed under addition because whenever the operation of addition is performed on members of this set the result is always a member of that same set.
   b. **The Commutative Axiom for Addition.** If a and b are any real numbers, then $a + b = b + a$. This axiom states that a sum is the same regardless of the order in which the numbers are added. Thus, $3 + 4 = 4 + 3 = 7$.
   c. **The Associative Axiom for Addition.** If a, b, and c are any real numbers, then $a + (b + c) = (a + b) + c$. Thus $(3 + 4) + 7 = 3 + (4 + 7)$ since $7 + 7 = 3 + 11 = 14$. This law states that when adding three or more numbers, they may be grouped or associated in any manner.
4. **The Axioms of Multiplication.**
   a. **The Axiom of Closure for Multiplication.** Multiplication, which may be considered as a process of repeated addition, may also be illustrated on the number line. For example, to multiply 2×3, we move two points to the right of zero three times. This is the same as adding $(2 + 2) + 2$. The set of real numbers is closed under the operation of multiplication because the product of any two real numbers will also be a real number.
   b. **The Commutative Axiom for Multiplication.** If a and b are any real numbers, then $ab = ba$. This axiom states that a product remains the same regardless of the order in which the factors are multiplied.
   c. **The Associative Axiom for Multiplication.** If a, b, and c are any real numbers, then $(ab)c$

$= a(bc)$. Thus, $(3 \cdot 4)8 = 3(4 \cdot 8)$ since $(12)8 = 3(32) = 96$. This axiom states that the product of three or more numbers is the same regardless of the manner in which the factors are grouped or associated.

d. **The Distributive Axiom for Multiplication with Respect to Addition.** If a, b, and c are any real numbers, then $a(b + c) = ab + ac$. This axiom means that the factor a is distributed over the addends included within the parentheses. Thus, $3(5 + 2) = 3 \cdot 5 + 3 \cdot 2 = 15 + 6 = 21$. Likewise, $(b + c + d)a = b \cdot a + c \cdot a + d \cdot a = ba + ca + da$.

Inverse Operations.

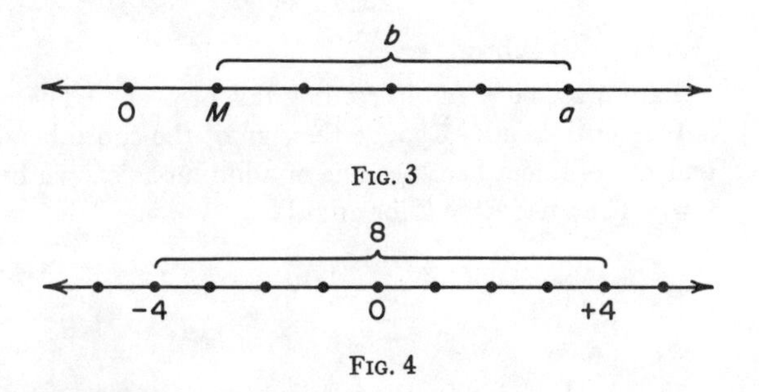

FIG. 3

FIG. 4

Subtraction is sometimes referred to as the inverse of addition. It can be described in terms of moves on the number line. Thus, if b is a natural number, to subtract b from a, we move b units to the left of a to the point M (see Fig. 3). Notice that if b is larger than a, we will have to continue the leftward movement beyond zero. For example, in order to subtract 8 from 4, we must move 8 units to the left of 4 (see Fig. 4). The answer to this operation is -4. Thus, the set of natural numbers is not closed under the operation of subtraction because subtraction introduces the use of negative numbers and zero. We can say, however, that the set of integers is closed under subtraction, and also that the set of real numbers is closed under subtraction. For any real number subtracted from another real number will result in a real number.

Note that if a negative number were being subtracted from some number, a, we would move to the right of a on the number line; if 0 were being subtracted, we would not move at all.

Division is defined as the inverse of multiplication. In other words, $\frac{a}{b} = c$ if and only if $c \cdot b = a$. Division by zero is not permitted in the real number system; for if $b = 0$, there would have to be a number c which when multiplied by 0 would equal a. However, any number multiplied by 0 equals 0. Similarly, if we let both a and b equal 0, the problem becomes $\frac{0}{0} = c$ for which there is no unique answer.

Zero is the only number in the set of real numbers that cannot be used as a divisor; and, because of it, the set of real numbers is not closed under the operation of division. Note also that when certain combinations of integers are used, such as $\frac{15}{2}$ and $\frac{2}{7}$, the result of dividing is a fraction or a repeating decimal. Thus, the set of integers is not closed under the operation of division.

A number is said to be the **reciprocal** of a given number if the product of the number and its reciprocal is equal to **1**. Thus, the reciprocal of d is $\frac{1}{d}$ because $d \cdot \frac{1}{d} = 1$. Likewise, the reciprocal of $\frac{1}{2}$ is 2 since $\frac{1}{2} \cdot 2 = 1$.

To divide by a number is the same as to multiply by its reciprocal. Thus, for example,

$$a \div a = a \times \frac{1}{a}.$$

$$1 = 1.$$

ADDITION AND SUBTRACTION

Rules for Addition and Subtraction:

1. Only similar terms may be combined by addition or subtraction. Similar terms are terms whose literal parts are identical.
2. When adding expressions whose signs are alike, add their absolute values and prefix the common sign.

Add:	
$3x$	$-4x$
$2x$	$-2x$
$4x$	$-2x$
$9x$	$-8x$

3. When adding two expressions whose signs are unlike, subtract the smaller absolute value from the larger absolute value and prefix to the answer the sign of the expression whose absolute value is larger.

Add:	
$8x$	$-9x$
$-3x$	$3x$
$5x$	$-6x$

4. In a subtraction problem, change the signs of all terms in the subtrahend; the problem thus becomes one of addition. Proceed with the addition, observing the above rules.

Subtract:
$3m - 4n + 5p$
$m + 2n - 3p$
$2m - 6n + 8p$

1. Add $3ab$ **and** $7ab$.

Use Rule 2.

$$\begin{array}{r} 3ab \\ \underline{7ab} \\ 10ab \end{array}$$

2. Add $-5xy$ and $-8xy$.

Use Rule 2.

$$\begin{array}{r} -5xy \\ \underline{-8xy} \\ -13xy \end{array}$$

3. Add $4mn$ and $-2mn$.

Use Rule 3.

$$\begin{array}{r} 4mn \\ \underline{-2mn} \\ 2mn \end{array}$$

4. Add:

$$\begin{array}{r} 5xy \\ 3xy \\ \underline{-2xy} \end{array}$$

Combine $5xy$ and $3xy$, using Rule 2. Then use Rule 3 to combine that answer with $-2xy$. This is an application of the associative axiom for addition.

$$\begin{array}{r} 5xy \\ 3xy \\ \underline{-2xy} \\ 6xy \end{array}$$

5. Add:

$$\begin{array}{r} -9cd \\ -3cd \\ \underline{4cd} \end{array}$$

Combine $-9cd$ and $-3cd$, using Rule 2. Then use Rule 3 to combine that answer with $4cd$.

$$\begin{array}{r} -9cd \\ -3cd \\ \underline{4cd} \\ -8cd \end{array}$$

6. Add:

$$\begin{array}{r} 5st \\ -3st \\ \underline{7st} \end{array}$$

Combine $5st$ and $7st$, using Rule 2. Then use Rule 3 to combine that answer with $-3st$. This is an application of both the commutative and associative axioms for addition.

$$\begin{array}{r} 5st \\ -3st \\ \underline{7st} \\ 9st \end{array}$$

7. Subtract $3xy$ from $8xy$.

Applying Rule 4, change the sign of the subtrahend, and the problem becomes one of addition as shown below. Then use Rule 3 for addition.

$$\begin{array}{r} 8xy \\ \underline{-3xy} \\ 5xy \end{array}$$

8. Subtract $-3ab$ from $7ab$.

Change the sign of the subtrahend, and the problem becomes one of addition as shown below. Then use Rule 2 for addition.

$$\begin{array}{r} 7ab \\ \underline{3ab} \\ 10ab \end{array}$$

9. From $-7xy$ subtract xy.

Care must be taken in reading this problem. The xy is the subtrahend. Change the sign of the subtrahend, and the problem becomes one of addition as shown below. Then use Rule 2 for addition.

$$\begin{array}{r} -7xy \\ \underline{-\ \ xy} \\ -8xy \end{array}$$

It must be kept in mind that when no coefficient is shown for the literal factors, that coefficient is 1.

10. Subtract $-7ab$ from $-4ab$.

Change the sign of the subtrahend, and the problem becomes one of addition as shown below. Then use Rule 3 for addition.

$$\begin{array}{r} -4ab \\ \underline{7ab} \\ 3ab \end{array}$$

11. Add $4xy + 3ab$ and $3xy - 2ab$.

When adding polynomials, add each column independently of the other columns. Care must be taken to observe the rules for addition as each column is added.

$$\begin{array}{rr} 4xy & +\ 3ab \\ \underline{3xy} & \underline{-\ 2ab} \\ 7xy & +\ \ ab \end{array}$$

12. Add $14mn - 3ab$ and $-9mn + 6ab$.

$$\begin{array}{rr} 14mn & -\ 3ab \\ \underline{-9mn} & \underline{+\ 6ab} \\ 5mn & +\ 3ab \end{array}$$

13. Add $xy + 3mn - 6ab$ and $4xy - 3mn + 9ab$.

$$\begin{array}{rrr} xy & +\ 3mn & -\ 6ab \\ \underline{4xy} & \underline{-\ 3mn} & \underline{+\ 9ab} \\ 5xy & & +\ 3ab \end{array}$$

14. Add $12rst + 7abc - 5mn$ and $3mn - 4abc - 3rst$.

These two expressions must be rearranged so that similar terms are in the same columns. In rearranging terms, the sign that immediately precedes a term goes with that term.

$$\begin{array}{r} 12rst + 7abc - 5mn \\ \underline{-3rst - 4abc + 3mn} \\ 9rst + 3abc - 2mn \end{array}$$

15. Add $3a^2 + 4a - 5c$ and $6a^2 - 2a + 3c$.

It must be kept in mind, from the first rule for addition and subtraction (page 5), that only terms with the same exponent can be added or subtracted.

$$\begin{array}{r} 3a^2 + 4a - 5c \\ \underline{6a^2 - 2a + 3c} \\ 9a^2 + 2a - 2c \end{array}$$

16. Add $7a^3 - 4a^2 + 5a$ and $7a^2 - 2a$.

$$\begin{array}{r} 7a^3 - 4a^2 + 5a \\ \underline{7a^2 - 2a} \\ 7a^3 + 3a^2 + 3a \end{array}$$

17. Add $4ab - 5b^2 + 8ac$ and $b^2 + ac - 4a^2$.

$$\begin{array}{r} 4ab - 5b^2 + 8ac \\ \underline{b^2 + ac - 4a^2} \\ 4ab - 4b^2 + 9ac - 4a^2 \end{array}$$

18. Subtract $3xy - ab$ from $7xy + 8ab$.

After all signs in the subtrahend have been changed, the problem becomes one of addition as shown below.

$$\begin{array}{r} 7xy + 8ab \\ \underline{-3xy + ab} \\ 4xy + 9ab \end{array}$$

19. Subtract $8t - 4b$ from $11t + 5b$.

The $8t - 4b$ is the subtrahend in this problem, and, after the signs have been changed, the problem becomes one of addition as shown below.

$$\begin{array}{r} 11t + 5b \\ \underline{-8t + 4b} \\ 3t + 9b \end{array}$$

20. Subtract $-3mn + 3st$ from $-5mn - 4st$.

This problem becomes one of addition as shown below.

$$\begin{array}{r} -5mn - 4st \\ \underline{3mn - 3st} \\ -2mn - 7st \end{array}$$

21. From $8ab + 7xy$ subtract $5ab - 2xy$.

Care must be taken in reading this problem. The $5ab - 2xy$ is the subtrahend.

$$\begin{array}{r} 8ab + 7xy \\ \underline{-5ab + 2xy} \\ 3ab + 9xy \end{array}$$

22. From $9a^2 + 7a - 6c$ subtract $4a - 5a^2 - 3c$.

Rearrange the terms in columns that consist of similar terms and change the signs of the subtrahend.

$$\begin{array}{r} 9a^2 + 7a - 6c \\ \underline{5a^2 - 4a + 3c} \\ 14a^2 + 3a - 3c \end{array}$$

23. Subtract $9a^2b - 7ab^2 + 3ab$ from $7ab + 5a^2b - 2ab^2$.

$$\begin{array}{r} 7ab + 5a^2b - 2ab^2 \\ \underline{-3ab - 9a^2b + 7ab^2} \\ 4ab - 4a^2b + 5ab^2 \end{array}$$

24. Subtract $7a^2b^3$ from $9a^3b^2$.

Since neither of these terms has similar exponents, they are not similar terms and cannot be combined into one column. The only way to show that the subtrahend is to be subtracted from the minuend is to use a minus sign. The answer is shown below.

$$9a^3b^2 - 7a^2b^3.$$

25. From the sum of $9a - 6c$ and $5a + 9c$, subtract $3a - c$.

$$\begin{array}{r} 9a - 6c \\ \underline{5a + 9c} \\ 14a + 3c \end{array}$$

Subtract $3a - c$ from this sum.

$$\begin{array}{r} 14a + 3c \\ \underline{-3a + c} \\ 11a + 4c \end{array}$$

26. To the difference obtained when $2ab - 2xy$ is subtracted from $5ab + 7xy$, add $7ab - 6xy$.

$$\begin{array}{r} 5ab + 7xy \\ \underline{-2ab + 2xy} \\ 3ab + 9xy \end{array}$$

$$\begin{array}{r} 3ab + 9xy \\ \underline{7ab - 6xy} \\ 10ab + 3xy \end{array}$$

27. From the difference obtained when $4mn - 5st$ is subtracted from $11mn - 3st$, subtract $mn - 3st$.

$$\begin{array}{r} 11mn - 3st \\ \underline{-4mn + 5st} \\ 7mn + 2st \end{array}$$

$$\begin{array}{r} 7mn + 2st \\ \underline{-mn + 3st} \\ 6mn + 5st \end{array}$$

28. To the sum of $6ab + 7xy - 2rs$ and $3ab + xy + 6rs$, add the sum of $2ab - 3xy + 5rs$ and $5ab + 4xy - 3rs$.

$$\begin{array}{l} 6ab + 7xy - 2rs \\ \underline{3ab + xy + 6rs} \\ 9ab + 8xy + 4rs \quad \text{(sum of the first group)} \end{array}$$

$$\begin{array}{l} 2ab - 3xy + 5rs \\ \underline{5ab + 4xy - 3rs} \\ 7ab + xy + 2rs \quad \text{(sum of the second group)} \end{array}$$

Add these two sums.

$$\begin{array}{r} 9ab + 8xy + 4rs \\ \underline{7ab + xy + 2rs} \\ 16ab + 9xy + 6rs \end{array}$$

29. To the sum of $4cd - 5xy + 6ab$ and $3cd + 2xy - 8ab$, add the difference obtained when $2cd + 3xy + 4ab$ is subtracted from $5cd - xy - 3ab$.

$$\begin{array}{l} 4cd - 5xy + 6ab \\ \underline{3cd + 2xy - 8ab} \\ 7cd - 3xy - 2ab \quad \text{(sum of the first group)} \end{array}$$

$$\begin{array}{l} 5cd - xy - 3ab \\ \underline{-2cd - 3xy - 4ab} \\ 3cd - 4xy - 7ab \quad \text{(difference of the second group)} \end{array}$$

Add these two answers.

$$\begin{array}{r} 7cd - 3xy - 2ab \\ \underline{3cd - 4xy - 7ab} \\ 10cd - 7xy - 9ab \end{array}$$

30. Subtract the difference that is obtained when $mn - 8st - 5pq$ is subtracted from $3mn - 4st + 2pq$ from the difference that is obtained when $3mn - st + 2pq$ is subtracted from $10mn + 3st - 6pq$.

$$\begin{array}{l} 3mn - 4st + 2pq \\ \underline{-\ mn + 8st + 5pq} \\ 2mn + 4st + 7pq \quad \text{(difference of the first group)} \end{array}$$

$$\begin{array}{l} 10mn + 3st - 6pq \\ \underline{-3mn + st - 2pq} \\ 7mn + 4st - 8pq \quad \text{(difference of the second group)} \end{array}$$

Subtract the first difference from the second difference.

$$\begin{array}{r} 7mn + 4st - 8pq \\ \underline{-2mn - 4st - 7pq} \\ 5mn - 15pq \end{array}$$

MULTIPLICATION AND DIVISION

Rules for Multiplication and Division:

1. If both numbers have the same sign, the product or quotient is a positive expression.

$$(a)(b) = ab. \qquad a \div b = \frac{a}{b}.$$
$$(-a)(-b) = ab. \qquad (-a) \div (-b) = \frac{a}{b}.$$

2. If the signs of the two numbers are not alike (if one is positive and the other negative), the product or quotient is a negative expression.

$$(-a)(b) = -ab. \qquad (-a) \div (b) = -\frac{a}{b}.$$
$$(a)(-b) = -ab. \qquad (a) \div (-b) = -\frac{a}{b}.$$

3. When multiplying one expression by another, add the exponents of similar factors.

$$(a^2b)(a^3b^2) = a^5b^3.$$
$$(3x^4)(2x^2) = 6x^6.$$

4. When dividing one expression by another, subtract the exponents of similar factors.

$$x^5 \div x^3 = x^2.$$

1. Multiply 5 by 4.
   Use Rule 1.
   $$(5)(4) = 20.$$

2. Multiply -5 by -4.
   Use Rule 1.
   $$(-5)(-4) = 20.$$

3. Multiply -5 by 4.
   Use Rule 2.
   $$(-5)(4) = -20.$$

4. Multiply 5 by -4.
   Use Rule 2.
   $$(5)(-4) = -20.$$

5. Multiply x by $-y$.
   $$(x)(-y) = -xy.$$

6. Multiply $-x$ by $-y$.
   $$(-x)(-y) = xy.$$

7. Divide 5 by 4.
   Use Rule 1.
   $$5 \div 4 = \tfrac{5}{4}.$$

8. Divide -5 by 4.
   Use Rule 2.
   $$-5 \div 4 = -\tfrac{5}{4}.$$

9. Divide -5 by -4.
Use Rule 1.
$$-5 \div -4 = \tfrac{5}{4}.$$
10. Divide 5 by -4.
Use Rule 2.
$$5 \div -4 = -\tfrac{5}{4}.$$
11. Divide $6x$ by $-3y$.
$$\frac{6x}{-3y} = -\frac{2x}{y}.$$
12. Divide $-8a$ by -2.
$$\frac{-8a}{-2} = 4a.$$

Meaning of Exponents. An **exponent** is a symbol placed at the upper right-hand corner of an expression to indicate how many times the expression is to be multiplied by itself. Evaluate the given expressions in Problems 13–16 so that the multiplication indicated by the exponent is carried out.

13. $3^4 = (3)(3)(3)(3) = 81.$
14. $y^3 = (y)(y)(y).$
15. $4a^3 = (4)(a)(a)(a).$
16. $(4a)^3 = (4a)(4a)(4a) = (4)(4)(4)(a)(a)(a)$
$= (64)(a)(a)(a).$
Observe that the exponent affects all factors enclosed within the parentheses.

Perform the indicated operations in Problems 17–30.

17. $(x^3)(x^2) = x^5.$
Rule 3 is applied here (together with Rule 1).
18. $(-3a^2)(4a^4) = -12a^6.$
Rule 3 is applied here (together with Rule 2).
19. $(-2a^2b^3)(3a^3b) = -6a^5b^4.$
20. $(3xy)^2(4xy^2).$
The exponent 2 associated with the first factor means that the entire factor is to be used as a factor twice.
$$(3xy)^2(4xy^2) = (3xy)(3xy)(4xy^2) = 36x^3y^4.$$
21. $(4x^2)(2x) = 8x^3.$
22. $(-3a^3)(4a^2) = -12a^5.$
23. $(2m^2)(-4m^2) = -8m^4.$
24. $(-5xy)(-3x^2y^3)(2xy^2) = 30x^4y^6.$
25. $(3r^2s)(-2rs^2)(rs) = -6r^4s^4.$
26. $\dfrac{8a^3}{2} = 4a^3.$
Rule 4 is applied here.
27. $\dfrac{-10x^4}{2x} = -5x^3.$
28. $\dfrac{-6m^5}{3m^2} = -2m^3.$
29. $\dfrac{-x^2y^3}{-xy^2} = xy.$
30. $\dfrac{12a^6b^2}{-4b} = -3a^6b.$

Denominator Greater than Numerator. If the exponent of the denominator is larger than the exponent of the numerator, one still subtracts exponents according to Rule 4, but the difference is placed in the denominator.

$$\frac{x^2}{x^5} = \frac{1}{x^{5-2}} = \frac{1}{x^3}.$$

Perform the indicated divisions in Problems 31–35.

31. $\dfrac{10a}{a^2} = \dfrac{10}{a}.$
32. $\dfrac{6xy}{2x^3} = \dfrac{3y}{x^2}.$
33. $\dfrac{-8ab}{2a^2b^2} = -\dfrac{4}{ab}.$
34. $\dfrac{-9m^2n^4}{-3m^4n^2} = \dfrac{3n^2}{m^2}.$
35. $\dfrac{2xy}{-12x^2y^3} = -\dfrac{1}{6xy^2}.$

Polynomial by Monomial. In multiplying a polynomial by a monomial, each term of the polynomial is multiplied by the monomial.

$$2a(3a - a^2 + 2a^3) = 6a^2 - 2a^3 + 4a^4.$$

Perform the indicated multiplications in Problems 36–38.

36. $4b(3b - 1) = 12b^2 - 4b.$
37. $-2ac(3c - 4a) = -6ac^2 + 8a^2c.$
38. $3x^2y^3(xy^2 - 2x^2y + 1) = 3x^3y^5 - 6x^4y^4 + 3x^2y^3.$

Polynomial by Polynomial. In multiplying a polynomial by a polynomial, each term of the multiplicand is multiplied by each term of the multiplier. Similar terms are placed in the same column and are added. The product consists of the algebraic sum of these columns.

$$\begin{array}{r} a + b \\ a + b \\ \hline a^2 + ab \\ ab + b^2 \\ \hline a^2 + 2ab + b^2 \end{array}$$

39. Multiply $a^2 + ab + b^2$ by $a + b$.

$$\begin{array}{r} a^2 + ab + b^2 \\ a + b \\ \hline a^3 + a^2b + ab^2 \\ a^2b + ab^2 + b^3 \\ \hline a^3 + 2a^2b + 2ab^2 + b^3 \end{array}$$

40. Multiply $2x^2 - 3x + 5$ by $x + 1$.

$$\begin{array}{r} 2x^2 - 3x + 5 \\ \underline{x + 1} \\ 2x^3 - 3x^2 + 5x \\ \underline{2x^2 - 3x + 5} \\ 2x^3 - x^2 + 2x + 5 \end{array}$$

41. Multiply $x + y$ by $a + b$.

$$\begin{array}{l} x + y \\ \underline{a + b} \\ ax + ay \\ \underline{\qquad + bx + by} \\ ax + ay + bx + by \end{array}$$

42. Multiply $4m^3 - 3m^2 + m - 1$ by $m^2 + 3m - 4$.

$$\begin{array}{r} 4m^3 - 3m^2 + m - 1 \\ \underline{m^2 + 3m - 4} \\ 4m^5 - 3m^4 + m^3 - m^2 \qquad\qquad \\ 12m^4 - 9m^3 + 3m^2 - 3m \qquad \\ \underline{- 16m^3 + 12m^2 - 4m + 4} \\ 4m^5 + 9m^4 - 24m^3 + 14m^2 - 7m + 4 \end{array}$$

43. Multiply $2m^2 + m^3 + 1 - m$ by $m^2 - m - 1$.

If both the multiplicand and the multiplier are arranged in either ascending or descending powers of the unknown, every term of each partial product will automatically fall into proper order in the columns to be added. The commutative axiom for addition permits our making such a rearrangement.

$$\begin{array}{l} m^3 + 2m^2 - m + 1 \\ \underline{m^2 - m - 1} \\ m^5 + 2m^4 - m^3 + m^2 \\ \quad - m^4 - 2m^3 + m^2 - m \\ \underline{\qquad\quad - m^3 - 2m^2 + m - 1} \\ m^5 + m^4 - 4m^3 \qquad\qquad\quad - 1 \end{array}$$

Dividing Polynomials. In dividing a polynomial by a monomial, each term of the polynomial is divided by the monomial.

$$\boxed{\frac{2a^4 + 6a^3 - 4a^2}{2a} = a^3 + 3a^2 - 2a.}$$

44. Divide $9x^2 - 3x^3 - 12x^4$ by $-3x^2$.

$$\frac{9x^2 - 3x^3 - 12x^4}{-3x^2} = -3 + x + 4x^2.$$

45. Divide $2bc + 6b^2c$ by $4bc$.

$$\frac{2bc + 6b^2c}{4bc} = \frac{1 + 3b}{2}.$$

46. Divide $-12m^2n^2 + 8m^3n^3 + 4m^4n^5$ by $-4m^2n^2$.

$$\frac{-12m^2n^2 + 8m^3n^3 + 4m^4n^5}{-4m^2n^2} = 3 - 2mn - m^2n^3.$$

47. Divide $x^2 + 12x + 32$ by $x + 4$.

This type of division is similar to long division in arithmetic. When solving problems in long division, one should arrange both the dividend and divisor in either ascending or descending powers of the unknown.

To find the first term of the quotient, divide the first term of the dividend by the first term of the divisor. Multiply each term of the divisor by the first term of the quotient and subtract the result from the dividend.

$$\begin{array}{r|l} & x \\ \hline x + 4 & x^2 + 12x + 32 \\ & \underline{x^2 + 4x} \\ & \qquad 8x + 32 \end{array}$$

This difference becomes the new dividend. The next term of the quotient is obtained by dividing the first term of this new dividend by the first term of the divisor. Each term of the divisor is multiplied by this next term of the quotient and the result is subtracted from the dividend.

$$\begin{array}{r|l} & x + 8 \\ \hline x + 4 & x^2 + 12x + 32 \\ & \underline{x^2 + 4x} \\ & \qquad 8x + 32 \\ & \qquad \underline{8x + 32} \end{array}$$

48. Divide $x^3 + 4x^2 + 5x + 2$ by $x + 2$.

$$\begin{array}{r|l} & x^2 + 2x + 1 \\ \hline x + 2 & x^3 + 4x^2 + 5x + 2 \\ & \underline{x^3 + 2x^2} \\ & \qquad 2x^2 + 5x \\ & \qquad \underline{2x^2 + 4x} \\ & \qquad\qquad x + 2 \\ & \qquad\qquad \underline{x + 2} \end{array}$$

49. Divide $2a^4 + 4a^3 - 7a^2 - 20a - 15$ by $a^2 - 5$.

$$\begin{array}{r|l} & 2a^2 + 4a + 3 \\ \hline a^2 - 5 & 2a^4 + 4a^3 - 7a^2 - 20a - 15 \\ & \underline{2a^4 \qquad\quad - 10a^2} \\ & \qquad 4a^3 + 3a^2 - 20a \\ & \qquad \underline{4a^3 \qquad\quad - 20a} \\ & \qquad\qquad 3a^2 \qquad\quad - 15 \\ & \qquad\qquad \underline{3a^2 \qquad\quad - 15} \end{array}$$

50. Divide $x^3 - y^3$ by $x - y$.

Frequently certain terms in descending order of the powers of the unknown are missing in the dividend, but they appear in the quotient. Space should be left for them as shown below.

$$\begin{array}{r|l} & x^2 + xy + y^2 \\ \hline x - y & x^3 \qquad\qquad\qquad - y^3 \\ & \underline{x^3 - x^2y} \\ & \qquad x^2y \\ & \qquad \underline{x^2y - xy^2} \\ & \qquad\qquad xy^2 - y^3 \\ & \qquad\qquad \underline{xy^2 - y^3} \end{array}$$

51. Divide $4m^3 - 6m^2 - 2m + 4$ by $2m^2 - 1$.

$$\begin{array}{r|l} & 2m - 3 \\ 2m^2 - 1 & 4m^3 - 6m^2 - 2m + 4 \\ & 4m^3 \qquad - 2m \\ & - 6m^2 \qquad + 4 \\ & - 6m^2 \qquad + 3 \\ & +1 \end{array}$$

Since the first term of the divisor is not contained in the 1 in the new dividend, the answer is $2m - 3$ with a remainder of 1.

As was done in arithmetic, the remainder may be shown as a fraction of the divisor, and the answer can be written: $2m - 3 + \frac{1}{2m^2 - 1}$.

GROUPING SYMBOLS

Types of Symbols. Grouping symbols are found throughout the study of algebra. Their purpose is to group more than one expression into a single unit. The following grouping symbols are the most frequently used.

parentheses	$(a + b + c)$;
brackets	$[a + b + c]$;
braces	$\{a + b + c\}$;
vinculum	$\overline{a + b + c}$.

Rules for the Use of Grouping Symbols:

1. When removing any of the symbols used for grouping, observe the sign that immediately precedes the expression. If the sign is positive, merely remove the symbols and write the enclosed expression as it appeared originally. If the sign is negative, remove the grouping symbols and change the signs of all the terms enclosed within them.

$$(a + b + c) = a + b + c.$$
$$-[a + b - c] = -a - b + c.$$

2. If more than one type of symbol is used in one expression, remove the innermost symbols first, observing Rule 1 above. Then continue until all symbols have been removed, one type in each operation.

$$-[-x - (a + b) + 3] = -[-x - a - b + 3]$$
$$= x + a + b - 3.$$

1. Remove the parentheses from $4a + (b - c)$.

Use Rule 1.

$$4a + (b - c) = 4a + b - c.$$

2. Remove the brackets from $6x - [y + 3z]$.

Use Rule 1.

$$6x - [y + 3z] = 6x - y - 3z.$$

3. Remove the vinculum from $m + \overline{n - p} + q$.

$$m + \overline{n - p} + q = m + n - p + q.$$

4. Remove the braces from $3 - \{a + b - c\}$.

$$3 - \{a + b - c\} = 3 - a - b + c.$$

5. Remove the parentheses from

$$a + (b - c) - (a + 2b - 3c).$$

Use Rule 1 and remove both sets of parentheses at the same time.

$$a + (b - c) - (a + 2b - 3c) = a + b - c - a - 2b + 3c$$

Combine similar terms for the final answer.

$$= 2c - b.$$

6. Simplify $5x - [-(x + y) + (x - 3y) + 2y] - y + x$.

Both sets of parentheses are considered innermost and may be removed in the same operation.

$$5x - [-(x + y) + (x - 3y) + 2y] - y + x$$
$$= 5x - [-x - y + x - 3y + 2y] - y + x$$

Remove the brackets, observing Rule 1.

$$= 5x + x + y - x + 3y - 2y - y + x$$

Combine similar terms for the final answer.

$$= 6x + y.$$

7. Simplify $3a - [2b + c - (a + b + c) + 4b]$.

Since the parentheses are the innermost pair of symbols, remove them first, observing Rule 1.

$$3a - [2b + c - (a + b + c) + 4b]$$
$$= 3a - [2b + c - a - b - c + 4b]$$

Remove the brackets, observing Rule 1.

$$= 3a - 2b - c + a + b + c - 4b$$
$$= 4a - 5b.$$

8. Simplify $3 - b - \{2b - (2 + 5b)\} - 4$.

Remove the parentheses first.

$$3 - b - \{2b - (2 + 5b)\} - 4 = 3 - b - \{2b - 2 - 5b\} - 4$$

Remove the braces, observing Rule 1.

$$= 3 - b - 2b + 2 + 5b - 4$$
$$= 1 + 2b.$$

9. Simplify $5x - (3y + z - [-4x + y - 5z] + 2y)$.

Remove the brackets first since they are innermost.

$$5x - (3y + z - [-4x + y - 5z] + 2y)$$
$$= 5x - (3y + z + 4x - y + 5z + 2y)$$

Remove the parentheses, observing Rule 1.

$$= 5x - 3y - z - 4x + y - 5z - 2y$$
$$= x - 4y - 6z.$$

10. Simplify $-(-a + 3b - \overline{c + 2b} + 3c - d)$.

Remove the vinculum, observing the minus sign immediately before it.

$$-(-a + 3b - \overline{c + 2b} + 3c - d)$$
$$= -(-a + 3b - c - 2b + 3c - d)$$

Remove the parentheses, observing the minus sign immediately before it.

$$= a - 3b + c + 2b - 3c + d$$
$$= a - b - 2c + d.$$

11. Simplify $4m - [m + n - (-m + \overline{m + 2n}) - m - 3n]$.

Remove the vinculum first.

$$4m - [m + n - (-m + \overline{m + 2n}) - m - 3n]$$
$$= 4m - [m + n - (-m + m + 2n) - m - 3n]$$

Remove the parentheses, observing the minus sign.

$$= 4m - [m + n + m - m - 2n - m - 3n]$$

Remove the brackets and combine similar terms.

$$= 4m - m - n - m + m + 2n + m + 3n$$
$$= 4m + 4n.$$

12. Simplify $4x - [y + z - \overline{x - (y + z) + 3y + x}] - y$.

Remove the parentheses first as they are innermost.

$$4x - [y + z - \overline{x - (y + z) + 3y + x}] - y$$
$$= 4x - [y + z - \overline{x - y - z + 3y + x}] - y$$

Remove the vinculum, observing the minus sign that precedes it.

$$= 4x - [y + z - x + y + z - 3y + x] - y$$

Remove the brackets, observing the minus sign.

$$= 4x - y - z + x - y - z + 3y - x - y$$
$$= 4x - 2z.$$

Numerical Coefficients. In the twelve problems shown thus far, you may have noticed that there was no coefficient for any grouped expression. As with other algebraic expressions, when no coefficient is shown it is assumed to be 1. Thus, as those grouping symbols were removed, each enclosed term was multiplied by 1 or -1 by use of the distributive axiom.

Should there appear a coefficient other than 1, each term included within the grouping symbols must be multiplied by that coefficient in accord with Rule 1.

13. Simplify $x + 3(x - y)$.

Multiply each term included within the parentheses by 3 and make no changes in the signs since the 3 is preceded by a plus sign.

$$x + 3(x - y) = x + 3x - 3y$$
$$= 4x - 3y.$$

14. Simplify $m - 2(m + n) + 5(n - m)$.

Multiply each term in the first parentheses by 2 and change all signs included within those parentheses since the 2 is preceded by a minus sign. This means that each term within the parentheses is multiplied by -2. Multiply each term included within the second set of parentheses by 5 and change no signs.

$$m - 2(m + n) + 5(n - m) = m - 2m - 2n + 5n - 5m$$
$$= 3n - 6m.$$

15. Simplify $a - 2[a + 3b - 3(a - b) - b]$.

Remove the parentheses first, since they are innermost, observing the minus sign that precedes the 3.

$$a - 2[a + 3b - 3(a - b) - b]$$
$$= a - 2[a + 3b - 3a + 3b - b]$$

Remove the brackets, observing the -2.

$$= a - 2a - 6b + 6a - 6b + 2b$$
$$= 5a - 10b.$$

16. Simplify $5x - 2[y - \overline{3z + x} - y]$.

Remove the vinculum first, observing the minus sign that precedes it.

$$5x - 2[y - \overline{3z + x} - y] = 5x - 2[y - 3z - x - y]$$
$$= 5x - 2y + 6z + 2x + 2y$$
$$= 7x + 6z.$$

17. Simplify $-\{3 + 2m - 3(1 + m) - 5\}$.

$$-\{3 + 2m - 3(1 + m) - 5\} = -\{3 + 2m - 3 - 3m - 5\}$$
$$= -3 - 2m + 3 + 3m + 5$$
$$= m + 5.$$

18. Simplify $2[-(a + 3\{2 - 3a + b\}) - 1]$.

$$2[-(a + 3\{2 - 3a + b\}) - 1] = 2[-(a + 6 - 9a + 3b) - 1]$$
$$= 2[-a - 6 + 9a - 3b - 1]$$
$$= -2a - 12 + 18a - 6b - 2$$
$$= 16a - 6b - 14.$$

Literal Coefficients. If the coefficient is literal, each term within the grouping symbols must be multiplied by that literal coefficient. The rules for multiplying exponents must be followed as usual in such problems.

19. Simplify $a(a + b) - 2b$.

$$a(a + b) - 2b = a^2 + ab - 2b.$$

20. Simplify $x - y(x + y) + x(y - x)$.

$$x - y(x + y) + x(y - x) = x - xy - y^2 + xy - x^2$$
$$= x - y^2 - x^2.$$

21. Simplify $2a - a(a + [3b - 2a])$.

$$2a - a(a + [3b - 2a]) = 2a - a(a + 3b - 2a)$$
$$= 2a - a^2 - 3ab + 2a^2$$
$$= a^2 + 2a - 3ab.$$

Composite Coefficients. The coefficient for a set of grouping symbols may consist of both numerical and literal factors. Here again the distributive axiom is used and the rules for multiplication apply.

22. Simplify $2x - 3x(x + 2y)$.

$$2x - 3x(x + 2y) = 2x - 3x^2 - 6xy.$$

23. Simplify $-[2a - 3ab(a + b)]$.

$$-[2a - 3ab(a + b)] = -[2a - 3a^2b - 3ab^2]$$
$$= -2a + 3a^2b + 3ab^2.$$

24. Simplify $4x - 2[x - (y - \overline{x - z} + 2y)] - 5$.

$$4x - 2[x - (y - \overline{x - z} + 2y)] - 5$$
$$= 4x - 2[x - (y - x + z + 2y)] - 5$$
$$= 4x - 2[x - y + x - z - 2y] - 5$$
$$= 4x - 2x + 2y - 2x + 2z + 4y - 5$$
$$= 6y + 2z - 5.$$

25. Simplify $-2\{-3[x + 2(x + y) - 3y]\}$.

$$-2\{-3[x + 2(x + y) - 3y]\} = -2\{-3[x + 2x + 2y - 3y]\}$$
$$= -2\{-3x - 6x - 6y + 9y\}$$
$$= 6x + 12x + 12y - 18y$$
$$= 18x - 6y.$$

Enclosing Expressions. Frequently problems arise in which it is necessary or desirable to enclose certain expressions within grouping symbols. In such cases we place either a plus or a minus sign before the grouping symbols to accomplish the given conditions. If a plus sign is inserted, the enclosed terms are preceded by the same sign as appeared in the original form of the expression. If a minus sign is inserted, the sign preceding each term is changed.

26. In the expression $a + b - c$, enclose the $b - c$ within parentheses preceded by a plus sign.

$$a + b - c = a + (b - c).$$

27. In the expression $3x - 4y - z$, enclose the $-4y - z$ within parentheses preceded by a minus sign.

$$3x - 4y - z = 3x - (4y + z).$$

28. By use of parentheses, show that the sum of a and b is to be subtracted from c.

$$c - (a + b).$$

29. By use of parentheses, show that three times the sum of m and n is to be subtracted from five times the sum of x and y.

$$5(x + y) - 3(m + n).$$

30. By use of parentheses, show that the difference obtained by subtracting p from 3 is to be added to q.

$$q + (3 - p).$$

31. By use of parentheses, show that the sum of c and d is to be subtracted from the difference obtained by subtracting n from m.

$$(m - n) - (c + d).$$

32. By use of parentheses, show that three times the sum of m and n is to be subtracted from three times the difference obtained by subtracting t from 5.

$$3(5 - t) - 3(m + n).$$

33. By use of parentheses, show that one-half the sum of a and b is to be added to x.

$$x + \tfrac{1}{2}(a + b).$$

34. By use of parentheses, indicate that one-third of the difference obtained by subtracting 3 from r is to be added to two-thirds of the sum of r and 3.

$$\tfrac{2}{3}(r + 3) + \tfrac{1}{3}(r - 3).$$

35. By use of parentheses, show that one-half the sum of a and b is to be subtracted from four times the sum of three times a and five times b.

$$4(3a + 5b) - \tfrac{1}{2}(a + b).$$

SUPPLEMENTARY PROBLEMS

The supplementary problems listed at the end of each chapter are provided for further study and practice. Answers to these problems will be found in the back of the book.

1. Describe the following set: $A = \{-2, -1, 0, 1, 2\}$.
2. Describe the following set: $B = \{3, 5, 7, 11, 13\}$.
3. List the elements of the following set: A is the set of all the proper fractions whose numerator and denominator are taken from the universe, $U = \{1, 2, 3\}$.
4. List all of the subsets of the following set:
$$X = \{-1, 0, 1\}.$$
5. List all of the subsets of the following set:
$$B = \{a, b, c, d\}.$$
6. Add $7x^3y + 5x^2y^2 - 8xy^3 + 3$ and $2xy^3 - 5 - 7x^2y^2 + x^3y$.
7. Subtract $3m^2 - 7n^3 + 5$ from $7m^2 - 3m + 2n^3$.
8. From $a^3 - 3a^2b + 4ab^2 + b^3$ subtract $-a^3 + 2ab^2 - 6$.
9. From zero subtract the sum of $3ax + 4ac - 1$ and $2ax + ac - ad + 3$.
10. From the sum of $7x^2 - 3xy + 4y^2$ and $4xy - 2y^2 + 4$ subtract the difference obtained by subtracting $x^2 + y^2 - 3$ from $2xy + 5y^2 - 5$.
11. Multiply $2x^3 - 7x^2 + 3x - 4$ by $x^2 - 2x + 3$.
12. Multiply $-4d^2x + 3dx^2 - 3$ by $2 + d - x$.
13. Divide $2x^4 - 10x + 11x^2 + 3 - 7x^3$ by $2x - 1$.
14. Divide $8a^7 - 4a^6 - 14a^3 - 2 - 3a^2 + 28a^4 - 28a^5 + 3a$ by $1 + 4a^3 - 2a^2$.
15. Divide $2x^4y^4 - 9x^3y^3 + 22x^2y^2 - 6xy + 3$ by $x^2y^2 - 2xy + 3$.
16. Simplify $2 - \{3m - [2n + m + (3m - 4n) + 2]\}$.
17. Simplify
$$-\{-3 - 4b - [2 + b - (5 + 6b) - 4 + b] - 5\}.$$
18. Simplify $2[-3 + m - 2(3m + 6) + 7m + 3]$.
19. Simplify $\{4 + 3d - 2d - (1 + d) - 2d - 5\}$.
20. Simplify $-2\{x - 3y + x[x - y(x + 3)] + 2\}$.

Chapter 2

Special Products and Factoring

Definitions. Expressions which yield a product when multiplied are known as **factors.** Examples of factors and their products are $(4)(3) = 12$, $(7)(8) = 56$, $(2x)(4) = 8x$, and $(3x)(5x^2) = 15x^3$.

A **prime factor** is one that has no factors except itself and 1. If we multiply 3 by 4 we get 12, so 3 and 4 are factors of 12. The 3 is a prime factor. However, the 4 can be broken down to $(2)(2)$. Hence, 4 is not a prime factor. For most uses in algebra it is desirable to have prime factors.

Factoring is the process of changing a product into its component multipliers. If we factor 12 into its prime factors, we get $(2)(2)(3)$. The prime factors of $8x^2$ are $(2)(2)(2)(x)(x)$ and of $10x^3y^2$ are $(5)(2)(x)(x)(x)(y)(y)$.

1. Find the product of 4 and $(x + y)$.

   Multiply each term within the parentheses by the 4.

$$4(x + y) = 4x + 4y.$$

2. Find the product of a and $(b - c + d)$.

   Multiply each term within the parentheses by a.

$$a(b - c + d) = ab - ac + ad.$$

3. Find the product of $3x$ and $(2 + y - z)$.

   Multiply each term within the parentheses by $3x$.

$$3x(2 + y - z) = 6x + 3xy - 3xz.$$

Simple Factoring. In each of the above problems, the product was found by multiplying one factor by the other. The coefficient of the enclosed expression was one factor and the enclosed expression was the second factor. If one started with the product and expressed it in terms of its factors, he would have performed the simplest form of factoring. Simple factoring is merely dividing each term by a common factor found in each term. That common factor is said to be the first factor and the resulting quotient is the second factor.

$$\boxed{mx + nx + px = x(m + n + p).}$$

4. Factor $4x + 6y - 10z$.

   Each term is divisible by 2. Therefore 2 is the first factor.

$$4x + 6y - 10z = 2(2x + 3y - 5z).$$

5. Factor $ax - bx - cx$.

   Each term is divisible by x. Therefore x is the first factor.

$$ax - bx - cx = x(a - b - c).$$

6. Factor $9abx + 3acx - 12adx + 6aex$.

   Each term is divisible by $3ax$. Hence, $3ax$ is the first factor.

$$9abx + 3acx - 12adx + 6aex = 3ax(3b + c - 4d + 2e).$$

7. Multiply $a - b$ by $a + b$.

$$\begin{array}{l} a - b \\ \underline{a + b} \\ a^2 - ab \\ \underline{\quad\ + ab - b^2} \\ a^2 \qquad\quad - b^2 \end{array}$$

8. Multiply $m - 2n$ by $m + 2n$.

$$\begin{array}{l} m - 2n \\ \underline{m + 2n} \\ m^2 - 2mn \\ \underline{\qquad + 2mn - 4n^2} \\ m^2 \qquad\qquad - 4n^2 \end{array}$$

9. Multiply $4x + 3y$ by $4x - 3y$.

$$\begin{array}{l} 4x + 3y \\ \underline{4x - 3y} \\ 16x^2 + 12xy \\ \underline{\qquad - 12xy - 9y^2} \\ 16x^2 \qquad\qquad - 9y^2 \end{array}$$

Difference of Two Perfect Squares. In Problems 7, 8, and 9, two binomials were multiplied together. In each case the expression was the product of the sum and the difference of two quantities. Each time, the second term of the product was eliminated. This holds true every time the sum and difference of two expressions are multiplied. Consider the products in Problems 7, 8, and 9. In each case the product is the difference of two perfect squares. This leads to a second type of factoring, that of factoring the difference of two perfect squares.

The rule for this type of factoring is as follows: *Write the square root of each term and place a positive sign between them for one factor and a negative sign between them for the second factor.*

$$x^2 - y^2 = (x + y)(x - y).$$

10. Factor $a^2 - m^2$.

$$a^2 - m^2 = (a + m)(a - m).$$

11. Factor $b^2 - 4$.

$$b^2 - 4 = (b + 2)(b - 2).$$

12. Factor $25 - 9y^2$.

$$25 - 9y^2 = (5 + 3y)(5 - 3y).$$

13. Factor $x^2y^2 - m^2n^2$.

$$x^2y^2 - m^2n^2 = (xy + mn)(xy - mn).$$

14. Factor $ax^2 - 4ay^2$.

In this problem it will be observed that neither term is a perfect square. However, it can be seen that there is a common factor in each of the two terms. When factoring, it is always necessary to look at all terms first to see whether there is a common factor. If so, remove that factor.

$$ax^2 - 4ay^2 = (a)(x^2 - 4y^2).$$

The second factor is the difference of two perfect squares. To factor $ax^2 - 4ay^2$ completely, $x^2 - 4y^2$ must be factored.

$$\begin{aligned} ax^2 - 4ay^2 &= a(x^2 - 4y^2) \\ &= a(x + 2y)(x - 2y). \end{aligned}$$

15. Factor $18m^2 - 8$.

$$\begin{aligned} 18m^2 - 8 &= 2(9m^2 - 4) \\ &= 2(3m + 2)(3m - 2). \end{aligned}$$

16. Factor $75ab^2 - 27ac^2$.

$$\begin{aligned} 75ab^2 - 27ac^2 &= 3a(25b^2 - 9c^2) \\ &= 3a(5b + 3c)(5b - 3c). \end{aligned}$$

17. Multiply $a - b$ by $a - b$.

$$\begin{array}{l} a - b \\ \underline{a - b} \\ a^2 - ab \\ \underline{\quad - ab + b^2} \\ a^2 - 2ab + b^2 \end{array}$$

18. Multiply $x + y$ by $x + y$.

$$\begin{array}{l} x + y \\ \underline{x + y} \\ x^2 + xy \\ \underline{\quad + xy + y^2} \\ x^2 + 2xy + y^2 \end{array}$$

19. Multiply $m + 2n$ by $m + 2n$.

$$\begin{array}{l} m + 2n \\ \underline{m + 2n} \\ m^2 + 2mn \\ \underline{\quad + 2mn + 4n^2} \\ m^2 + 4mn + 4n^2 \end{array}$$

20. Multiply $3ab - 5c$ by $3ab - 5c$.

$$\begin{array}{l} 3ab - 5c \\ \underline{3ab - 5c} \\ 9a^2b^2 - 15abc \\ \underline{\quad - 15abc + 25c^2} \\ 9a^2b^2 - 30abc + 25c^2 \end{array}$$

Squaring Binomials. In Problems 17–20 the two factors to be multiplied were exactly alike in each case. Those problems might have been written as follows: $(a - b)^2$; $(x + y)^2$; $(m + 2n)^2$; and $(3ab - 5c)^2$. This type of problem is often referred to as the square of a binomial.

The square of a binomial is equal to the square of the first term plus twice the product of the first and second terms plus the square of the second term. The first and last terms are positive. The second term has the same sign as the sign between the two terms in the binomial form.

21. Square the binomial $c - 2d$.

$$(c - 2d)^2 = c^2 - 4cd + 4d^2.$$

22. Square the binomial $2a + 5b$.

$$(2a + 5b)^2 = 4a^2 + 20ab + 25b^2.$$

23. Square the binomial $ab + 3cd$.

$$(ab + 3cd)^2 = a^2b^2 + 6abcd + 9c^2d^2.$$

24. Square the binomial $ab^2 - 4c^2$.

$$(ab^2 - 4c^2)^2 = a^2b^4 - 8ab^2c^2 + 16c^4.$$

Perfect-Square Trinomials. One should learn to identify perfect-square trinomials and remember that the two factors of such expressions are identical. If the first and third terms of the trinomial are positive perfect squares, the expression *may* be of this type. If it is, the second term must be twice the product of the square roots of the first and third terms.

Thus, the rule for factoring a perfect square trinomial is: *Write the square root of the first and third term and place the sign of the second term between them. Repeat this process since there will be two identical factors.*

$$d^2 + 6d + 9 = (d + 3)(d + 3).$$

25. Factor $x^2 + 2x + 1$.

The square root of the first term is x and of the third term is 1. Twice the product of these square roots is $2x$. Since this is the same as the second term of the trinomial, the original expression is the product of a squared binomial.

$$x^2 + 2x + 1 = (x + 1)(x + 1).$$

26. Factor $a^2 + 4a + 4$.

$$a^2 + 4a + 4 = (a + 2)(a + 2).$$

27. Factor $9 - 6y + y^2$.

$$9 - 6y + y^2 = (3 - y)(3 - y).$$

28. Factor $25m^2 - 40mn + 16n^2$.

$$25m^2 - 40mn + 16n^2 = (5m - 4n)(5m - 4n).$$

29. Factor $x^2y^2 + 8xyz + 16z^2$.

$$x^2y^2 + 8xyz + 16z^2 = (xy + 4z)(xy + 4z).$$

30. Factor $3s^2 - 6s + 3$.

Remove the common factor 3.

$$3s^2 - 6s + 3 = 3(s^2 - 2s + 1)$$

Now the second factor can be converted into prime factors.

$$= 3(s - 1)(s - 1).$$

31. Factor $5x^2 - 30x + 45$.

$$5x^2 - 30x + 45 = 5(x^2 - 6x + 9)$$
$$= 5(x - 3)(x - 3).$$

32. Factor $x^4 - 8x^2 + 16$.

$$x^4 - 8x^2 + 16 = (x^2 - 4)(x^2 - 4).$$

Both factors are the difference of two perfect squares and can be factored again.

$$= (x + 2)(x - 2)(x + 2)(x - 2).$$

33. Multiply $x + 3$ by $x + 2$.

$$\begin{array}{r} x + 3 \\ \underline{x + 2} \\ x^2 + 3x \\ \underline{+ 2x + 6} \\ x^2 + 5x + 6 \end{array}$$

34. Multiply $y - 4$ by $y - 2$.

$$\begin{array}{r} y - 4 \\ \underline{y - 2} \\ y^2 - 4y \\ \underline{- 2y + 8} \\ y^2 - 6y + 8 \end{array}$$

35. Multiply $m + 5$ by $2m + 3$.

$$\begin{array}{r} m + 5 \\ \underline{2m + 3} \\ 2m^2 + 10m \\ \underline{+ 3m + 15} \\ 2m^2 + 13m + 15 \end{array}$$

36. Multiply $x + 7$ by $x - 3$.

$$\begin{array}{r} x + 7 \\ \underline{x - 3} \\ x^2 + 7x \\ \underline{- 3x - 21} \\ x^2 + 4x - 21 \end{array}$$

37. Multiply $3a - 1$ by $2a + 3$.

$$\begin{array}{r} 3a - 1 \\ \underline{2a + 3} \\ 6a^2 - 2a \\ \underline{+ 9a - 3} \\ 6a^2 + 7a - 3 \end{array}$$

Factoring Trinomials. In Problems 33–37, the product is a trinomial. The binomials multiplied to get those trinomials are the factors of the trinomials. It will be observed that the two factors are not identical. (They were identical in Problems 25–32.)

When factoring any trinomial, the following procedure is used.

Factor $x^2 + 9x + 18$.

Set up two sets of parentheses, and in the first position within the sets of parentheses place factors of the first term in the trinomial.

$$(x \quad\quad)(x \quad\quad)$$

Then place the proper sign between the terms of each binomial factor according to the rule,

If the third term of the trinomial is positive, the signs will be alike and will be the same as the sign of the second term in the trinomial. If the sign of the third term of the trinomial is negative, the two signs will be unlike.

Thus, the above problem becomes

$$(x + ?)(x + ?)$$

The second terms of the binomial factors must be factors of the third term of the trinomial. Insert such a pair of factors in place of the question marks and the problem continues

outer

$$(x + 6)(x + 3)$$

inner

The product of the first terms of the factors is x^2, the first term in the trinomial. The product of the last terms of the factors is 18, the last term in the trinomial. The final test to determine whether or not the chosen factors are correct is to find the algebraic sum of the products of the *outer terms* and the *inner terms* (as shown above). If this sum is the same as the second term in the trinomial, the factors are correct.

If that sum is not the same as the second term in the trinomial, other trials must be made. Sometimes by finding another pair of factors for either the first or the third term, the desired result may be obtained. In the above problem, other factors of the 18 are 9 and 2 or 18 and 1. If the signs between the binomial factors are unlike, it may be necessary to use the opposite sign in each case. This type of factoring requires considerable practice, after which the correct procedure will become routine.

$$x^2 + 9x + 18 = (x + 6)(x + 3).$$
$$x^2 - 2x - 15 = (x - 5)(x + 3).$$

38. Factor $x^2 - 9x + 20$.

$$x^2 - 9x + 20 = (x - 4)(x - 5).$$

39. Factor $a^2 + 11a + 24$.

$$a^2 + 11a + 24 = (a + 3)(a + 8).$$

40. Factor $2a^2 + a - 3$.

$$2a^2 + a - 3 = (2a + 3)(a - 1).$$

41. Factor $10x^2 - 11x - 6$.

$$10x^2 - 11x - 6 = (5x + 2)(2x - 3).$$

42. Factor $2x^2y^2 + 5xy - 3$.

$$2x^2y^2 + 5xy - 3 = (2xy - 1)(xy + 3).$$

43. Factor $6x^2 + 7xy - 3y^2$.

$$6x^2 + 7xy - 3y^2 = (3x - y)(2x + 3y).$$

44. Factor $6a^2x - 21abx + 9b^2x$.

Factor out the common factor $3x$ from each term.

$$6a^2x - 21abx + 9b^2x = 3x(2a^2 - 7ab + 3b^2)$$

Then the trinomial is factorable.

$$= 3x(a - 3b)(2a - b).$$

45. Factor $4m^2 + 2mn - 6n^2$.

$$\begin{aligned} 4m^2 + 2mn - 6n^2 &= 2(2m^2 + mn - 3n^2) \\ &= 2(m - n)(2m + 3n). \end{aligned}$$

46. Factor $x^2(a - b) + 5x(a - b) + 6(a - b)$.

Each term of the trinomial has the common factor $(a - b)$.

$$\begin{aligned} x^2(a - b) + 5x(a - b) + 6(a - b) &= (a - b)(x^2 + 5x + 6) \\ &= (a - b)(x + 2)(x + 3). \end{aligned}$$

47. Factor $2m^2(x + 2y) - m(x + 2y) - x - 2y$.

Place parentheses around the $-x - 2y$ to form one term, inserting a negative sign before the parentheses.

$$2m^2(x + 2y) - m(x + 2y) - (x + 2y)$$

We have the common factor $(x + 2y)$ in each term of the trinomial. Thus

$$\begin{aligned} &2m^2(x + 2y) - m(x + 2y) - x - 2y \\ &= 2m^2(x + 2y) - m(x + 2y) - (x + 2y) \\ &= (x + 2y)(2m^2 - m - 1) \\ &= (x + 2y)(2m + 1)(m - 1). \end{aligned}$$

48. Factor $mx^2 - 2nx^2 - 3mx + 6nx - 10m + 20n$.

Enclose the first two terms with parentheses, the third and fourth terms with parentheses preceded by a negative sign, and the last two terms with parentheses preceded by a negative sign.

$$\begin{aligned} &mx^2 - 2nx^2 - 3mx + 6nx - 10m + 20n \\ &= (mx^2 - 2nx^2) - (3mx - 6nx) - (10m - 20n) \end{aligned}$$

Factor x^2 from the first binomial within parentheses, $3x$ from the second binomial, and 10 from the last binomial.

$$\begin{aligned} &= x^2(m - 2n) - 3x(m - 2n) - 10(m - 2n) \\ &= (m - 2n)(x^2 - 3x - 10) \\ &= (m - 2n)(x - 5)(x + 2). \end{aligned}$$

Sum or Difference of Two Cubes. A special case of factoring is that of the sum of two cubes or the difference of two cubes. Examples of those types are $x^3 + y^3$ and $x^3 - y^3$. These two expressions are factored as follows.

$$x^3 + y^3 = (x + y)(x^2 - xy + y^2).$$
$$x^3 - y^3 = (x - y)(x^2 + xy + y^2).$$

There are two factors, one a binomial and the other a trinomial. *The binomial is formed by taking the cube root of each term in the original expression and placing between them the same sign that is between the two terms in the original expression.*

The trinomial factor is formed from the binomial factor and can be determined by the following rule: *Square the two terms in the binomial factor to form the first and third terms in the trinomial factor, both of which will be positive. The second term of the trinomial factor is the negative of the product of the two terms in the binomial factor.* The sign preceding the second term in the trinomial factor will therefore be the opposite of the sign between the two terms of the binomial factor.

49. Factor $a^3 - b^3$.

$$a^3 - b^3 = (a - b)(a^2 + ab + b^2).$$

50. Factor $m^3 + 8$.

This expression may be written as $m^3 + 2^3$.

$$\begin{aligned} m^3 + 8 &= m^3 + 2^3 \\ &= (m + 2)(m^2 - 2m + 4). \end{aligned}$$

51. Factor $64 - 125y^3$.

This expression may be written as $4^3 - (5y)^3$.

$$\begin{aligned} 64 - 125y^3 &= 4^3 - (5y)^3 \\ &= (4 - 5y)(16 + 20y + 25y^2). \end{aligned}$$

52. Factor $x^3y^3 + z^3$.

$$x^3y^3 + z^3 = (xy + z)(x^2y^2 - xyz + z^2).$$

53. Factor $16m^3 + 54n^3$.

Factor out the common factor 2.

$$\begin{aligned} 16m^3 + 54n^3 &= 2(8m^3 + 27n^3) \\ &= 2(2m + 3n)(4m^2 - 6mn + 9n^2). \end{aligned}$$

Grouping of Terms. In some types of expressions it is necessary to group certain terms together with parentheses and then use some of the other types of factoring on the new grouping. This is especially true if there are four or more terms in the expression.

$$\begin{aligned} mx + nx + my + ny &= (mx + nx) + (my + ny) \\ &= x(m + n) + y(m + n) \\ &= (m + n)(x + y). \end{aligned}$$

54. Factor $am + an + bm + bn$.

Group the first two terms together and the last two terms together.

$$am + an + bm + bn = (am + an) + (bm + bn)$$

In the first group, a is the common factor; in the second group, b is the common factor.

$$= a(m + n) + b(m + n)$$

Now we have two terms in the expression and $(m + n)$ is a common factor of both terms.

$$= (m + n)(a + b).$$

55. Factor $3x - 6 + ax - 2a$.

$$\begin{aligned} 3x - 6 + ax - 2a &= (3x - 6) + (ax - 2a) \\ &= 3(x - 2) + a(x - 2) \\ &= (x - 2)(3 + a). \end{aligned}$$

56. Factor $2ma - 2mx + 6a - 6x$.

Factor out the common factor 2.

$$\begin{aligned} 2ma - 2mx + 6a - 6x &= 2(ma - mx + 3a - 3x) \\ &= 2[(ma - mx) + (3a - 3x)] \\ &= 2[m(a - x) + 3(a - x)] \\ &= 2(a - x)(m + 3). \end{aligned}$$

57. Factor $3ax - 2by - 6ay + bx$.

Rearrange the terms of the expression.

$$\begin{aligned} 3ax - 2by - 6ay + bx &= 3ax - 6ay + bx - 2by \\ &= (3ax - 6ay) + (bx - 2by) \\ &= 3a(x - 2y) + b(x - 2y) \\ &= (x - 2y)(3a + b). \end{aligned}$$

58. Factor $2am - 4an - 2bn + bm$.

Rearrange the terms in the expression.

$$\begin{aligned} 2am - 4an - 2bn + bm &= 2am - 4an + bm - 2bn \\ &= (2am - 4an) + (bm - 2bn) \\ &= 2a(m - 2n) + b(m - 2n) \\ &= (m - 2n)(2a + b). \end{aligned}$$

59. Factor $x^2 - a^2 - 4a - 4$.

Group this expression as shown below. Remember that a negative sign placed before the parentheses changes all signs included within the parentheses.

$$\begin{aligned} x^2 - a^2 - 4a - 4 &= (x^2) - (a^2 + 4a + 4) \\ &= (x)^2 - (a + 2)^2 \end{aligned}$$

Now we have the difference of two perfect squares.

$$\begin{aligned} &= (x - [a + 2])(x + [a + 2]) \\ &= (x - a - 2)(x + a + 2). \end{aligned}$$

60. Factor $y^2 + 6y + 9 - x^2$.

$$\begin{aligned} y^2 + 6y + 9 - x^2 &= (y^2 + 6y + 9) - (x^2) \\ &= (y + 3)^2 - (x)^2 \\ &= (y + 3 + x)(y + 3 - x). \end{aligned}$$

61. Factor $x^2 - 4x + 4 - y^2 - 10y - 25$.

$$\begin{aligned} &x^2 - 4x + 4 - y^2 - 10y - 25 \\ &= (x^2 - 4x + 4) - (y^2 + 10y + 25) \\ &= (x - 2)^2 - (y + 5)^2 \\ &= ([x - 2] + [y + 5])([x - 2] - [y + 5]) \\ &= (x - 2 + y + 5)(x - 2 - y - 5) \end{aligned}$$

Combine similar terms.

$$= (x + y + 3)(x - y - 7).$$

Addition and Subtraction of Terms. It is usually assumed that the sum of two perfect squares cannot be factored. This is true in most cases. However, there are some cases in which, by a special device, such expressions can be factored.

Frequently in mathematics, in order to change an expression into a form that can be factored, we add to and subtract from the expression the same amount. If the amount to be added to and subtracted from the expression is a perfect square, the sum of two squares can be factored. The amount to be added to the expression must make it a perfect square trinomial. The second term of that perfect square trinomial is twice the product of the square roots of the two terms that form the sum of two squares.

$$\begin{aligned} x^4 + 4 &= x^4 + 4x^2 + 4 - 4x^2 \\ &= (x^2 + 2)^2 - (2x)^2 \\ &= [(x^2 + 2) + 2x][(x^2 + 2) - 2x] \\ &= (x^2 + 2x + 2)(x^2 - 2x + 2). \end{aligned}$$

62. Factor x^4+64.

Add $16x^2$ (twice the product of the square roots of the two terms in the given binomial) to the expression, using it as the second term of a trinomial. Then subtract $16x^2$ from the expression. Since the same amount has been both added to and subtracted from the expression, its value remains the same, though its form has changed.

$$\begin{aligned} x^4+64 &= x^4+16x^2+64-16x^2 \\ &= (x^4+16x^2+64)-(16x^2) \\ &= (x^2+8)^2-(4x)^2 \\ &= ([x^2+8]+4x)([x^2+8]-4x) \\ &= (x^2+8+4x)(x^2+8-4x) \\ &= (x^2+4x+8)(x^2-4x+8). \end{aligned}$$

63. Factor $x^4y^4+4z^4$.

$$\begin{aligned} x^4y^4+4z^4 &= x^4y^4+4x^2y^2z^2+4z^4-4x^2y^2z^2 \\ &= (x^4y^4+4x^2y^2z^2+4z^4)-(4x^2y^2z^2) \\ &= (x^2y^2+2z^2)^2-(2xyz)^2 \\ &= (x^2y^2+2xyz+2z^2)(x^2y^2-2xyz+2z^2). \end{aligned}$$

64. Factor x^8+4.

$$\begin{aligned} x^8+4 &= x^8+4x^4+4-4x^4 \\ &= (x^4+2)^2-(2x^2)^2 \\ &= (x^4+2+2x^2)(x^4+2-2x^2) \\ &= (x^4+2x^2+2)(x^4-2x^2+2). \end{aligned}$$

65. Factor $3x^4+12$.

$$\begin{aligned} 3x^4+12 &= 3(x^4+4) \\ &= 3(x^4+4x^2+4-4x^2) \\ &= 3([x^4+4x^2+4]-[4x^2]) \\ &= 3([x^2+2]^2-[2x]^2) \\ &= 3(x^2+2+2x)(x^2+2-2x) \\ &= 3(x^2+2x+2)(x^2-2x+2). \end{aligned}$$

It can be observed from the above problems that this device for factoring will work only when the amount to be added to and subtracted from the original expression is a perfect square.

SUPPLEMENTARY PROBLEMS

Find the product of the following.

1. $m(a-2b+c)$
2. $3c(4+3a-5b)$
3. $(d+m)(d-m)$
4. $(2b-3c)(2b+3c)$
5. $(2x+y)(2x+y)$
6. $(5k-3)(5k-3)$
7. $(x+3z)(2x+z)$
8. $(4m-5n)(3m+2n)$
9. $(m-2n)(m^2+2mn+4n^2)$
10. $(3+4d)(9-12d+16d^2)$
11. $(d-a)(m+n)$
12. $(2-z)(a-3y)$

Find the prime factors of the following.

13. $4ax-4ay+8az$
14. $-3bcm+6bcn-9bcp$
15. $50ax^2-98ay^2$
16. $12c^3d-75cd^3$
17. $x^2-8xy+16y^2$
18. $4x^2+20xz+25z^2$
19. $3m^2-24mp+48p^2$
20. $6p^2+7py-3y^2$
21. $12-d-6d^2$
22. $12m^2-29am+15a^2$
23. $6a^2+3ax-63x^2$
24. $2a^4b^3-a^3b^2c-6a^2bc^2$
25. $mn-dn+mc-dc$
26. $3b+3c-2ab-2ac$
27. $2pr-2px+2qr-2qx$
28. d^4+4c^4
29. x^4+4x^2+16
30. $4adm+16ad^5m$

Chapter 3

Fractions

Reducing Fractions. A fraction is in simplest form when it is reduced to lowest terms. This is done, as in arithmetic, by factoring fractions which are equal to 1 out of the original expressions.

$$\begin{aligned}\frac{a^2-b^2}{a+b} &= \frac{(a+b)(a-b)}{(a+b)}\\ &= \frac{a+b}{a+b}\cdot(a-b)\\ &= a-b.\end{aligned}$$

1. Reduce $\frac{4}{12}$ to lowest terms.

$$\frac{4}{12}=\frac{\overset{1}{\cancel{(2)}}\overset{1}{\cancel{(2)}}}{\underset{1}{\cancel{(2)}}\underset{1}{\cancel{(2)}}(3)}$$
$$=\frac{1}{3}.$$

2. Reduce $\frac{4x}{6x}$ to lowest terms.

$$\frac{4x}{6x}=\frac{\overset{1}{\cancel{(2)}}(2)\overset{1}{\cancel{(x)}}}{\underset{1}{\cancel{(2)}}(3)\underset{1}{\cancel{(x)}}}$$
$$=\frac{2}{3}.$$

3. Reduce $\frac{3x^2}{x}$ to lowest terms.

$$\frac{3x^2}{x}=\frac{(3)(x)\overset{1}{\cancel{(x)}}}{\underset{1}{\cancel{(x)}}}$$
$$=3x.$$

4. Reduce $\frac{15a^3c^2}{5abc}$ to lowest terms.

$$\frac{15a^3c^2}{5abc}=\frac{\overset{1}{\cancel{(5)}}(3)\overset{1}{\cancel{(a)}}(a)(a)(c)\overset{1}{\cancel{(c)}}}{\underset{1}{\cancel{(5)}}\underset{1}{\cancel{(a)}}(b)\underset{1}{\cancel{(c)}}}$$
$$\frac{3a^2c}{b}$$

It is not necessary to place the 1 above those factors that have been divided out in the numerator nor below those divided out in the denominator. However, if all factors in a numerator or denominator have been divided out, it must be remembered that the value of that numerator or denominator is 1.

5. Reduce $\frac{18mn^2}{10mn^3}$ to lowest terms.

$$\frac{18mn^2}{10mn^3}=\frac{(3)(3)\cancel{(2)}\cancel{(m)}\cancel{(n)}\cancel{(n)}}{(5)\underset{1}{\cancel{(2)}}\underset{1}{\cancel{(m)}}\underset{1}{\cancel{(n)}}\underset{1}{\cancel{(n)}}(n)}$$
$$=\frac{9}{5n}.$$

6. Reduce $\frac{a^2+a}{a+1}$ to lowest terms.

$$\frac{a^2+a}{a+1}=\frac{a\cancel{(a+1)}}{\cancel{(a+1)}}$$
$$=a.$$

7. Reduce $\frac{m^2-n^2}{m+n}$ to lowest terms.

$$\frac{m^2-n^2}{m+n}=\frac{\cancel{(m+n)}(m-n)}{\cancel{(m+n)}}$$
$$=m-n.$$

8. Reduce $\frac{4(x+b)}{2x+2b}$ to lowest terms.

$$\frac{4(x+b)}{2x+2b}=\frac{(2)\cancel{(2)}\cancel{(x+b)}}{\cancel{(2)}\cancel{(x+b)}}$$
$$=2.$$

9. Reduce $\frac{m^4-n^4}{m^2+n^2}$ to lowest terms.

$$\frac{m^4-n^4}{m^2+n^2}=\frac{\cancel{(m^2+n^2)}(m^2-n^2)}{\cancel{(m^2+n^2)}}$$
$$=m^2-n^2$$
$$=(m+n)(m-n).$$

10. Reduce $\frac{a^2+4a+4}{a+2}$ to lowest terms.

$$\frac{a^2+4a+4}{a+2}=\frac{(a+2)\cancel{(a+2)}}{\cancel{(a+2)}}$$
$$=a+2.$$

11. Reduce $\frac{x^3 + y^3}{x + y}$ to lowest terms.

$$\frac{x^3 + y^3}{x + y} = \frac{\cancel{(x + y)}(x^2 - xy + y^2)}{\cancel{(x + y)}}$$
$$= x^2 - xy + y^2.$$

12. Reduce $\frac{3x^2 - 15x + 18}{3x - 9}$ to lowest terms.

$$\frac{3x^2 - 15x + 18}{3x - 9} = \frac{3(x^2 - 5x + 6)}{3(x - 3)}$$
$$= \frac{\cancel{3(x - 3)}(x - 2)}{\cancel{3(x - 3)}}$$
$$= x - 2.$$

13. Reduce $\frac{m^2 - 4m - 21}{2m + 6}$ to lowest terms.

$$\frac{m^2 - 4m - 21}{2m + 6} = \frac{\cancel{(m + 3)}(m - 7)}{2\cancel{(m + 3)}}$$
$$= \frac{m - 7}{2}.$$

14. Reduce $\frac{6a^2 + 7a - 3}{3ax - x}$ to lowest terms.

$$\frac{6a^2 + 7a - 3}{3ax - x} = \frac{\cancel{(3a - 1)}(2a + 3)}{x\cancel{(3a - 1)}}$$
$$= \frac{2a + 3}{x}.$$

15. Reduce $\frac{2m^2 + 13m + 15}{m^2 + 6m + 5}$ to lowest terms.

$$\frac{2m^2 + 13m + 15}{m^2 + 6m + 5} = \frac{\cancel{(m + 5)}(2m + 3)}{\cancel{(m + 5)}(m + 1)}$$
$$= \frac{2m + 3}{m + 1}.$$

16. Reduce $\frac{2x^2y^2 + 5xy - 3}{x^2y^2 - 9}$ to lowest terms.

$$\frac{2x^2y^2 + 5xy - 3}{x^2y^2 - 9} = \frac{(2xy - 1)\cancel{(xy + 3)}}{(xy - 3)\cancel{(xy + 3)}}$$
$$= \frac{2xy - 1}{xy - 3}.$$

17. Reduce $\frac{4m^2 + 2mn - 6n^2}{4m^2 - 4n^2}$ to lowest terms.

$$\frac{4m^2 + 2mn - 6n^2}{4m^2 - 4n^2} = \frac{2(2m^2 + mn - 3n^2)}{4(m^2 - n^2)}$$
$$= \frac{\cancel{2(m - n)}(2m + 3n)}{\underset{2}{\cancel{4(m - n)}}(m + n)}$$
$$= \frac{2m + 3n}{2(m + n)}.$$

18. Reduce $\frac{a^2 + 11a + 24}{a^3 + 27}$ to lowest terms.

$$\frac{a^2 + 11a + 24}{a^3 + 27} = \frac{\cancel{(a + 3)}(a + 8)}{\cancel{(a + 3)}(a^2 - 3a + 9)}$$
$$= \frac{a + 8}{a^2 - 3a + 9}.$$

19. Reduce $\frac{ax + ay + bx + by}{x + y}$ to lowest terms.

$$\frac{ax + ay + bx + by}{x + y} = \frac{a(x + y) + b(x + y)}{x + y}$$
$$= \frac{\cancel{(x + y)}(a + b)}{\cancel{(x + y)}}$$
$$= a + b.$$

20. Reduce $\frac{3x - 6 + ax - 2a}{x - 2}$ to lowest terms.

$$\frac{3x - 6 + ax - 2a}{x - 2} = \frac{3(x - 2) + a(x - 2)}{(x - 2)}$$
$$= \frac{\cancel{(x - 2)}(3 + a)}{\cancel{(x - 2)}}$$
$$= 3 + a.$$

Multiplication of Fractions. The multiplication of algebraic fractions is similar to the multiplication of arithmetic fractions. The numerators are multiplied to form the numerator of the product and the denominators are multiplied to form the denominator of the product. The product must be stated in simplest form. Often the product becomes so complex that its simplification is difficult. In order to simplify the entire process of multiplication, we convert each numerator and each denominator into factored form. Then when similar factors appear in any denominator and any numerator, we have a fraction which is equal to one, and we may "cancel". This process leads to a product in lowest terms.

$$\frac{a^2 - b^2}{ab^2} \cdot \frac{b}{a + b} = \frac{\cancel{(a + b)}(a - b)}{(a)(b)\cancel{(b)}} \cdot \frac{\cancel{b}}{\cancel{(a + b)}}$$
$$= \frac{a - b}{ab}.$$

21. Multiply $\frac{x^2 + 2x + 1}{x + 2}$ by $\frac{x^2 + 4x + 4}{x + 1}$.

$$\frac{x^2 + 2x + 1}{x + 2} \cdot \frac{x^2 + 4x + 4}{x + 1} = \frac{\cancel{(x + 1)}(x + 1)}{\cancel{(x + 2)}} \cdot \frac{\cancel{(x + 2)}(x + 2)}{\cancel{(x + 1)}}$$
$$= (x + 1)(x + 2)$$
$$= x^2 + 3x + 2.$$

22. Multiply $\frac{x^2+3x}{x^2+5x+6}$ by $\frac{x+2}{x+1}$.

$$\frac{x^2+3x}{x^2+5x+6}\cdot\frac{x+2}{x+1}=\frac{x\cancel{(x+3)}}{\cancel{(x+2)}\cancel{(x+3)}}\cdot\frac{\cancel{(x+2)}}{(x+1)}$$
$$=\frac{x}{x+1}.$$

23. Multiply $\frac{a^3-y^3}{a-y}$ by $\frac{a+y}{a^2+ay+y^2}$.

$$\frac{a^3-y^3}{a-y}\cdot\frac{a+y}{a^2+ay+y^2}=\frac{\cancel{(a-y)}\cancel{(a^2+ay+y^2)}}{\cancel{(a-y)}}\cdot\frac{(a+y)}{\cancel{(a^2+ay+y^2)}}$$
$$=a+y.$$

24. Multiply $\frac{ax+ay}{x^2+2x+1}$ by $\frac{ax+a}{x+y}$.

$$\frac{ax+ay}{x^2+2x+1}\cdot\frac{ax+a}{x+y}=\frac{a\cancel{(x+y)}}{(x+1)\cancel{(x+1)}}\cdot\frac{a\cancel{(x+1)}}{\cancel{(x+y)}}$$
$$=\frac{a^2}{x+1}.$$

25. Multiply $\frac{ax+ay-bx-by}{x+y}$ by $\frac{5x}{a-b}$.

$$\frac{ax+ay-bx-by}{x+y}\cdot\frac{5x}{a-b}=\frac{a(x+y)-b(x+y)}{x+y}\cdot\frac{5x}{a-b}$$
$$=\frac{\cancel{(x+y)}\cancel{(a-b)}}{\cancel{(x+y)}}\cdot\frac{5x}{\cancel{(a-b)}}$$
$$=5x.$$

26. Multiply $\frac{x^2-5x-24}{x+3}$ by $\frac{2}{8-x}$.

$$\frac{x^2-5x-24}{x+3}\cdot\frac{2}{8-x}=\frac{(x+3)(x-8)}{(x+3)}\cdot\frac{2}{8-x}$$

Change all signs in both the numerator and denominator of the second factor.

$$=\frac{\cancel{(x+3)}\cancel{(x-8)}}{\cancel{(x+3)}}\cdot\frac{-2}{\cancel{(x-8)}}$$
$$=-2.$$

27. Multiply $\frac{x^3-y^3}{x^2-2x-15}$ by $\frac{x+3}{y-x}$.

$$\frac{x^3-y^3}{x^2-2x-15}\cdot\frac{x+3}{y-x}=\frac{(x-y)(x^2+xy+y^2)}{(x-5)(x+3)}\cdot\frac{(x+3)}{(y-x)}$$

Change the sign of the second factor and the signs of all terms in the denominator of the second factor.

$$=\frac{\cancel{(x-y)}(x^2+xy+y^2)}{(x-5)\cancel{(x+3)}}\cdot(-)\frac{\cancel{(x+3)}}{\cancel{(x-y)}}$$
$$=-\frac{x^2+xy+y^2}{x-5}\text{ or }\frac{x^2+xy+y^2}{5-x}.$$

28. Multiply $\frac{3ab^2-27ax^2}{3}$ by $\frac{1}{ab+3ax}$.

$$\frac{3ab^2-27ax^2}{3}\cdot\frac{1}{ab+3ax}=\frac{\cancel{3a}(b-3x)\cancel{(b+3x)}}{\cancel{3}}\cdot\frac{1}{\cancel{a(b+3x)}}$$
$$=b-3x.$$

29. Multiply $\frac{x^2-144}{x^2-14x+24}$ by $\frac{x-2}{x^2+14x+24}$.

$$\frac{x^2-144}{x^2-14x+24}\cdot\frac{x-2}{x^2+14x+24}$$
$$=\frac{\cancel{(x-12)}\cancel{(x+12)}}{\cancel{(x-12)}\cancel{(x-2)}}\cdot\frac{\cancel{(x-2)}}{\cancel{(x+12)}(x+2)}$$
$$=\frac{1}{x+2}.$$

30. Multiply $\frac{3m^2-8mn-3n^2}{m^2-n^2}$ by $\frac{m^2+mn-2n^2}{3m^2+7mn+2n^2}$.

$$\frac{3m^2-8mn-3n^2}{m^2-n^2}\cdot\frac{m^2+mn-2n^2}{3m^2+7mn+2n^2}$$
$$=\frac{\cancel{(3m+n)}(m-3n)}{\cancel{(m-n)}(m+n)}\cdot\frac{\cancel{(m+2n)}\cancel{(m-n)}}{\cancel{(3m+n)}\cancel{(m+2n)}}$$
$$=\frac{m-3n}{m+n}.$$

Division of Fractions. When dividing fractions in arithmetic, the divisor is inverted and then the problem becomes one of multiplication; for division by a number is the same as multiplication by its reciprocal. The same procedure is used in the division of algebraic fractions. After the divisor has been inverted, the problem becomes one that is solved exactly as those in the preceding group.

$$\frac{x^2-y^2}{x+1}\div\frac{x-y}{x^2-1}=\frac{x^2-y^2}{x+1}\cdot\frac{x^2-1}{x-y}$$
$$=\frac{(x+y)\cancel{(x-y)}}{\cancel{(x+1)}}\cdot\frac{\cancel{(x+1)}(x-1)}{\cancel{(x-y)}}$$
$$=(x+y)(x-1).$$

31. Divide $\frac{15y}{4}$ by $\frac{3y}{x}$.

$$\frac{15y}{4}\div\frac{3y}{x}=\frac{\overset{5}{\cancel{15y}}}{4}\cdot\frac{x}{\cancel{3y}}$$
$$=\frac{5x}{4}.$$

32. Divide $-\frac{12x^2}{3a}$ by $\frac{6x}{a^2}$.

$$-\frac{12x^2}{3a}\div\frac{6x}{a^2}=-\frac{\overset{2x}{\cancel{12x^2}}}{\underset{3}{\cancel{3a}}}\cdot\frac{\overset{a}{\cancel{a^2}}}{\cancel{6x}}$$
$$=-\frac{2ax}{3}.$$

33. Divide $\dfrac{x^2 - y^2}{x + y}$ by $\dfrac{x - y}{a}$.

$$\frac{x^2 - y^2}{x + y} \div \frac{x - y}{a} = \frac{x^2 - y^2}{x + y} \cdot \frac{a}{x - y}$$
$$= \frac{(x - y)(x + y)}{(x + y)} \cdot \frac{a}{(x - y)}$$
$$= a.$$

34. Divide $\dfrac{x^2 + 7x + 12}{x^2 + x - 12}$ by $\dfrac{x^2 + 3x + 9}{x^3 - 27}$.

$$\frac{x^2 + 7x + 12}{x^2 + x - 12} \div \frac{x^2 + 3x + 9}{x^3 - 27}$$
$$= \frac{(x + 4)(x + 3)}{(x + 4)(x - 3)} \div \frac{(x^2 + 3x + 9)}{(x - 3)(x^2 + 3x + 9)}$$
$$= \frac{(x + 4)(x + 3)}{(x + 4)(x - 3)} \cdot \frac{(x - 3)(x^2 + 3x + 9)}{(x^2 + 3x + 9)}$$
$$= x + 3.$$

35. Divide $\dfrac{am^2 - an^2}{bx + by}$ by $\dfrac{am + an}{b^2 + cb}$.

$$\frac{am^2 - an^2}{bx + by} \div \frac{am + an}{b^2 + cb} = \frac{a(m + n)(m - n)}{b(x + y)} \div \frac{a(m + n)}{b(b + c)}$$
$$= \frac{a(m + n)(m - n)}{b(x + y)} \cdot \frac{b(b + c)}{a(m + n)}$$
$$= \frac{(m - n)(b + c)}{x + y}.$$

36. Divide $\dfrac{8x^2 - 2xy - y^2}{8x^2 + 14xy + 3y^2}$ by $\dfrac{4x - 2y}{6x^2 + 13xy + 6y^2}$.

$$\frac{8x^2 - 2xy - y^2}{8x^2 + 14xy + 3y^2} \div \frac{4x - 2y}{6x^2 + 13xy + 6y^2}$$
$$= \frac{(4x + y)(2x - y)}{(4x + y)(2x + 3y)} \cdot \frac{(2x + 3y)(3x + 2y)}{2(2x - y)}$$
$$= \frac{3x + 2y}{2}$$

37. Divide $\dfrac{a^2 - 3ab + 2b^2}{ax + ay - bx - by}$ by $\dfrac{a - 2b}{x + y}$.

$$\frac{a^2 - 3ab + 2b^2}{ax + ay - bx - by} \div \frac{a - 2b}{x + y} = \frac{(a - 2b)(a - b)}{(x + y)(a - b)} \cdot \frac{(x + y)}{(a - 2b)}$$
$$= 1.$$

38. Divide $\dfrac{x^2 - y^2}{a + b} \div \dfrac{x + y}{a - b}$ by $\dfrac{x - y}{a + b}$.

In this problem, two fractions must be inverted.

$$\left[\frac{x^2 - y^2}{a + b} \div \frac{x + y}{a - b}\right] \div \frac{x - y}{a + b}$$
$$= \left[\frac{(x - y)(x + y)}{(a + b)} \div \frac{(x + y)}{(a - b)}\right] \div \frac{(x - y)}{(a + b)}$$
$$= \left[\frac{(x - y)(x + y)}{(a + b)} \cdot \frac{(a - b)}{(x + y)}\right] \cdot \frac{(a + b)}{(x - y)}$$
$$= a - b.$$

Lowest Common Multiple. The lowest common multiple (abbreviated L.C.M.) of a group of algebraic expressions with integral coefficients is the polynomial of lowest degree of which each of those given expressions is a factor. To find the L.C.M. of a group of algebraic expressions we first convert each of them into a product of its prime factors. Then the L.C.M. is the product of all the different prime factors, using each one the greatest number of times it occurs in any one of the given expressions.

Find the L.C.M. of $x - y$, $x^2 - xy$, and $x^2 - y^2$.

$$x - y = (x - y).$$
$$x^2 - xy = x(x - y).$$
$$x^2 - y^2 = (x - y)(x + y).$$
$$\text{L.C.M.} = x(x - y)(x + y).$$

39. Find the L.C.M. of $a + b$, $a^2 + 2ab + b^2$, and $4a + 4b$.

$$a + b = (a + b).$$
$$a^2 + 2ab + b^2 = (a + b)(a + b).$$
$$4a + 4b = 4(a + b).$$
$$\text{L.C.M.} = 4(a + b)(a + b).$$

40. Find the L.C.M. of $ax + bx - ay - by$, $a^2 + ab$, $a^3 + b^3$, and $x^2 - y^2$.

$$ax + bx - ay - by = (a + b)(x - y).$$
$$a^2 + ab = a(a + b).$$
$$a^3 + b^3 = (a + b)(a^2 - ab + b^2).$$
$$x^2 - y^2 = (x - y)(x + y).$$
$$\text{L.C.M.} = a(a + b)(a^2 - ab + b^2)(x - y)(x + y).$$

41. Find the L.C.M. of $x^2 - x - 12$, $x^2 + 6x + 9$, and $x^2 - 6x + 8$.

$$x^2 - x - 12 = (x - 4)(x + 3).$$
$$x^2 + 6x + 9 = (x + 3)(x + 3).$$
$$x^2 - 6x + 8 = (x - 4)(x - 2).$$
$$\text{L.C.M.} = (x - 4)(x - 2)(x + 3)(x + 3).$$

Addition and Subtraction of Fractions. When adding or subtracting fractions the denominator of all such fractions must be identical. When such is not the case, we find the lowest common denominator (the L.C.M. of all the given denominators) and change each of the given fractions to a new fraction whose denominator is that lowest common multiple. Any additional factor in the denominator of each of those new fractions must also be a new factor in the numerator of that fraction. This does not change the value of the original fraction; for to multiply both the numerator and the denominator of a fraction by the same number is to multiply the fraction by one. Thus, for example, $\dfrac{x}{y} \cdot \dfrac{a}{a} = \dfrac{x}{y} \cdot 1$. The sum or difference (as the case may be) of the fractions thus obtained will be a single fraction whose numerator is the sum or difference of all of the numerators, and whose denominator is the lowest common denominator. If the fraction is reducible it should be converted into its simplest form.

Add: $\frac{1}{x-y}+\frac{2}{x+y}$.

$x-y=(x-y)$.

$x+y=(x+y)$.

L.C.M. $=(x-y)(x+y)$.

$$\frac{1}{x-y}+\frac{2}{x+y}=\frac{1(x+y)}{(x-y)(x+y)}+\frac{2(x-y)}{(x-y)(x+y)}$$

$$=\frac{x+y+2x-2y}{(x-y)(x+y)}$$

$$=\frac{3x-y}{(x-y)(x+y)}.$$

42. Add: $\frac{x-y}{x+y}+\frac{x+y}{x-y}$.

$$\frac{x-y}{x+y}+\frac{x+y}{x-y}=\frac{(x-y)(x-y)}{(x+y)(x-y)}+\frac{(x+y)(x+y)}{(x+y)(x-y)}$$

$$=\frac{x^2-2xy+y^2+x^2+2xy+y^2}{(x+y)(x-y)}$$

$$=\frac{2x^2+2y^2}{x^2-y^2}.$$

43. Subtract: $\frac{x+y}{x-y}-\frac{x-y}{x+y}$.

$$\frac{x+y}{x-y}-\frac{x-y}{x+y}=\frac{(x+y)(x+y)}{(x-y)(x+y)}-\frac{(x-y)(x-y)}{(x-y)(x+y)}$$

$$=\frac{x^2+2xy+y^2-x^2+2xy-y^2}{(x-y)(x+y)}$$

Note that the signs of the terms in the product $(x-y)(x-y)$ were changed. This is consistent with the rule which requires that the signs of the subtrahend be changed when subtracting.

$$=\frac{4xy}{(x-y)(x+y)}.$$

44. Subtract $\frac{4x}{x^2+x}$ from $\frac{7x}{x+1}$.

$x^2+x=x(x+1)$.

$x+1=(x+1)$.

L.C.M. $=x(x+1)$.

$$\frac{7x}{x+1}-\frac{4x}{x^2+x}=\frac{(7x)(x)}{x(x+1)}-\frac{4x}{x(x+1)}$$

$$=\frac{7x^2-4x}{x(x+1)}$$

$$=\frac{\cancel{x}(7x-4)}{\cancel{x}(x+1)}$$

$$=\frac{7x-4}{x+1}.$$

45. Simplify $\frac{3}{m-n}+\frac{1}{m+n}-\frac{m+n}{m^2-2mn+n^2}$.

$m-n=(m-n)$.

$m+n=(m+n)$.

$m^2-2mn+n^2=(m-n)(m-n)$.

L.C.M. $=(m-n)(m-n)(m+n)$.

$$\frac{3}{m-n}+\frac{1}{m+n}-\frac{m+n}{m^2-2mn+n^2}$$

$$=\frac{3(m-n)(m+n)}{(m-n)(m-n)(m+n)}+\frac{1(m-n)(m-n)}{(m-n)(m-n)(m+n)}-\frac{(m+n)(m+n)}{(m-n)(m-n)(m+n)}$$

$$=\frac{3m^2-3n^2+m^2-2mn+n^2-m^2-2mn-n^2}{(m-n)(m-n)(m+n)}$$

$$=\frac{3m^2-4mn-3n^2}{(m-n)(m-n)(m+n)}.$$

46. Simplify $\frac{x+1}{x-1}+\frac{3}{x^2-2x+1}-\frac{3x+4}{x^2+x-2}$.

$x-1=(x-1)$.

$x^2-2x+1=(x-1)(x-1)$.

$x^2+x-2=(x-1)(x+2)$.

L.C.M. $=(x-1)(x-1)(x+2)$.

$$=\frac{x+1}{x-1}+\frac{3}{x^2-2x+1}-\frac{3x+4}{x^2+x-2}$$

$$=\frac{(x+1)(x-1)(x+2)}{(x-1)(x-1)(x+2)}+\frac{3(x+2)}{(x-1)(x-1)(x+2)}-\frac{(3x+4)(x-1)}{(x-1)(x-1)(x+2)}$$

$$=\frac{x^3+2x^2-x-2+3x+6-3x^2-x+4}{(x-1)(x-1)(x+2)}$$

$$=\frac{x^3-x^2+x+8}{(x-1)(x-1)(x+2)}.$$

Complex Fractions. A complex fraction is a fraction whose numerator or denominator (or both) is a fraction. To simplify a complex fraction, divide the fractional numerator by the fractional denominator.

$$\frac{\frac{a+b}{a-b}}{\frac{a+b}{a}}=\frac{a+b}{a-b}\div\frac{a+b}{a}$$

$$=\frac{\cancel{(a+b)}}{(a-b)}\cdot\frac{a}{\cancel{(a+b)}}$$

$$=\frac{a}{a-b}.$$

If either the numerator or denominator consists of more than one term it must be converted into a form involving one fraction. This requires finding a common denominator and making a single fraction out of the numerous terms involved.

47. Simplify $\frac{1-\frac{1}{\frac{1}{x}}}{1-x^2}$.

Change the second term of the numerator.

$$\frac{1}{\frac{1}{x}}=1\cdot\frac{x}{1}=x.$$

$$\frac{1-\frac{1}{\frac{1}{x}}}{1-x^2} = \frac{1-x}{1-x^2}$$

$$= \frac{\cancel{(1-x)}}{\cancel{(1-x)}(1+x)}$$

$$= \frac{1}{1+x}.$$

48. Simplify $\dfrac{3+\dfrac{a}{a+b}}{\dfrac{a}{a-b}-3}$.

$$\frac{3+\dfrac{a}{a+b}}{\dfrac{a}{a-b}-3} = \frac{\dfrac{3(a+b)}{a+b}+\dfrac{a}{a+b}}{\dfrac{a}{a-b}-\dfrac{3(a-b)}{a-b}}$$

$$= \frac{\dfrac{3a+3b+a}{a+b}}{\dfrac{a-3a+3b}{a-b}}$$

$$= \frac{4a+3b}{a+b}\cdot\frac{a-b}{3b-2a}$$

$$= \frac{(4a+3b)(a-b)}{(a+b)(3b-2a)}$$

$$= \frac{4a^2-4ab+3ab-3b^2}{3ab-2a^2+3b^2-2ab}$$

$$= \frac{4a^2-ab-3b^2}{-2a^2+ab+3b^2}$$

$$= -\frac{4a^2-ab-3b^2}{2a^2-ab-3b^2}.$$

49. Simplify $\dfrac{1-\dfrac{1}{a-\dfrac{1}{b}}}{1+\dfrac{1}{a-\dfrac{1}{b}}}$.

Combine the $\left(a-\frac{1}{b}\right)$ in each part into a single fraction.

$$\frac{1-\dfrac{1}{a-\dfrac{1}{b}}}{1+\dfrac{1}{a-\dfrac{1}{b}}} = \frac{1-\dfrac{1}{\dfrac{ab}{b}-\dfrac{1}{b}}}{1+\dfrac{1}{\dfrac{ab}{b}-\dfrac{1}{b}}}$$

$$= \frac{1-\dfrac{1}{\dfrac{ab-1}{b}}}{1+\dfrac{1}{\dfrac{ab-1}{b}}}$$

$$= \frac{1-\dfrac{b}{ab-1}}{1+\dfrac{b}{ab-1}}$$

$$= \frac{\dfrac{ab-1}{ab-1}-\dfrac{b}{ab-1}}{\dfrac{ab-1}{ab-1}+\dfrac{b}{ab-1}}$$

$$= \frac{\dfrac{ab-1-b}{ab-1}}{\dfrac{ab-1+b}{ab-1}}$$

$$= \frac{ab-1-b}{\cancel{(ab-1)}}\cdot\frac{\cancel{(ab-1)}}{ab-1+b}$$

$$= \frac{ab-1-b}{ab-1+b}.$$

50. Simplify $\dfrac{\dfrac{1}{m}-\dfrac{1}{m+\dfrac{1}{m}}}{\dfrac{1}{m+\dfrac{1}{m}}-\dfrac{1}{m}}$.

$$\frac{\dfrac{1}{m}-\dfrac{1}{m+\dfrac{1}{m}}}{\dfrac{1}{m+\dfrac{1}{m}}-\dfrac{1}{m}} = \frac{\dfrac{1}{m}-\dfrac{1}{\dfrac{m^2+1}{m}}}{\dfrac{1}{\dfrac{m^2+1}{m}}-\dfrac{1}{m}}$$

$$= \frac{\dfrac{1}{m}-\dfrac{m}{m^2+1}}{\dfrac{m}{m^2+1}-\dfrac{1}{m}}$$

$$= \frac{\dfrac{1(m^2+1)}{m(m^2+1)}-\dfrac{m^2}{m(m^2+1)}}{\dfrac{m^2}{m(m^2+1)}-\dfrac{1(m^2+1)}{m(m^2+1)}}$$

$$= \frac{\dfrac{m^2+1-m^2}{m(m^2+1)}}{\dfrac{m^2-m^2-1}{m(m^2+1)}}$$

$$= \frac{\dfrac{1}{m(m^2+1)}}{\dfrac{-1}{m(m^2+1)}}$$

$$= \frac{1}{\cancel{m(m^2+1)}}\cdot\frac{\cancel{m(m^2+1)}}{-1}$$

$$= -1.$$

SUPPLEMENTARY PROBLEMS

Reduce the following fractions.

1. $\dfrac{4x^2 - 5x - 6}{2x^2 - 4x}$

2. $\dfrac{3m^2 + 25m + 28}{m + 7}$

3. $\dfrac{6r^2 + 13rs + 6s^2}{2r + 3s}$

4. $\dfrac{36x^2 + 38xy - 24y^2}{4x + 6y}$

5. $\dfrac{x^3 + 27}{x + 3}$

6. $\dfrac{am - bm + 2an - 2bn}{2a - 2b}$

Find the lowest common multiple for each of the following groups of expressions.

7. $x^2 - y^2,\ x^2 - 2xy + y^2,\ 3x + 3y$
8. $x^2 - x - 20,\ x^3 + 4x^2,\ x^2 - 7x + 10$
9. $am - bm - an + bn,\ m^2 - 2mn + n^2,\ a^2 - 2ab + b^2$
10. $s^3 + 125,\ s^2 + 4s - 5,\ s^2 - 2s + 1$

Perform the indicated operations and simplify.

11. $\dfrac{t^2 + t - 20}{t^2 + 8t + 15} \cdot \dfrac{2t^2 + 9t + 9}{2t^2 - 5t - 12}$

12. $\dfrac{6p^2 + 25p - 9}{12p^2 + 41p - 15} \cdot \dfrac{4p^2 + 3p - 45}{12p^2 + 64p + 45}$

13. $\dfrac{6m^2 - m - 15}{4m^2 - 9} \div \dfrac{6m^2 - 19m + 15}{4m^2 - 12m + 9}$

14. $\dfrac{3ac^2 - 12ad^2}{12ac^2 - 18acd - 12ad^2} \div \dfrac{c^2 - 4d^2}{8c^2 - 2cd - 3d^2}$

15. $\dfrac{1 - b^3}{4b + 12} \div \dfrac{1 + b + b^2}{2b + 3} \cdot \dfrac{b^2 + 2b - 3}{b^2 - 2b + 1}$

16. $\dfrac{x + y}{x - y} + \dfrac{2x + 3y}{x + y}$

17. $\dfrac{r^2 - 2r + 1}{r^2 - 1} + \dfrac{r^2 - 1}{1 - r}$

18. $\dfrac{2c + d}{c + d} - \dfrac{3c - 2d}{c - 2d}$

19. $\dfrac{3m - 3n}{m^2 + mn} - \dfrac{6}{m - n}$

20. $\dfrac{a + b}{a - b} + \dfrac{a - b}{a + b} - \dfrac{2a^3 - 2b^3}{a^3 - ab^2}$

21. $\dfrac{\dfrac{1}{x - y}}{1 - \dfrac{1}{x - y}}$

22. $\dfrac{a - \dfrac{b}{a}}{\dfrac{b}{a} - 3}$

23. $\dfrac{\dfrac{1}{x}}{\dfrac{\dfrac{1}{x - y}}{\dfrac{x - y}{x}}}$

24. $\dfrac{a - b + \dfrac{1}{b}}{\dfrac{1}{a} + b - \dfrac{b^2}{a}}$

25. $\dfrac{\dfrac{a^2}{a - b} - a - b}{\dfrac{b^2}{a - b}}$

Chapter 4

Linear Equations

Equations. An equation is a statement that two numerals or symbols represent the same number. Equality is shown by the use of the equals symbol (=). That part of the equation lying to the left of the equals symbol is called the **left member** of the equation while the part to the right of the equals symbol is called the **right member** of the equation. For example, in the equation $\frac{1}{2} = \frac{4}{8}$, $\frac{1}{2}$ is the left member, $\frac{4}{8}$ is the right member, and '$\frac{1}{2}$' and '$\frac{4}{8}$' are names for the same number.

Variables. In some equations, the number or numbers named by the terms' literal coefficients are not known. Thus, for example, in $4x = 20$, or in $4x < 20$, we do not know what number(s) are named by x. Therefore, x is called a variable; for it can have a variety of values. And, depending on the value assigned to x, the mathematical sentence will be true or false. Thus, for example, if x is given the value 5, $4x = 20$ (or, $4 \times 5 = 20$) is true.

Solution Sets. The set of values of a variable which make a mathematical statement true is known as the solution set of that statement. Thus, for example, the solution set of $4x = 20$ is $\{5\}$; and the solution set of $4x < 20$ is $\{\ldots, -1, 0, 1, 2, 3, 4\}$ if the integers constitute the **replacement set** (that universal set from which you are to find values of a variable which will make a given statement true).

Linear Equations in One Variable. A **linear equation** is an equation in which the variable or variables are raised to the first power. For example, $x = 2$ and $x + y + 3 = 0$ are linear equations. A linear equation in one variable may be solved for the value of that variable that makes the statement of equality true. That value is known as the **root** of the equation, and is the solution set of that equation. In the example, $2x = 8$, the solution set is $\{4\}$.

The following axioms will be used in solving all types of equations.

1. **The Addition Axiom.** If equals are added to equals, the sums are equal.
2. **The Subtraction Axiom.** If equals are subtracted from equals, the remainders are equal.
3. **The Multiplication Axiom.** If equals are multiplied by equals, the products are equal.
4. **The Division Axiom.** If equals are divided by equals, the quotients are equal (provided the divisor is not zero).
5. Quantities equal to the same or equal quantities are equal to each other.
6. Like powers of equals are equal.
7. Like roots of equals are equal.

To employ set notation in the solution of equations, we use the set-builder form explained in Chapter 1. Thus, for example, if we wish to find the solution set (or, the root) of the equation $x + 3 = 4$, we would proceed as follows.

$$\{x \mid x + 3 = 4\}.$$

By use of the subtraction axiom, 3 can be subtracted from both members of the equation. Therefore,

$$\{x \mid x + 3 = 4\} = \{x \mid x = 1\}.$$

Therefore, $\{1\}$ is the solution set of the equation. We can check the accuracy of our work by substituting the element of the solution set for the variable; and, in this way, determine whether or not the value satisfies the equation. Thus,

$$x + 3 = 4.$$
$$1 + 3 = 4.$$
$$4 = 4.$$

In solving the following equation, we must make use of several of the axioms mentioned above.

$$\left\{x \mid 3(x - 6) = \frac{3x}{4} + 9\right\}.$$

Remove the parentheses, and we have

$$\left\{x \mid 3(x - 6) = \frac{3x}{4} + 9\right\} = \left\{x \mid 3x - 18 = \frac{3x}{4} + 9\right\}.$$

By use of the subtraction axiom this becomes

$$\left\{x \mid 3x - \frac{3x}{4} - 18 = 9\right\}.$$

Using the addition axiom we get

$$\left\{x \mid 3x - \frac{3x}{4} = 27\right\}.$$

Combining the terms in the left member of the equation gives us

$$\left\{x \mid \frac{9x}{4} = 27\right\}.$$

By use of the multiplication axiom this becomes

$$\{x \mid 9x = 108\}.$$

Finally, using the division axiom, we get

$$\{x \mid x = 12\}.$$

The solution set is $\{12\}$.

To check our solution we substitute the member of the solution set for the variable each time it is found in the original equation. Thus,

$$3(x - 6) = \frac{3x}{4} + 9.$$

$$3(12 - 6) = \frac{3(12)}{4} + 9.$$

$$3(6) = \frac{36}{4} + 9.$$

$$18 = 9 + 9.$$

$$18 = 18.$$

These and similar equations may also be solved without the use of set structure as shown in the problems that follow.

Solve for m: $2m = 3 - m$.

Add m to both members of the equation (addition axiom).

$$3m = 3.$$

Divide both members of the equation by 3 (division axiom).

$$m = 1.$$

Solve for a: $\frac{2}{3}a = 1 + \frac{1}{3}a$.

Subtract $\frac{1}{3}a$ from each member of the equation (subtraction axiom).

$$\tfrac{1}{3}a = 1.$$

Multiply both members of the equation by 3 (multiplication axiom).

$$a = 3.$$

1. Solve for x: $x + x = 16$.

$$2x = 16.$$

Divide both members of the equation by 2.

$$x = 8.$$

Check the solution by substituting the root (the 8 in this case) in the original equation.

Check:

$$x + x = 16.$$
$$8 + 8 = 16.$$
$$16 = 16.$$

2. Solve for x: $3x - x = 12$.

$$2x = 12.$$

Divide both members of the equation by 2.

$$x = 6.$$

Check:

$$3x - x = 12.$$
$$3(6) - 6 = 12.$$
$$18 - 6 = 12.$$
$$12 = 12.$$

3. Solve for y: $3y = y + 6$.

Since the right member of the equation is to be free of the unknown for which we have been asked to solve, subtract y from each member of the equation.

$$2y = 6.$$
$$y = 3.$$

Check:

$$3y = y + 6.$$
$$3(3) = 3 + 6.$$
$$9 = 9.$$

4. Solve for m: $m - 4 = 12$.

Add 4 to both members of the equation.

$$m = 16.$$

Check:

$$m - 4 = 12.$$
$$16 - 4 = 12.$$
$$12 = 12.$$

5. Solve for a: $4a + 2 = a - 1$.

Subtract a from both members of the equation.

$$3a + 2 = -1.$$

Subtract 2 from both members of the equation.

$$3a = -3.$$

Divide both members of the equation by 3.

$$a = -1.$$

Check:

$$4a + 2 = a - 1.$$
$$4(-1) + 2 = -1 - 1.$$
$$-4 + 2 = -2.$$
$$-2 = -2.$$

6. Solve for x: $x - 5 = 1$.

Add 5 to each member of the equation.

$$x = 6.$$

Check:

$$x - 5 = 1.$$
$$6 - 5 = 1.$$
$$1 = 1.$$

7. Solve for c: $2c - 3 = 3 - c$.

$$3c = 6.$$

Divide both members of the equation by 3.

$$c = 2.$$

Check:

$$2c - 3 = 3 - c.$$
$$2(2) - 3 = 3 - 2.$$
$$4 - 3 = 3 - 2.$$
$$1 = 1.$$

8. Solve for x: $\frac{1}{3}x = 2$.

Multiply both members of the equation by 3.

$$x = 6.$$

Check:

$$\frac{1}{3}x = 2.$$
$$\frac{1}{3}(6) = 2.$$
$$2 = 2.$$

9. Solve for y: $\frac{1}{2}y - 2 = \frac{1}{4}y + 1$.

$$\frac{1}{4}y = 3.$$

Multiply both members of the equation by 4.

$$y = 12.$$

Check:

$$\frac{1}{2}y - 2 = \frac{1}{4}y + 1.$$
$$\frac{1}{2}(12) - 2 = \frac{1}{4}(12) + 1.$$
$$6 - 2 = 3 + 1.$$
$$4 = 4.$$

10. Solve for m: $3m - 2 + 2m = m + 10$.

$$4m = 12.$$
$$m = 3.$$

Check:

$$3m - 2 + 2m = m + 10.$$
$$3(3) - 2 + 2(3) = 3 + 10.$$
$$9 - 2 + 6 = 13.$$
$$13 = 13.$$

11. Solve for a: $a + 2 + (a - 3) = a - (4a - 14)$.

First remove the parentheses, observing the rules for such removal. Then solve as in the above problems.

$$a + 2 + a - 3 = a - 4a + 14.$$
$$5a = 15.$$
$$a = 3.$$

Check:

$$a + 2 + (a - 3) = a - (4a - 14).$$
$$3 + 2 + (3 - 3) = 3 - (4[3] - 14).$$
$$3 + 2 + 0 = 3 - (12 - 14).$$
$$5 = 3 - (-2).$$
$$5 = 3 + 2.$$
$$5 = 5.$$

12. Solve for b: $3b - (3 + b) = -(2b - 9)$.

$$3b - 3 - b = -2b + 9.$$
$$4b = 12.$$
$$b = 3.$$

Check:

$$3b - (3 + b) = -(2b - 9).$$
$$3(3) - (3 + 3) = -(2[3] - 9).$$
$$9 - 6 = -(6 - 9).$$
$$3 = -(-3).$$
$$3 = 3.$$

13. Solve for t: $t - [-2t - (t - 1)] = t - (3t - 11)$.

$$t - [-2t - t + 1] = t - 3t + 11.$$
$$t + 2t + t - 1 = t - 3t + 11.$$
$$6t = 12.$$
$$t = 2.$$

Check:

$$t - [-2t - (t - 1)] = t - (3t - 11).$$
$$2 - [-2(2) - (2 - 1)] = 2 - (3[2] - 11).$$
$$2 - [-4 - 1] = 2 - (6 - 11).$$
$$2 - [-4 - 1] = 2 - (-5).$$
$$2 - [-5] = 2 + 5.$$
$$2 + 5 = 7.$$
$$7 = 7.$$

14. Solve for x: $2x - [3 - \overline{x + 4} + x] = x + 5$.

$$2x - [3 - x - 4 + x] = x + 5.$$
$$2x - 3 + x + 4 - x = x + 5.$$
$$x = 4.$$

Check:

$$2x - [3 - \overline{x + 4} + x] = x + 5.$$
$$2(4) - [3 - \overline{4 + 4} + 4] = 4 + 5.$$
$$8 - [3 - 8 + 4] = 9.$$
$$8 - [-1] = 9.$$
$$8 + 1 = 9.$$
$$9 = 9.$$

15. Solve for m: $4 - m = (m - [3m - 1] + 2)$.

$$4 - m = (m - 3m + 1 + 2).$$
$$4 - m = m - 3m + 1 + 2.$$
$$m = -1.$$

Check:

$$4 - m = (m - [3m - 1] + 2).$$
$$4 - (-1) = (-1 - [3(-1) - 1] + 2).$$
$$4 + 1 = (-1 - [-3 - 1] + 2).$$
$$5 = (-1 - [-4] + 2).$$
$$5 = (-1 + 4 + 2).$$
$$5 = 5.$$

16. Solve for y: $y - 3(2 - y) = 6$.

$$y - 6 + 3y = 6.$$
$$4y = 12.$$
$$y = 3.$$

Check:

$$y - 3(2 - y) = 6.$$
$$3 - 3(2 - 3) = 6.$$
$$3 - 3(-1) = 6.$$
$$3 + 3 = 6.$$
$$6 = 6.$$

17. Solve for a: $a - 2[a - (3a + 4)] = a - (3 + a) + 6$.

$$a - 2[a - 3a - 4] = a - 3 - a + 6.$$
$$a - 2a + 6a + 8 = 3.$$
$$5a = -5.$$
$$a = -1.$$

Check:

$$a - 2[a - (3a + 4)] = a - (3 + a) + 6.$$
$$-1 - 2[-1 - (3[-1] + 4)] = -1 - (3 - 1) + 6.$$
$$-1 - 2[-1 - (-3 + 4)] = -1 - (2) + 6.$$
$$-1 - 2[-1 - (1)] = -1 - 2 + 6.$$
$$-1 - 2[-2] = 3.$$
$$-1 + 4 = 3.$$
$$3 = 3.$$

18. Solve for x: $x - x(x - 2) = 6 + x(-x)$.

$$x - x^2 + 2x = 6 - x^2.$$
$$3x = 6.$$
$$x = 2.$$

Check:

$$x - x(x - 2) = 6 + x(-x).$$
$$2 - 2(2 - 2) = 6 + 2(-2).$$
$$2 - 2(0) = 6 - 4.$$
$$2 - 0 = 2.$$
$$2 = 2.$$

19. Solve for b: $5 + [b - b(4 - b)] = b^2 - (b + 3)$.

$$5 + [b - 4b + b^2] = b^2 - b - 3.$$
$$5 + b - 4b + b^2 = b^2 - b - 3.$$
$$-2b = -8.$$
$$b = 4.$$

Check:

$$5 + [b - b(4 - b)] = b^2 - (b + 3).$$
$$5 + [4 - 4(4 - 4)] = 4^2 - (4 + 3).$$
$$5 + [4 - 4(0)] = 16 - 4 - 3.$$
$$5 + 4 = 9.$$
$$9 = 9.$$

20. Solve for n: $n^2 - n(n + 3) = 2 - n$.

$$n^2 - n^2 - 3n = 2 - n.$$
$$-2n = 2.$$
$$n = -1.$$

Check:

$$n^2 - n(n + 3) = 2 - n.$$
$$(-1)^2 - (-1)(-1 + 3) = 2 - (-1).$$
$$1 - (-1)(2) = 2 + 1.$$
$$1 - (-2) = 3.$$
$$1 + 2 = 3.$$
$$3 = 3.$$

21. Solve for x: $2x - a = x + a$.

The letter a will appear in the answer.

$$x = 2a.$$

Check:

$$2x - a = x + a.$$
$$2(2a) - a = 2a + a.$$
$$4a - a = 3a.$$
$$3a = 3a.$$

22. Solve for y: $3y - (y + b) = b + 1$.

$$3y - y - b = b + 1.$$
$$2y = 2b + 1.$$
$$y = \frac{2b + 1}{2}.$$

Check:

$$3y - (y + b) = b + 1.$$
$$3\left(\frac{2b + 1}{2}\right) - \left(\frac{2b + 1}{2} + b\right) = b + 1.$$
$$\frac{6b + 3}{2} - \frac{2b + 1}{2} - b = b + 1.$$

Multiply both members of the equation by 2.

$$6b + 3 - 2b - 1 - 2b = 2b + 2.$$
$$2b + 2 = 2b + 2.$$

23. Solve for m: $-[4 - (m + 1)] = m(-1) - m$.

$$-[4 - m - 1] = -m - m.$$
$$-4 + m + 1 = -m - m.$$
$$3m = 3.$$
$$m = 1.$$

Check:

$$-[4 - (m + 1)] = m(-1) - m.$$
$$-[4 - (1 + 1)] = 1(-1) - 1.$$
$$-[4 - 2] = -1 - 1.$$
$$-2 = -2.$$

24. Solve for y: $y - [3 + 2y] = 4 - (2y - 6)$.

$$y - 3 - 2y = 4 - 2y + 6.$$
$$y = 13.$$

Check:

$$y - [3 + 2y] = 4 - (2y - 6).$$
$$13 - [3 + 2 \cdot 13] = 4 - (2 \cdot 13 - 6).$$
$$13 - [3 + 26] = 4 - (26 - 6).$$
$$13 - 29 = 4 - 20.$$
$$-16 = -16.$$

25. Solve for a: $4 - 3(-a + 2) = a - (a - 1)$.

$$4 + 3a - 6 = a - a + 1.$$
$$3a = 3.$$
$$a = 1.$$

Check:

$$4 - 3(-a + 2) = a - (a - 1).$$
$$4 - 3(-1 + 2) = 1 - (1 - 1).$$
$$4 - 3(1) = 1 - 0.$$
$$4 - 3 = 1.$$
$$1 = 1.$$

26. Solve for x: $ax + bx = 1$.

$$x(a + b) = 1.$$
$$x = \frac{1}{a + b}.$$

Check:

$$ax + bx = 1.$$
$$a\left(\frac{1}{a + b}\right) + b\left(\frac{1}{a + b}\right) = 1.$$

Multiply both members of the equation by $a + b$.

$$\frac{a}{a + b} + \frac{b}{a + b} = \frac{a + b}{a + b}.$$
$$a + b = a + b.$$

27. Solve for y: $by + cy = b + c$.

$$y(b + c) = b + c.$$
$$y = \frac{b + c}{b + c}.$$
$$y = 1.$$

Check:

$$by + cy = b + c.$$
$$b(1) + c(1) = b + c.$$
$$b + c = b + c.$$

28. Solve for m: $am + bm + cm = ax + bx + cx$.

$$m(a + b + c) = x(a + b + c).$$
$$m = \frac{x(a + b + c)}{(a + b + c)}.$$
$$m = x.$$

Check:

$$am + bm + cm = ax + bx + cx.$$
$$a(x) + b(x) + c(x) = ax + bx + cx.$$
$$ax + bx + cx = ax + bx + cx.$$

29. Solve for y: $4y - 2(a + 2y) = ay - ax$.

$$4y - 2a - 4y = ay - ax.$$
$$-ay = -ax + 2a.$$
$$ay = ax - 2a.$$
$$ay = a(x - 2).$$

Divide both members of the equation by a.

$$y = x - 2.$$

Check:

$$4y - 2(a + 2y) = ay - ax.$$
$$4(x - 2) - 2(a + 2[x - 2]) = a(x - 2) - ax.$$
$$4x - 8 - 2(a + 2x - 4) = ax - 2a - ax.$$
$$4x - 8 - 2a - 4x + 8 = -2a.$$
$$-2a = -2a.$$

30. Solve for m: $a - b = 2am - 2bm$.

$$2bm - 2am = b - a.$$
$$2m(b - a) = b - a.$$
$$2m = \frac{b - a}{b - a}.$$
$$2m = 1.$$
$$m = \tfrac{1}{2}.$$

Check:

$$a - b = 2am - 2bm.$$
$$a - b = 2a(\tfrac{1}{2}) - 2b(\tfrac{1}{2}).$$
$$a - b = a - b.$$

Verbal Problems. The solution of verbal problems involves the translation of words into symbols. When the proper equation has been set up, it can be solved for the required unknown as in the preceding problems.

31. If one number is three more than a second number and their sum is 61, what are the numbers?

Let x = the second number;
$x + 3$ = the first number;
$2x + 3$ = their sum.

$$2x + 3 = 61.$$
$$2x = 58.$$
$$x = 29.$$
$$x + 3 = 32.$$

Check:

$$x + (x + 3) = 61.$$
$$29 + (29 + 3) = 61.$$
$$29 + 32 = 61.$$
$$61 = 61.$$

32. If the sum of two consecutive numbers is 77, what are the numbers?

Let x = the smaller number;
$x + 1$ = the larger number;
$2x + 1$ = their sum.

$$2x + 1 = 77.$$
$$2x = 76.$$
$$x = 38.$$
$$x + 1 = 39.$$

Check:

$$x + (x + 1) = 77.$$
$$38 + (38 + 1) = 77.$$
$$38 + 39 = 77.$$
$$77 = 77.$$

33. The sum of two consecutive even numbers is 118. What are the numbers?

The general form for consecutive even numbers is $2x$, $2x + 2$, $2x + 4$, ...

Let $2y$ = the smaller number;
$2y + 2$ = the larger number;
$4y + 2$ = their sum.

$$4y + 2 = 118.$$
$$4y = 116.$$
$$2y = 58.$$
$$2y + 2 = 60.$$

Check:

$$2y + (2y + 2) = 118.$$
$$58 + (58 + 2) = 118.$$
$$58 + 60 = 118.$$
$$118 = 118.$$

34. The sum of two consecutive odd numbers is 136. What are the numbers?

The general form for consecutive odd numbers is $2x - 1$, $2x + 1$, $2x + 3$, $2x + 5$, ...

Let $2x - 1$ = the smaller number;
$2x + 1$ = the larger number;
$4x$ = their sum.

$$4x = 136.$$
$$2x = 68.$$
$$2x - 1 = 67.$$
$$2x + 1 = 69.$$

Check:

$$(2x - 1) + (2x + 1) = 136.$$
$$67 + 69 = 136.$$
$$136 = 136.$$

35. The sum of three consecutive odd numbers is 69. What are the numbers?

Let $2x - 1 =$ the smallest number;
$2x + 1 =$ the second number;
$2x + 3 =$ the largest number;
$6x + 3 =$ their sum.
$6x + 3 = 69.$
$6x = 66.$
$2x = 22.$
$2x - 1 = 21.$
$2x + 1 = 23.$
$2x + 3 = 25.$

Check:
$(2x - 1) + (2x + 1) + (2x + 3) = 69.$
$21 + 23 + 25 = 69.$
$69 = 69.$

36. One number is three more than twice another number. Their sum is 54. What are the numbers?

Let $b =$ the smaller number;
$2b + 3 =$ the larger number;
$3b + 3 =$ their sum.
$3b + 3 = 54.$
$3b = 51.$
$b = 17.$
$2b + 3 = 37.$

Check:
$b + (2b + 3) = 54.$
$17 + (2[17] + 3) = 54.$
$17 + (34 + 3) = 54.$
$17 + 37 = 54.$
$54 = 54.$

37. A second number is three less than twice another number. Their sum is 51. What are the numbers?

Let $x =$ the first number;
$2x - 3 =$ the second number;
$3x - 3 =$ their sum.
$3x - 3 = 51.$
$3x = 54.$
$x = 18.$
$2x - 3 = 33.$

Check:
$x + (2x - 3) = 51.$
$18 + (2[18] - 3) = 51.$
$18 + (36 - 3) = 51.$
$18 + 33 = 51.$
$51 = 51.$

38. Of three given numbers, the second is three times the first, and the third is four more than one-half the sum of the first two. The sum of the three numbers is 52. What are the three numbers?

Let $y =$ the first number;
$3y =$ the second number;
$2y + 4 =$ the third number;
$6y + 4 =$ their sum.
$6y + 4 = 52.$
$6y = 48.$
$y = 8.$
$3y = 24.$
$2y + 4 = 20.$

Check:
$y + 3y + 2y + 4 = 52.$
$8 + 3(8) + 2(8) + 4 = 52.$
$8 + 24 + 16 + 4 = 52.$
$52 = 52.$

39. Alice is three times as old as Joan. In eight years, the sum of their ages will be 28. What are their present ages?

Let $x =$ Joan's present age;
$3x =$ Alice's present age;
$x + 8 =$ Joan's age 8 years hence;
$3x + 8 =$ Alice's age 8 years hence;
$4x + 16 =$ sum of their ages 8 years hence.
$4x + 16 = 28.$
$4x = 12.$
$x = 3$ years.
$3x = 9$ years.

Check:
$x + 8 + 3x + 8 = 28.$
$3 + 8 + 3(3) + 8 = 28.$
$3 + 8 + 9 + 8 = 28.$
$28 = 28.$

40. Jack is twice as old as John. In four years the sum of the ages of Jack and John will be 20 years. How old is each at the present time?

Let $x =$ John's present age;
$2x =$ Jack's present age;
$x + 4 =$ John's age 4 years hence;
$2x + 4 =$ Jack's age 4 years hence;
$3x + 8 =$ sum of their ages 4 years hence.
$3x + 8 = 20.$
$3x = 12.$
$x = 4$ years.
$2x = 8$ years.

Check:
$4 + 8 = 12.$ (1)
$12 = 12.$

$(4 + 4) + (8 + 4) = 20.$ (2)
$8 + 12 = 20.$
$20 = 20.$

41. The length of a rectangle is 3 feet less than twice its width, and its perimeter is 48 feet. What are the dimensions of the rectangle?

The perimeter is equal to twice the length plus twice the width.

$$\begin{aligned} \text{Let } x &= \text{width;} \\ 2x - 3 &= \text{length;} \\ 2(x) + 2(2x - 3) &= \text{perimeter.} \\ 2x + 4x - 6 &= 48. \\ 6x &= 54. \\ x &= 9 \text{ feet.} \\ 2x - 3 &= 15 \text{ feet.} \end{aligned}$$

Check:

$$\begin{aligned} 2(9) + 2(15) &= 48. \\ 18 + 30 &= 48. \\ 48 &= 48. \end{aligned}$$

42. A farmer has 60 feet of wire fence with which he expects to build a chicken pen, using the side of the barn for one side of the length of the pen. If the length is to be three times the width, what are the dimensions of the chicken pen?

It is often a good idea to draw a diagram when a physical problem is to be solved. This practice will be a great aid in visualizing the situation and determining how to go about solving the problem. A simple diagram such as Fig. 5 is usually quite adequate.

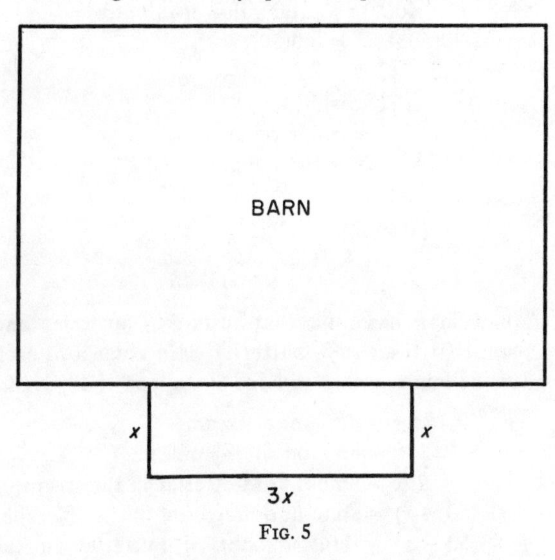

FIG. 5

$$\begin{aligned} \text{Let } x &= \text{width;} \\ 3x &= \text{length.} \end{aligned}$$

The wire will be used to enclose one length and both widths.

$$\begin{aligned} 2x + 3x &= \text{sum of two widths and one length.} \\ 5x &= 60. \\ x &= 12 \text{ feet.} \\ 3x &= 36 \text{ feet.} \end{aligned}$$

Check:

$$\begin{aligned} x + x + 3x &= 60. \\ 12 + 12 + 3(12) &= 60. \\ 12 + 12 + 36 &= 60. \\ 60 &= 60. \end{aligned}$$

43. Two automobiles start from the same place, the second one an hour later than the first. The first one was traveling 40 miles per hour and the second 50 miles per hour. How long will it take the second one to overtake the first one?

$$\begin{aligned} \text{Let } t &= \text{time for second automobile;} \\ t + 1 &= \text{time for first automobile.} \end{aligned}$$

Use the formula: *Distance equals rate times time.*

$$\begin{aligned} d_1 &= 40(t + 1). \\ d_2 &= 50t. \end{aligned}$$

Since the distance for both automobiles is the same, set one distance equal to the other.

$$\begin{aligned} d_2 &= d_1. \\ 50t &= 40 + 40t. \\ 10t &= 40. \\ t &= 4 \text{ hours.} \end{aligned}$$

Check:

$$\begin{aligned} 50t &= 40t + 40. \\ 50(4) &= 40(4) + 40. \\ 200 &= 160 + 40. \\ 200 &= 200. \end{aligned}$$

44. How far did each of the automobiles in Problem 43 travel?

$$\begin{aligned} d &= rt. \\ d_2 &= 50(4) \\ &= 200 \text{ miles.} \\ d_1 &= 40(4 + 1) \\ &= 40(5) \\ &= 200 \text{ miles.} \end{aligned}$$

45. Two automobiles 425 miles apart start driving toward each other at the same time. The first one averages 45 miles per hour; the second one averages 40 miles per hour. How far will each have traveled when they meet?

$$\begin{aligned} \text{Let } x &= \text{distance first one travels;} \\ 425 - x &= \text{distance second one travels.} \end{aligned}$$

Use the formula: $d = rt$, and solve for t.

$$\begin{aligned} t &= \frac{d}{r}. \\ t_1 &= \frac{x}{45}. \\ t_2 &= \frac{425 - x}{40}. \end{aligned}$$

Since the time is the same for both automobiles, set the two times equal to each other.

$$\begin{aligned} t_1 &= t_2. \\ \frac{x}{45} &= \frac{425 - x}{40}. \\ 40(x) &= 45(425 - x). \\ 40x &= 19{,}125 - 45x. \end{aligned}$$

$$85x = 19{,}125.$$
$$x = 225 \text{ miles.}$$
$$425 - x = 200 \text{ miles.}$$

Check:

$$225 + 200 = 425.$$
$$425 = 425.$$

46. After changing a dollar bill on a bus, a man has a total of \$1.08 in pennies, nickels, and dimes. He finds that he has twice as many dimes as pennies and three times as many nickels as pennies. How many pennies, nickels, and dimes does he have?

Let x = number of pennies;
$3x$ = number of nickels;
$2x$ = number of dimes.

In this problem each of the items has to be converted into its value in terms of cents.

$$1(x) = x;$$
$$5(3x) = 15x;$$
$$10(2x) = 20x;$$
$36x$ = total value in cents.
$$36x = 108.$$
$$x = 3.$$
$$3x = 9.$$
$$2x = 6.$$

Check:

$$3(0.01) + 9(0.05) + 6(0.10) = \$1.08.$$
$$0.03 + 0.45 + 0.60 = \$1.08.$$
$$\$1.08 = \$1.08.$$

47. A grocer counted the coins in his cash register. He found that he had twice as many quarters as half dollars, twice as many nickels as half dollars, three times as many dimes as half dollars, and twice as many pennies as nickels. The value of all the coins was \$8.64. How many coins of each denomination did he have?

Decide first which coin you have no information about and let x equal the number of those coins.

Let x = number of half dollars;
$2x$ = number of quarters;
$2x$ = number of nickels;
$3x$ = number of dimes;
$4x$ = number of pennies.

Find the value in cents for each denomination of the coins.

$$50(x) = 50x;$$
$$25(2x) = 50x;$$
$$5(2x) = 10x;$$
$$10(3x) = 30x;$$
$$1(4x) = 4x;$$
$144x$ = total value in cents.

$$144x = 864.$$
$$x = 6.$$
$$2x = 12.$$
$$2x = 12.$$
$$3x = 18.$$
$$4x = 24.$$

Check:

$$6(0.50) + 12(0.25) + 12(0.05) + 18(0.10) + 24(0.01) = \$8.64.$$
$$\$3.00 + \$3.00 + \$0.60 + \$1.80 + \$0.24 = \$8.64.$$
$$\$8.64 = \$8.64.$$

48. A storekeeper has walnuts that sell for 45 cents per pound and peanuts that sell for 30 cents per pound. He wishes to make a mixture of 50 pounds of the two kinds of nuts so they can sell for 40 cents per pound. How many pounds of each will he have to put into the mixture?

Let x = weight of walnuts;
$50 - x$ = weight of peanuts;
$.45x$ = value of walnuts;
$.30(50 - x)$ = value of peanuts;
$.45x + .30(50 - x)$ = total value of nuts to be mixed;
$.40(50)$ = total value of mixture.

$$.45x + .30(50 - x) = .40(50).$$
$$.45x + 15 - .30x = 20.$$
$$.15x = 5.$$
$$x = 33\tfrac{1}{3} \text{ pounds.}$$
$$50 - x = 16\tfrac{2}{3} \text{ pounds.}$$

Check:

$$0.45(33\tfrac{1}{3}) + 0.30(16\tfrac{2}{3}) = \$20.$$
$$\$15.00 + \$5.00 = \$20.$$
$$\$20.00 = \$20.$$

49. A dairyman has milk that tests 3% butterfat and cream that tests 20% butterfat. How much of each must he mix together to have 50 gallons of 4% milk?

Let x = amount of cream;
$50 - x$ = amount of 3% milk;
$.20x$ = amount of butterfat in the cream;
$.03(50 - x)$ = amount of butterfat in the 3% milk;
$.20x + .03(50 - x)$ = total amount of butterfat in the cream and milk;
$.04(50)$ = total amount of butterfat in the 4% mixture.

$$.20x + .03(50 - x) = .04(50).$$
$$.20x + 1.5 - .03x = 2.$$
$$.17x = .5.$$
$x = 2.9$ or approximately 3 gallons.
$50 - x = 47.1$ or approximately 47 gallons.

Check:

$$0.20(3) + 0.03(47) = 0.04(50).$$
$$0.60 + 1.41 = 2.$$
$$2.01 = 2.$$

This answer does not check out exactly since the amounts of each were approximate values.

50. A motorist has an 18-quart radiator in his automobile. The radiator is filled with a 35% solution of alcohol in water. If the motorist decides to make the solution 45%, how many quarts will he have to drain from the radiator and replace with pure alcohol?

Let x = amount of pure alcohol to be added;
x = amount of 35% solution to be drained;
$18 - x$ = amount of 35% solution left after draining;
$.35(18 - x)$ = amount of pure alcohol left after draining;
$x + .35(18 - x)$ = amount of pure alcohol present after x quarts are added;
$.45(18)$ = amount of pure alcohol in desired mixture.

$$x + .35(18 - x) = .45(18).$$
$$x + 6.3 - .35x = 8.1.$$
$$.65x = 1.8.$$
$$x = 2.77 \text{ quarts}.$$

Check:

$$2.77 + .35(18 - 2.77) = .45(18).$$
$$2.77 + 6.30 - .97 = 8.1.$$
$$8.1 = 8.1.$$

51. A man invests $18,000 in two separate investments. The first one pays 5% interest and the second $4\frac{1}{2}$% interest. His annual income from the two investments is $857.50. How much has he in each investment?

Let x = amount invested at $4\frac{1}{2}$%;
$18,000 − x = amount invested at 5%;
$.045x$ = interest on $4\frac{1}{2}$% investment;
$.05(\$18,000 - x)$ = interest on 5% investment;
$.045x + .05(\$18,000 - x)$ = total interest on both investments.

$$.045x + .05(\$18,000 - x) = \$857.50.$$
$$.045x + \$900 - .05x = \$857.50.$$
$$-.005x = -\$42.50.$$
$$x = \$8,500.$$
$$\$18,000 - x = \$9,500.$$

Check:

$$0.045(\$8,500) + 0.05(\$9,500) = \$857.50.$$
$$\$382.50 + \$475.00 = \$857.50.$$
$$\$857.50 = \$857.50.$$

SUPPLEMENTARY PROBLEMS

Using the set-builder form, find the solution set for the following equations.

1. $3a - 4 = a + 8$.
2. $m + 3 = 4m - 3$.
3. $c + 13 = 1 - 3c$.
4. $2(x - 3) = -2(x - 5)$.
5. $13 = 8 - d$.
6. $y - 2 = 2y + 3$.
7. $x - 2(x - 3) = 3(2 - x)$.
8. $3 - 2(x + 3) = 3[x - (2x - 4)]$.
9. $2x + 3(x - 1) = 2(x - 2)$.
10. $2[-(3 + x) - 1] = 4 + 3(1 - x)$.

Without using the set-builder form, find the roots of the following equations.

11. $m - 6 = 3m - (m + 1)$.
12. $2(p + 3) = 3p - 2(p - 5)$.
13. $3 - g = 7 + 2(3g - 1)$.
14. $3b - 2(3b + 4) = b - 1$.
15. $2[1 - (3x + 5) + 3] = 2x - 3$.
16. $y + [1 - (2 + 3y)] = 4 - y$.

Solve for x in each of the following.

17. $2x - 5 = 2a + 3$.
18. $x - a = 4a - 3x$.
19. $1 - 3y = 2x + 3$.
20. $2x - 3 = 4 - 2[3m + 1]$.

Solve the following verbal problems.

21. A man took a bus at eight o'clock in the morning and rode for three hours at which time he left the bus to change to another one which was to leave an hour later. The second bus traveled an average of 10 miles per hour faster than the first one, and reached the man's destination at 5 P.M. the same day. He had traveled a total of 410 miles. What was the average speed of the second bus?
22. A man made two investments using his total inheritance of $15,000. One paid him $3\frac{1}{2}$% per annum and the second one $4\frac{1}{2}$%. This gave him an annual income of $605 on the two investments. How much did he invest at each rate?
23. A merchant mixed a quantity of 80-cent coffee with a quantity of 95-cent coffee to make 100 pounds that sold for 86 cents per pound. How many pounds of each grade of coffee did he use to make the mixture?
24. A small boy took the coins, consisting of nickels, dimes and quarters, from his toy bank. He counted the money and found that he had a total of $4.95. The number of dimes was one more than three times the number of nickels, and the number of quarters was one less than twice the number of nickels. How many of each type of coin did he have?
25. Students were admitted to a high school football game for 25 cents if they had a student-body card, but had to pay $1.00 if they did not have such a card. Non-students were charged $1.50. At a given game there were an equal number of students and non-students, and there were five times as many students with cards as without them. The total receipts were $450.00. How many students without cards attended the game?

Chapter 5

Fractional Equations

Clearing of Fractions. When working with equations involving fractions, it is necessary to clear the equation of fractions before solving for the unknown. This is done by multiplying each member of the equation by the lowest common multiple of the denominators.

$$\frac{x}{2} + \frac{x}{4} = 3.$$

$$4\left(\frac{x}{2} + \frac{x}{4}\right) = 4(3). \text{ (multiplication axiom)}$$

$$4\left(\frac{x}{2}\right) + 4\left(\frac{x}{4}\right) = 4(3). \text{ (distributive axiom)}$$

$$2x + x = 12.$$

$$3x = 12.$$

$$x = 4.$$

Check:

$$\frac{x}{2} + \frac{x}{4} = 3.$$

$$\frac{4}{2} + \frac{4}{4} = 3.$$

$$2 + 1 = 3.$$

$$3 = 3.$$

1. Solve for x: $\frac{x}{3} + \frac{2x}{6} = \frac{4}{3}$.

$$\frac{2x}{6} + \frac{2x}{6} = \frac{8}{6}.$$

Multiply each member of the equation by 6.

$$\frac{(6)(2x)}{6} + \frac{(6)(2x)}{6} = \frac{(6)(8)}{6}.$$

$$2x + 2x = 8.$$

$$4x = 8.$$

$$x = 2.$$

Check:

$$\frac{x}{3} + \frac{2x}{6} = \frac{4}{3}.$$

$$\frac{2}{3} + \frac{2(2)}{6} = \frac{4}{3}.$$

$$\frac{2}{3} + \frac{4}{6} = \frac{4}{3}.$$

$$\frac{2}{3} + \frac{2}{3} = \frac{4}{3}.$$

$$\frac{4}{3} = \frac{4}{3}.$$

2. Solve for y: $\frac{2y}{21} + \frac{y}{7} = 15$.

$$\frac{2y}{21} + \frac{(3)(y)}{21} = \frac{(21)(15)}{21}.$$

$$2y + 3y = 315.$$

$$5y = 315.$$

$$y = 63.$$

Check:

$$\frac{2y}{21} + \frac{y}{7} = 15.$$

$$\frac{(2)(63)}{21} + \frac{63}{7} = 15.$$

$$6 + 9 = 15.$$

$$15 = 15.$$

3. Solve for m: $\frac{3m}{5} = \frac{15}{2}$.

$$\frac{(2)(3m)}{10} = \frac{(5)(15)}{10}.$$

$$6m = 75.$$

$$m = 12\tfrac{1}{2}.$$

Check:

$$\frac{3m}{5} = \frac{15}{2}.$$

$$\frac{(3)(12\frac{1}{2})}{5} = \frac{15}{2}.$$

$$\frac{37\frac{1}{2}}{5} = \frac{15}{2}.$$

$$7\tfrac{1}{2} = 7\tfrac{1}{2}.$$

4. Solve for a: $\frac{a+3}{2} + \frac{a-3}{3} = \frac{4}{3}$.

$$\frac{3(a+3)}{6} + \frac{2(a-3)}{6} = \frac{8}{6}.$$

$$3a + 9 + 2a - 6 = 8.$$

$$5a = 5.$$

$$a = 1.$$

Check:

$$\frac{a+3}{2} + \frac{a-3}{3} = \frac{4}{3}.$$

$$\frac{1+3}{2} + \frac{1-3}{3} = \frac{4}{3}.$$

$$\frac{4}{2} + \frac{-2}{3} = \frac{4}{3}.$$

$$\frac{6}{3} - \frac{2}{3} = \frac{4}{3}.$$

$$\frac{4}{3} = \frac{4}{3}.$$

5. Solve for x: $\frac{x}{2} - \frac{x+3}{5} = \frac{3}{2}$.

$$\frac{5(x)}{10} - \frac{2(x+3)}{10} = \frac{15}{10}.$$

$$5x - 2(x+3) = 15.$$

$$5x - 2x - 6 = 15.$$

$$3x = 21.$$

$$x = 7.$$

Check:

$$\frac{x}{2} - \frac{x+3}{5} = \frac{3}{2}.$$
$$\frac{7}{2} - \frac{7+3}{5} = \frac{3}{2}.$$
$$\tfrac{7}{2} - 2 = \tfrac{3}{2}.$$
$$\tfrac{7}{2} - \tfrac{4}{2} = \tfrac{3}{2}.$$
$$\tfrac{3}{2} = \tfrac{3}{2}.$$

6. Solve for a: $\dfrac{12}{a} + \dfrac{2}{3a} = \dfrac{2}{3}$

$$\frac{36}{3a} + \frac{2}{3a} = \frac{2a}{3a}.$$
$$36 + 2 = 2a.$$
$$2a = 38.$$
$$a = 19.$$

Check:

$$\frac{12}{a} + \frac{2}{3a} = \frac{2}{3}.$$
$$\tfrac{12}{19} + \tfrac{2}{57} = \tfrac{2}{3}.$$
$$\tfrac{36}{57} + \tfrac{2}{57} = \tfrac{38}{57}.$$
$$\tfrac{38}{57} = \tfrac{38}{57}.$$

7. Solve for y: $1 + \dfrac{9}{y} = 2.$

$$\frac{y}{y} + \frac{9}{y} = \frac{2y}{y}.$$
$$y + 9 = 2y.$$
$$-y = -9.$$
$$y = 9.$$

Check:

$$1 + \frac{9}{y} = 2.$$
$$1 + \tfrac{9}{9} = 2.$$
$$1 + 1 = 2.$$
$$2 = 2.$$

8. Solve for b: $\dfrac{b+1}{2} + \dfrac{b-1}{7} = \dfrac{16}{7}.$

$$\frac{7(b+1)}{14} + \frac{2(b-1)}{14} = \frac{2(16)}{14}.$$
$$7b + 7 + 2b - 2 = 32.$$
$$9b = 27.$$
$$b = 3.$$

Check:

$$\frac{b+1}{2} + \frac{b-1}{7} = \frac{16}{7}.$$
$$\frac{3+1}{2} + \frac{3-1}{7} = \frac{16}{7}.$$
$$\tfrac{4}{2} + \tfrac{2}{7} = \tfrac{16}{7}.$$
$$\tfrac{28}{14} + \tfrac{4}{14} = \tfrac{32}{14}.$$
$$\tfrac{32}{14} = \tfrac{32}{14}.$$

9. Solve for w: $\dfrac{1}{w-1} - \dfrac{2}{w} = 0.$

$$\frac{w}{w(w-1)} - \frac{2(w-1)}{w(w-1)} = 0.$$
$$w - 2w + 2 = 0.$$
$$-w = -2.$$
$$w = 2.$$

Check:

$$\frac{1}{w-1} - \frac{2}{w} = 0.$$
$$\frac{1}{2-1} - \frac{2}{2} = 0.$$
$$\tfrac{1}{1} - \tfrac{2}{2} = 0.$$
$$1 - 1 = 0.$$
$$0 = 0.$$

10. Solve for x: $\dfrac{3x-1}{2} - \dfrac{x+3}{3} = \dfrac{x+2}{4} - \dfrac{3x+1}{8}.$

$$\frac{12(3x-1)}{24} - \frac{8(x+3)}{24} = \frac{6(x+2)}{24} - \frac{3(3x+1)}{24}.$$
$$36x - 12 - 8x - 24 = 6x + 12 - 9x - 3.$$
$$31x = 45.$$
$$x = \tfrac{45}{31}.$$

Check:

$$\frac{3x-1}{2} - \frac{x+3}{3} = \frac{x+2}{4} - \frac{3x+1}{8}.$$
$$\frac{3(\frac{45}{31}) - 1}{2} - \frac{\frac{45}{31} + 3}{3} = \frac{\frac{45}{31} + 2}{4} - \frac{3(\frac{45}{31}) + 1}{8}.$$
$$\frac{\frac{135}{31} - 1}{2} - \frac{\frac{45}{31} + 3}{3} = \frac{\frac{45}{31} + 2}{4} - \frac{\frac{135}{31} + 1}{8}.$$
$$\frac{\frac{135}{31} - \frac{31}{31}}{2} - \frac{\frac{45}{31} + \frac{93}{31}}{3} = \frac{\frac{45}{31} + \frac{62}{31}}{4} - \frac{\frac{135}{31} + \frac{31}{31}}{8}.$$
$$\frac{\frac{104}{31}}{2} - \frac{\frac{138}{31}}{3} = \frac{\frac{107}{31}}{4} - \frac{\frac{166}{31}}{8}.$$
$$\tfrac{52}{31} - \tfrac{46}{31} = \tfrac{107}{124} - \tfrac{166}{248}.$$
$$\tfrac{6}{31} = \tfrac{107}{124} - \tfrac{83}{124}.$$
$$\tfrac{6}{31} = \tfrac{24}{124}.$$
$$\tfrac{6}{31} = \tfrac{6}{31}.$$

11. Solve for y: $\dfrac{1}{y^2-1} + \dfrac{3}{y-1} = \dfrac{2}{y+1}.$

$$\frac{1}{(y-1)(y+1)} + \frac{3(y+1)}{(y-1)(y+1)} = \frac{2(y-1)}{(y-1)(y+1)}.$$
$$1 + 3y + 3 = 2y - 2.$$
$$y = -6.$$

Check:

$$\frac{1}{y^2-1} + \frac{3}{y-1} = \frac{2}{y+1}.$$
$$\frac{1}{36-1} + \frac{3}{-6-1} = \frac{2}{-6+1}.$$
$$\tfrac{1}{35} - \tfrac{3}{7} = -\tfrac{2}{5}.$$
$$\tfrac{1}{35} - \tfrac{15}{35} = -\tfrac{14}{35}.$$
$$-\tfrac{14}{35} = -\tfrac{14}{35}.$$

12. Solve for a: $\dfrac{5a+5}{a^2-1} - \dfrac{2a+3}{a-1} = 6.$

$$\frac{5(a+1)}{(a+1)(a-1)} - \frac{2a+3}{a-1} = 6.$$
$$\frac{5}{a-1} - \frac{2a+3}{a-1} = \frac{6(a-1)}{a-1}.$$
$$5 - 2a - 3 = 6a - 6.$$
$$-8a = -8.$$
$$a = 1.$$

Check:

$$\frac{5a+5}{a^2-1}-\frac{2a+3}{a-1}=6.$$
$$\frac{5(1)+5}{1-1}-\frac{2(1)+3}{1-1}=6.$$
$$\tfrac{10}{0}-\tfrac{5}{0}=6.$$

A root of an equation must satisfy the equation. Since division by zero is not permitted (see p. 5), the above solution will not satisfy the given equation. This tells us, then, that there is no value of a for which the equation is true. Hence, there is no solution. If set-builder form had been used, the solution set would be ϕ.

13. Solve for y: $\dfrac{15}{y^2+5y+6}-\dfrac{2}{y+3}=\dfrac{5}{y+2}$.

$$\frac{15}{(y+2)(y+3)}-\frac{2(y+2)}{(y+2)(y+3)}=\frac{5(y+3)}{(y+2)(y+3)}.$$
$$15-2y-4=5y+15.$$
$$-7y=4.$$
$$y=-\tfrac{4}{7}.$$

Check:

$$\frac{15}{y^2+5y+6}-\frac{2}{y+3}=\frac{5}{y+2}.$$
$$\frac{15}{\frac{16}{49}-\frac{20}{7}+6}-\frac{2}{-\frac{4}{7}+3}=\frac{5}{-\frac{4}{7}+2}.$$
$$\frac{15}{\frac{16}{49}-\frac{140}{49}+\frac{294}{49}}-\frac{2}{-\frac{4}{7}+\frac{21}{7}}=\frac{5}{-\frac{4}{7}+\frac{14}{7}}.$$
$$\frac{15}{\frac{170}{49}}-\frac{2}{\frac{17}{7}}=\frac{5}{\frac{10}{7}}.$$
$$\tfrac{735}{170}-\tfrac{14}{17}=\tfrac{35}{10}.$$
$$\tfrac{735}{170}-\tfrac{140}{170}=\tfrac{595}{170}.$$
$$\tfrac{595}{170}=\tfrac{595}{170}.$$

14. Solve for x: $\dfrac{2x+1}{x-1}-\dfrac{2x}{x+1}=\dfrac{11}{x^2-1}$.

$$\frac{(2x+1)(x+1)}{(x-1)(x+1)}-\frac{2x(x-1)}{(x-1)(x+1)}=\frac{11}{(x-1)(x+1)}.$$
$$2x^2+2x+x+1-2x^2+2x=11.$$
$$5x=10.$$
$$x=2.$$

Check:

$$\frac{2x+1}{x-1}-\frac{2x}{x+1}=\frac{11}{x^2-1}.$$
$$\frac{4+1}{2-1}-\frac{4}{2+1}=\frac{11}{4-1}.$$
$$\tfrac{5}{1}-\tfrac{4}{3}=\tfrac{11}{3}.$$
$$\tfrac{15}{3}-\tfrac{4}{3}=\tfrac{11}{3}.$$
$$\tfrac{11}{3}=\tfrac{11}{3}.$$

15. Solve for x: $\dfrac{3}{x^2-x-2}+\dfrac{3}{x-2}=\dfrac{1}{x+1}$.

$$\frac{3}{(x-2)(x+1)}+\frac{3(x+1)}{(x-2)(x+1)}=\frac{1(x-2)}{(x-2)(x+1)}.$$
$$3+3x+3=x-2.$$
$$2x=-8.$$
$$x=-4.$$

Check:

$$\frac{3}{x^2-x-2}+\frac{3}{x-2}=\frac{1}{x+1}.$$
$$\frac{3}{16+4-2}+\frac{3}{-4-2}=\frac{1}{-4+1}.$$
$$\frac{3}{18}+\frac{3}{-6}=\frac{1}{-3}.$$
$$\tfrac{1}{6}-\tfrac{3}{6}=-\tfrac{2}{6}.$$
$$-\tfrac{2}{6}=-\tfrac{2}{6}.$$

16. Solve for a: $\dfrac{a-1}{a}+\dfrac{a+1}{a-1}=2$.

$$\frac{(a-1)(a-1)}{a(a-1)}+\frac{a(a+1)}{a(a-1)}=\frac{2a(a-1)}{a(a-1)}.$$
$$a^2-2a+1+a^2+a=2a^2-2a.$$
$$a=-1.$$

Check:

$$\frac{a-1}{a}+\frac{a+1}{a-1}=2.$$
$$\frac{-1-1}{-1}+\frac{-1+1}{-1-1}=2.$$
$$\frac{-2}{-1}+\frac{0}{-2}=2.$$
$$2=2.$$

17. Solve for w: $\dfrac{w^2+w}{w^2-1}-\dfrac{3}{w+1}=1$.

$$\frac{w^2+w}{(w-1)(w+1)}-\frac{3(w-1)}{(w-1)(w+1)}=\frac{(w-1)(w+1)}{(w-1)(w+1)}.$$
$$w^2+w-3w+3=w^2-1.$$
$$-2w=-4.$$
$$w=2.$$

Check:

$$\frac{w^2+w}{w^2-1}-\frac{3}{w+1}=1.$$
$$\frac{4+2}{4-1}-\frac{3}{2+1}=1.$$
$$\tfrac{6}{3}-\tfrac{3}{3}=1.$$
$$2-1=1.$$
$$1=1.$$

18. Solve for b: $\dfrac{2b-3}{2b+1}=\dfrac{b-3}{b+1}$.

$$\frac{(2b-3)(b+1)}{(2b+1)(b+1)}=\frac{(b-3)(2b+1)}{(2b+1)(b+1)}.$$
$$2b^2-b-3=2b^2-5b-3.$$
$$4b=0.$$
$$b=0.$$

Check:

$$\frac{2b-3}{2b+1}=\frac{b-3}{b+1}.$$
$$\frac{2(0)-3}{2(0)+1}=\frac{0-3}{0+1}.$$
$$-3=-3.$$

19. Solve for m: $\dfrac{m^2+3}{m^2+8m+15}+\dfrac{4}{m+3}=1.$

$$\frac{m^2+3}{(m+3)(m+5)}+\frac{4(m+5)}{(m+3)(m+5)}=\frac{(m+3)(m+5)}{(m+3)(m+5)}.$$
$$m^2+3+4m+20=m^2+8m+15.$$
$$-4m=-8.$$
$$m=2.$$

Check:
$$\frac{m^2+3}{m^2+8m+15}+\frac{4}{m+3}=1.$$
$$\frac{4+3}{4+16+15}+\frac{4}{2+3}=1.$$
$$\tfrac{7}{35}+\tfrac{4}{5}=1.$$
$$\tfrac{1}{5}+\tfrac{4}{5}=1.$$
$$1=1.$$

20. Solve for y: $\dfrac{y^2-4}{y^2-5y+6}-\dfrac{4y-12}{y^2-6y+9}=2.$

Factor and reduce each fraction in the left member of the equation.

$$\frac{(y-2)(y+2)}{(y-2)(y-3)}-\frac{4(y-3)}{(y-3)(y-3)}=2.$$
$$\frac{y+2}{y-3}-\frac{4}{y-3}=2.$$
$$\frac{y+2}{y-3}-\frac{4}{y-3}=\frac{2(y-3)}{y-3}.$$
$$y+2-4=2y-6.$$
$$-y=-4.$$
$$y=4.$$

Check:
$$\frac{y^2-4}{y^2-5y+6}-\frac{4y-12}{y^2-6y+9}=2.$$
$$\frac{16-4}{16-20+6}-\frac{16-12}{16-24+9}=2.$$
$$\tfrac{12}{2}-\tfrac{4}{1}=2.$$
$$6-4=2.$$
$$2=2.$$

Verbal Problems. The equations set up from verbal problems frequently contain fractions. Such equations are solved in the same way as those in the preceding problems.

21. The width of a rectangle is two-thirds as long as its length. If the perimeter is 80 feet, what are the dimensions of the rectangle?

Let x = length;
$\tfrac{2}{3}x$ = width;
$2(x)+2(\tfrac{2}{3}x)$ = perimeter.
$$2x+\frac{4x}{3}=80.$$
$$\frac{6x}{3}+\frac{4x}{3}=\frac{240}{3}.$$
$$10x=240.$$
$x = 24$ feet.
$\tfrac{2}{3}x = 16$ feet.

Check:
$$2(24)+2(16)=80.$$
$$48+32=80.$$
$$80=80.$$

22. The second side of a triangle is three-fourths as long as the first side. The third side is 3 inches longer than one-half the second side. If the perimeter is 54 inches, what are the lengths of the three sides of the triangle?

Let x = the first side;
$\tfrac{3}{4}x$ = the second side;
$\tfrac{3}{8}x+3$ = the third side;
$x+\tfrac{3}{4}x+\tfrac{3}{8}x+3$ = perimeter.
$$x+\frac{3x}{4}+\frac{3x}{8}+3=54.$$
$$\frac{8x}{8}+\frac{6x}{8}+\frac{3x}{8}+\frac{24}{8}=\frac{432}{8}.$$
$$8x+6x+3x+24=432.$$
$$17x=408.$$
$x = 24$ inches.
$\tfrac{3}{4}x = 18$ inches.
$\tfrac{3}{8}x+3 = 12$ inches.

Check:
$$24+18+12=54.$$
$$54=54.$$

23. If a lot cost one-sixth as much as a house and together they cost \$18,200, what was the cost of each?

Let x = cost of the house;
$\dfrac{x}{6}$ = cost of the lot;
$x+\dfrac{x}{6}$ = total cost.
$$x+\frac{x}{6}=\$18{,}200.$$
$$\frac{6x}{6}+\frac{x}{6}=\frac{6(\$18{,}200)}{6}.$$
$$7x=\$109{,}200.$$
$$x=\$15{,}600.$$
$$\frac{x}{6}=\$2{,}600.$$

Check:
$$\$15{,}600+\$2{,}600=\$18{,}200.$$
$$\$18{,}200=\$18{,}200.$$

24. One number is five-sixths another number. If their sum is 66, what are the numbers?

Let x = the larger number;
$\tfrac{5}{6}x$ = the smaller number.
$$x+\frac{5x}{6}=66.$$
$$\frac{6x}{6}+\frac{5x}{6}=\frac{6(66)}{6}.$$
$$11x=396.$$
$$x=36.$$
$$\frac{5x}{6}=30.$$

Check:

$$36 + 30 = 66.$$
$$66 = 66.$$

25. A gardener had $29.00 to spend on improvements in the garden, but he spent only a part of it. The amount he had spent for seeds was one-third of the unspent amount; for shrubbery it was one-half of the unspent amount; and for new tools it was three times the unspent amount. How much did he have left?

Let x = amount left;

$\frac{x}{3}$ = amount spent for seeds;

$\frac{x}{2}$ = amount spent for shrubbery;

$3x$ = amount spent for new tools;

$x + 3x + \frac{x}{3} + \frac{x}{2}$ = amount left + amount spent.

$$4x + \frac{x}{3} + \frac{x}{2} = \$29.00.$$
$$\frac{6(4x)}{6} + \frac{2x}{6} + \frac{3x}{6} = \frac{6(\$29.00)}{6}.$$
$$24x + 2x + 3x = \$174.00.$$
$$29x = \$174.00.$$
$$x = \$6.00.$$

Check:

$$x + 3x + \frac{x}{3} + \frac{x}{2} = \$29.00.$$
$$\$6 + 3(\$6) + \frac{\$6}{3} + \frac{\$6}{2} = \$29.00.$$
$$\$6 + \$18 + \$2 + \$3 = \$29.00.$$
$$\$29 = \$29.00.$$

26. At a football game student tickets sold for $1.00 each and general admission tickets for $1.50 each. There were three-fifths as many student tickets sold as general admission tickets. If the total receipts were $1,512.00, how many of each type of ticket were sold?

Let x = number of general admission tickets sold;

$\frac{3x}{5}$ = number of student tickets sold;

$(1.50)(x)$ = value of general admission tickets;

$(1.00)\left(\frac{3x}{5}\right)$ = value of student tickets;

$1.5x + \frac{3x}{5}$ = total value of all tickets.

$$1.5x + \frac{3x}{5} = \$1{,}512.00.$$
$$\frac{5(1.5x)}{5} + \frac{3x}{5} = \frac{5(\$1{,}512)}{5}.$$
$$7.5x + 3x = \$7{,}560.$$
$$10.5x = \$7{,}560.$$
$$x = 720.$$
$$\frac{3x}{5} = 432.$$

Check:

$$720(\$1.50) + 432(\$1.00) = \$1{,}512.$$
$$\$1{,}080 + \$432 = \$1{,}512.$$
$$\$1{,}512 = \$1{,}512.$$

27. A man made two investments, the first one at 5% interest and the second one at 4%. The amount of the second was two-thirds as much as the first. If the two investments netted him an annual income of $966.00, how much was each investment?

Let x = the amount invested at 5%;

$\frac{2x}{3}$ = the amount invested at 4%;

$(0.05)(x)$ = income from 5% investment;

$(0.04)\left(\frac{2x}{3}\right)$ = income from 4% investment;

$0.05x + \frac{0.08x}{3}$ = total income.

$$0.05x + \frac{0.08x}{3} = \$966.00.$$
$$\frac{3(0.05x)}{3} + \frac{0.08x}{3} = \frac{3(\$966)}{3}.$$
$$0.15x + 0.08x = \$2{,}898.$$
$$0.23x = \$2{,}898.$$
$$x = \$12{,}600.$$
$$\frac{2x}{3} = \$8{,}400.$$

Check:

$$0.05(\$12{,}600) + 0.04(\$8{,}400) = \$966.00.$$
$$\$630.00 + \$336.00 = \$966.00.$$
$$\$966.00 = \$966.00.$$

28. Two pipes lead into an irrigation storage reservoir. If one of them can fill the reservoir in 15 hours and the second one can fill it in 18 hours, how long will it take to fill the reservoir if both are flowing at the same time?

Let x = time required if both pipes are flowing concurrently;

$\frac{1}{x}$ = part of full capacity both pipes can fill in one hour;

$\frac{1}{15}$ = part of full capacity first pipe can fill in one hour;

$\frac{1}{18}$ = part of full capacity second pipe can fill in one hour.

$$\frac{1}{15} + \frac{1}{18} = \frac{1}{x}.$$
$$\frac{18x}{x(15)(18)} + \frac{15x}{x(15)(18)} = \frac{(15)(18)}{x(15)(18)}.$$
$$18x + 15x = 270.$$
$$33x = 270.$$
$$x = 8\tfrac{2}{11} \text{ hours}$$
$$= 8 \text{ hours}, \tfrac{120}{11} \text{ minutes}$$

It will take approximately 8 hours, 11 minutes.

Check:

$$\frac{1}{15}+\frac{1}{18}=\frac{1}{8\frac{2}{11}}.$$
$$\frac{1}{15}+\frac{1}{18}=\frac{1}{\frac{90}{11}}.$$
$$\tfrac{1}{15}+\tfrac{1}{18}=\tfrac{11}{90}.$$
$$\tfrac{6}{90}+\tfrac{5}{90}=\tfrac{11}{90}.$$
$$\tfrac{11}{90}=\tfrac{11}{90}.$$

29. If the pipes in Problem 28 were set up so that the first pipe flowed into the reservoir and the second one was an outlet for irrigating the land, how long would it take to fill the tank if both pipes flowed continuously until the tank was full?

Let x = time required;

$\frac{1}{x}$ = part filled in one hour if both pipes are flowing;

$\frac{1}{15}$ = part filled in one hour if outlet is closed;

$\frac{1}{18}$ = part emptied in one hour if intake is closed.

$$\frac{1}{15}-\frac{1}{18}=\frac{1}{x}.$$
$$\frac{18x}{x(15)(18)}-\frac{15x}{x(15)(18)}=\frac{(15)(18)}{x(15)(18)}.$$
$$18x-15x=270.$$
$$3x=270.$$
$$x=90 \text{ hours.}$$

Check:

$$\tfrac{1}{15}-\tfrac{1}{18}=\tfrac{1}{90}.$$
$$\tfrac{6}{90}-\tfrac{5}{90}=\tfrac{1}{90}.$$
$$\tfrac{1}{90}=\tfrac{1}{90}.$$

SUPPLEMENTARY PROBLEMS

Solve the following problems for the variable involved.

1. $\dfrac{m+3}{2}+\dfrac{2m-1}{3}=7.$
2. $\dfrac{2(x+3)}{3}-\dfrac{5(2+x)}{4}=\dfrac{2}{3}.$
3. $\dfrac{2y+7}{7}+\dfrac{y-1}{6}=4.$
4. $\dfrac{3-a}{2}+\dfrac{3a-1}{8}=a-\dfrac{35}{4}.$
5. $\dfrac{12}{p}+\dfrac{4}{3p}=\dfrac{10}{p-1}.$
6. $\dfrac{x-1}{4}+\dfrac{2x+1}{7}=1+\dfrac{4-x}{2}.$
7. $\dfrac{5}{4c-1}+\dfrac{7}{3c+10}=0.$
8. $\dfrac{3}{1-x^2}+\dfrac{5}{1+x}=\dfrac{3}{1-x}.$
9. $\dfrac{3(x-1)}{1-x^2}+\dfrac{4}{x+1}=\dfrac{2}{1-x}.$
10. $4-\dfrac{3(a+1)}{a^2-1}=4.$

Solve each of the following equations for x.

11. $\dfrac{2x-a}{2}=\dfrac{a+1}{3}.$
12. $\dfrac{2(1-x)}{2}+3=\dfrac{4-x}{7}.$
13. $\dfrac{5x}{3}+\dfrac{5}{6}=\dfrac{3-y}{2}.$
14. $\dfrac{2(3+m)}{3}=\dfrac{2+2x}{5}.$
15. $\dfrac{1}{x}+\dfrac{2}{y}=\dfrac{3}{z}.$

Solve the following verbal problems:

16. The width of a rectangle is 2 feet more than three-fifths its length. If its perimeter is 52 feet what are the dimensions of the rectangle?
17. A boy has $5.05 consisting of quarters, dimes, and nickels. He has two-thirds as many dimes as nickels, and three more than one-fourth as many quarters as nickels. How many of each type of coin does he have?
18. One number is nine-sevenths of a given number. Their difference is 16. What are the numbers?
19. If 3 is subtracted from a given number and the remainder is divided by 3, the result will be the same as that obtained by adding 1 to the given number and dividing the sum by 5. What is the given number?
20. A housewife went to the market and spent $7.20 for a quantity of meat, some fresh vegetables, and certain canned goods. The canned goods cost two-thirds as much as the meat, and the fresh vegetables cost one-half as much as the canned goods. How much did she spend for each of the three types of food?

Chapter 6

Systems of Linear Equations

Ordered Pairs. If a and b represent real numbers then the expression (a, b) is called an ordered pair. An ordered pair, then, is two numbers separated by a comma and enclosed within parentheses. In the above example, a is the first member of the ordered pair and b the second member. The members of an ordered pair cannot be interchanged since "ordered" refers to the order in which they are stated.

Ordered pairs may be the elements of a set. As an example, consider the following problem: If $U = \{1, 2, 3, \ldots 9\}$, the set of all ordered pairs for which the second member is twice the first member would be $\{(1,2), (2,4), (3,6), (4,8)\}$. Thus, each element of the set is an ordered pair that satisfies the conditions stated in the problem.

Frequently, in algebra, we are given a problem in two variables in which two conditions involving those two variables are stated. In such problems we are confronted with the task of finding the values of those variables that will satisfy both conditions. The following illustration represents such a problem: If x and y are two numbers such that their sum is 12 and their difference is 2, what are the numbers? The conditions in this problem could be stated as a pair of equations which use the variables.

$$x + y = 12.$$
$$x - y = 2.$$

It can be seen easily that if $x = 7$ and $y = 5$ both equations will be satisfied. Thus, the ordered pair $\{(7,5)\}$ represents the solution set to the above pair of equations. Any such pair of equations in two variables is known as a **system of equations.**

It is customary to use the value of x as the first member of an ordered pair and the value of y as the second member. The reason for this will become clear when we graph algebraic equations (Chapter 11).

Intersection of Sets. A set C whose elements are those common to two or more given sets is the intersection of those given sets. Thus, if $A = \{4,6,8\}$ and $B = \{6,8,10\}$, then the intersection of A and B (symbolized $A \cap B$) would be $\{6,8\}$. Thus,

$$C = A \cap B = \{6,8\}.$$

The intersection of A and B may be a unique set, set A, set B, or the empty set.

Systems of equations, as well as individual equations, can be written in set-builder form. The following statements are set-builder formulations of a system of equations.

$$A = \{(x, y) \mid 2x + y = 7 \text{ and } 3x - y = 8\}.$$
$$A = \{(x, y) \mid 2x + y = 7\} \cap \{(x, y) \mid 3x - y = 8\}.$$

These would be read, respectively,

"A is the set of all x and y such that $2x + y = 7$ and $3x - y = 8$."

"A is the intersection of the set of all x and y such that $2x + y = 7$, and the set of all x and y such that $3x - y = 8$."

The second way of writing the set-builder form of a system of equations is illustrative of the fact that the solution set of the system of equations is the intersection of the solution sets of each equation in the system. Thus, for example, if we take the natural numbers as the universe, the solution set of $\{(x, y) \mid 2x + y = 7\}$ is $\{(1, 5), (2, 3), (3, 1)\}$; and the solution set of $\{(x, y) \mid 3x - y = 8\}$ is $\{(3, 1), (4, 4), (5, 7), (6, 10), (7, 13), \ldots\}$. The intersection of these solution sets is $\{(3, 1)\}$, the solution set of the system.

In set-builder form, the solution of a system of linear equations would be written in the following manner.

$$\begin{aligned} A &= \{(x, y) \mid x - y = 1 \text{ and } x + y = 9\} \\ &= \{(x, y) \mid x - y = 1\} \cap \{(x, y) \mid x + y = 9\} \\ &= \{(5, 4)\}. \end{aligned}$$

The ordered pair (5, 4), enclosed within braces, is the solution set.

The set-builder form need not be used in the solution of systems of linear equations, however, and the following problems do not employ it.

SOLVING SYSTEMS OF LINEAR EQUATIONS BY ELIMINATION

Definitions. As stated above, linear equations, equations in which the unknowns are first power expressions, sometimes involve two unknowns and are stated so that two equations may be formulated from the given facts. The solution requires the finding of values for the two unknowns, which, when substituted in the two equations, will satisfy both equations. Other problems may involve three or more unknowns. Such systems of equations are called

simultaneous equations if there is a common solution. In some cases there is no common solution and the equations are said to be **inconsistent.** In other systems of equations there may be an infinite number of solutions, in which case the equations are referred to as **dependent** equations.

Some methods used for solving systems of equations are called **solving by elimination.** We eliminate one of the variables by use of algebraic techniques and then determine the value of the remaining variable.

Three common methods for solving systems of linear equations by elimination are presented in the subsequent discussion. These are: solving by **addition,** solving by **subtraction,** and solving by **substitution.** All of the methods require that the equations be simplified so that each variable occurs only once in each equation and that all equations are in the same form.

Solving Systems of Equations by Addition. This method involves algebraically adding one equation to the other; and it has been pointed out that this is a valid operation since we are adding equals to equals (Chapter 4, Axiom 1). The method of solving by addition applies when a term containing an unknown in one of the equations is *the negative of* the term containing the same unknown in the other equation. In such a case, the unknown will be eliminated by adding the equations.

Solve for x and y:

$$x - y = 1 \quad (1)$$
$$x + y = 9 \quad (2)$$

Add equations (1) and (2).

$$\begin{array}{rl} x - y = 1 & (1) \\ x + y = 9 & (2) \\ \hline 2x \quad = 10 & \\ x = 5. & (3) \end{array}$$

Substitute the value of x from equation (3) in equation (2).

$$x + y = 9. \quad (2)$$
$$5 + y = 9.$$
$$y = 4.$$

Check:

$$x - y = 1. \quad (1)$$
$$5 - 4 = 1.$$
$$1 = 1.$$

$$x + y = 9. \quad (2)$$
$$5 + 4 = 9.$$
$$9 = 9.$$

1. Solve for x and y:

$$x + y = 3. \quad (1)$$
$$x - y = 1. \quad (2)$$

This is set up as a regular addition problem. The numbers in parentheses are given in order to make proper reference to previous parts of the problem.

$$\begin{array}{rl} x + y = 3 & (1) \\ x - y = 1 & (2) \\ \hline 2x \quad = 4 & \\ x = 2. & (3) \end{array}$$

Take the value of x as found in equation (3) and substitute it in either equation (1) or equation (2) to find the value of y.

$$x + y = 3. \quad (1)$$
$$2 + y = 3.$$
$$y = 1.$$

Check:

$$x + y = 3. \quad (1) \qquad x - y = 1. \quad (2)$$
$$2 + 1 = 3. \qquad 2 - 1 = 1.$$
$$3 = 3. \qquad 1 = 1.$$

Hence, $x = 2$ and $y = 1$.

2. Solve for x and y:

$$5x + y = 8. \quad (1)$$
$$2x - y = -1. \quad (2)$$

Add equations (1) and (2).

$$\begin{array}{rl} 5x + y = 8 & (1) \\ 2x - y = -1 & (2) \\ \hline 7x \quad = 7 & \\ x = 1. & (3) \end{array}$$

Substitute the value of x from equation (3) in equation (1).

$$5x + y = 8. \quad (1)$$
$$5(1) + y = 8.$$
$$5 + y = 8.$$
$$y = 3.$$

Check:

$$5x + y = 8. \quad (1) \qquad 2x - y = -1. \quad (2)$$
$$5(1) + 3 = 8. \qquad 2(1) - 3 = -1.$$
$$5 + 3 = 8. \qquad 2 - 3 = -1.$$
$$8 = 8. \qquad -1 = -1.$$

Hence, $x = 1$ and $y = 3$.

3. Solve for x and y:

$$-x - y = 0. \quad (1)$$
$$x - 3y = 4. \quad (2)$$

Add equations (1) and (2).

$$\begin{array}{rl} -x - y = 0 & (1) \\ x - 3y = 4 & (2) \\ \hline -4y = 4 & \\ y = -1. & (3) \end{array}$$

Substitute the value of y from equation (3) in equation (2).

$$x - 3y = 4. \quad (2)$$
$$x - 3(-1) = 4.$$
$$x + 3 = 4.$$
$$x = 1.$$

Check:

$$-x - y = 0. \quad (1) \qquad x - 3y = 4. \quad (2)$$
$$-1 - (-1) = 0. \qquad 1 - 3(-1) = 4.$$
$$-1 + 1 = 0. \qquad 1 + 3 = 4.$$
$$0 = 0. \qquad 4 = 4.$$

Hence, $x = 1$ and $y = -1$.

4. Solve for x and y:

$$2x + 2y = 10. \quad (1)$$
$$x - 2y = -4. \quad (2)$$

Add equations (1) and (2).

$$\begin{array}{rcl} 2x + 2y &=& 10 \quad (1) \\ x - 2y &=& -4 \quad (2) \\ \hline 3x \quad &=& 6 \end{array}$$
$$x = 2. \quad (3)$$

Substitute the value of x from equation (3) in equation (2).

$$x - 2y = -4. \quad (2)$$
$$2 - 2y = -4.$$
$$-2y = -6.$$
$$y = 3.$$

Check:

$$2x + 2y = 10. \ (1) \qquad x - 2y = -4. \ (2)$$
$$2(2) + 2(3) = 10. \qquad 2 - 2(3) = -4.$$
$$4 + 6 = 10. \qquad 2 - 6 = -4.$$
$$10 = 10. \qquad -4 = -4.$$

Hence, $x = 2$ and $y = 3$.

5. Solve for x and y:

$$x + y = 3. \quad (1)$$
$$x = y + 7. \quad (2)$$

Convert equation (2) to the form $x - y = 7$ and add it to equation (1).

$$\begin{array}{rcl} x + y &=& 3 \quad (1) \\ x - y &=& 7 \quad (2a) \\ \hline 2x \quad &=& 10 \end{array}$$
$$x = 5. \quad (3)$$

Substitute the value of x from equation (3) in equation (1).

$$x + y = 3. \quad (1)$$
$$5 + y = 3.$$
$$y = -2.$$

Check:

$$x + y = 3. \ (1) \qquad x = y + 7. \ (2)$$
$$5 + (-2) = 3. \qquad 5 = -2 + 7.$$
$$5 - 2 = 3. \qquad 5 = 5.$$
$$3 = 3.$$

Hence, $x = 5$ and $y = -2$.

Solving Systems of Equations by Subtraction. When a term containing an unknown in one of the equations is *the same as* the term containing the same unknown in the other equation, the unknown can be eliminated by subtraction.

Solve for x and y:

$$2x + y = 4. \quad (1)$$
$$3x + y = 5. \quad (2)$$

Subtract (2) from (1).

$$\begin{array}{rcl} 2x + y &=& 4 \quad (1) \\ 3x + y &=& 5 \quad (2) \\ \hline -x \quad &=& -1 \end{array}$$
$$x = 1. \quad (3)$$

Substitute the value of x from equation (3) in equation (1).

$$2x + y = 4. \quad (1)$$
$$2(1) + y = 4.$$
$$2 + y = 4.$$
$$y = 2.$$

Check:

$$2x + y = 4. \quad (1)$$
$$2(1) + 2 = 4.$$
$$2 + 2 = 4.$$
$$4 = 4.$$

$$3x + y = 5. \quad (2)$$
$$3(1) + 2 = 5.$$
$$3 + 2 = 5.$$
$$5 = 5.$$

6. Solve for x and y:

$$x + 3y = -4. \quad (1)$$
$$x + 2y = -3. \quad (2)$$

Subtract equation (2) from equation (1).

$$\begin{array}{rcl} x + 3y &=& -4 \quad (1) \\ x + 2y &=& -3 \quad (2) \\ \hline y &=& -1 \quad (3) \end{array}$$

Substitute the value of y from equation (3) in equation (1).

$$x + 3y = -4. \quad (1)$$
$$x + 3(-1) = -4.$$
$$x - 3 = -4.$$
$$x = -1.$$

Check:

$$x + 3y = -4. \ (1) \qquad x + 2y = -3. \ (2)$$
$$-1 + 3(-1) = -4. \qquad -1 + 2(-1) = -3.$$
$$-1 - 3 = -4. \qquad -1 - 2 = -3.$$
$$-4 = -4. \qquad -3 = -3.$$

Hence, $x = -1$ and $y = -1$.

7. Solve for x and y:

$$x + y = 9. \quad (1)$$
$$3x + y = 19. \quad (2)$$

Subtract equation (2) from equation (1).

$$\begin{array}{rcl} x + y &=& 9 \quad (1) \\ 3x + y &=& 19 \quad (2) \\ \hline -2x \quad &=& -10 \end{array}$$
$$x = 5. \quad (3)$$

Substitute the value of x from equation (3) in equation (1).

$$x + y = 9. \quad (1)$$
$$5 + y = 9.$$
$$y = 4.$$

Check:

$x + y = 9.$ (1) $3x + y = 19.$ (2)
$5 + 4 = 9.$ $3(5) + 4 = 19.$
$9 = 9.$ $15 + 4 = 19.$
$19 = 19.$

Hence, $x = 5$ and $y = 4$.

8. Solve for x and y:

$$x + y = 0. \quad (1)$$
$$5x = 12 - y. \quad (2)$$

Convert equation (2) to the form $5x + y = 12$ and subtract it from equation (1).

$$x + y = 0 \quad (1)$$
$$5x + y = 12 \quad (2a)$$
$$-4x = -12$$
$$x = 3. \quad (3)$$

Substitute the value of x from equation (3) in equation (1).

$$x + y = 0. \quad (1)$$
$$3 + y = 0.$$
$$y = -3.$$

Check:

$x + y = 0.$ (1) $5x = 12 - y.$ (2)
$3 + (-3) = 0.$ $5(3) = 12 - (-3).$
$3 - 3 = 0.$ $15 = 12 + 3.$
$0 = 0.$ $15 = 15.$

Hence, $x = 3$ and $y = -3$.

9. Solve for x and y:

$$3x = y + 6. \quad (1)$$
$$x = y + 2. \quad (2)$$

In this problem, the y can be left in the right member in both equations and the subtraction completed as the equations are given.

$$3x = y + 6 \quad (1)$$
$$x = y + 2 \quad (2)$$
$$2x = 4$$
$$x = 2. \quad (3)$$

Substitute the value of x from equation (3) in equation (2).

$$x = y + 2. \quad (2)$$
$$2 = y + 2.$$
$$-y = 2 - 2.$$
$$y = 0.$$

Check:

$3x = y + 6.$ (1) $x = y + 2.$ (2)
$3(2) = 0 + 6.$ $2 = 0 + 2.$
$6 = 6.$ $2 = 2.$

Hence, $x = 2$ and $y = 0$.

Solving Systems of Equations by Substitution. One of the equations can be solved for one variable in terms of the other variable. When this value of the variable is substituted in the other equation, the second variable is eliminated.

Solve for x and y:

$$x + 3y = 10. \quad (1)$$
$$x + y = 4. \quad (2)$$

Solve for y in terms of x in (2).

$$y = 4 - x. \quad (2a)$$

Substitute the value of y from equation (2a) in equation (1).

$$x + 3(4 - x) = 10.$$
$$x + 12 - 3x = 10.$$
$$-2x = -2.$$
$$x = 1. \quad (3)$$

Substitute the value of x from equation (3) in equation (2a).

$$y = 4 - x$$
$$= 4 - 1$$
$$= 3.$$

Check:

$x + 3y = 10.$ (1)
$1 + 3(3) = 10.$
$1 + 9 = 10.$
$10 = 10.$

$x + y = 4.$ (2)
$1 + 3 = 4.$
$4 = 4.$

10. Solve for x and y:

$$x + y = 1. \quad (1)$$
$$x = 2y + 7. \quad (2)$$

In equation (2), the value of x is stated in terms of y. Substitute that value in equation (1).

$$x + y = 1. \quad (1)$$
$$2y + 7 + y = 1.$$
$$3y = -6.$$
$$y = -2. \quad (3)$$

Substitute the value of y from equation (3) in equation (2) to find the value of x.

$$x = 2y + 7 \quad (2)$$
$$= 2(-2) + 7$$
$$= -4 + 7$$
$$= 3.$$

Check:

$x + y = 1.$ (1) $x = 2y + 7.$ (2)
$3 - 2 = 1.$ $3 = 2(-2) + 7.$
$1 = 1.$ $3 = -4 + 7.$
$3 = 3.$

Hence, $x = 3$ and $y = -2$.

11. Solve for x and y:

$$2x + y = 2. \quad (1)$$
$$y = x - 10. \quad (2)$$

Substitute the value of y from equation (2) in equation (1).

$$2x + y = 2. \quad (1)$$
$$2x + x - 10 = 2.$$
$$3x = 12.$$
$$x = 4. \quad (3)$$

Substitute the value of x from equation (3) in equation (2).

$$y = x - 10 \quad (2)$$
$$= 4 - 10$$
$$= -6.$$

Check:

$$2x + y = 2. \quad (1) \qquad y = x - 10. \quad (2)$$
$$2(4) + (-6) = 2. \qquad -6 = 4 - 10.$$
$$8 - 6 = 2. \qquad -6 = -6.$$
$$2 = 2.$$

Hence, $x = 4$ and $y = -6$.

12. Solve for x and y:

$$2x + 3y = 8. \quad (1)$$
$$2x = 10 - 2y. \quad (2)$$

Substitute the value of $2x$ as shown in equation (2) for the $2x$ in the equation (1).

$$2x + 3y = 8. \quad (1)$$
$$10 - 2y + 3y = 8.$$
$$y = -2. \quad (3)$$

Substitute the value of y from equation (3) in equation (2).

$$2x = 10 - 2y \quad (2)$$
$$= 10 - 2(-2)$$
$$= 10 + 4$$
$$= 14.$$
$$x = 7.$$

Check:

$$2x + 3y = 8. \quad (1) \qquad 2x = 10 - 2y. \quad (2)$$
$$2(7) + 3(-2) = 8. \qquad 2(7) = 10 - 2(-2).$$
$$14 - 6 = 8. \qquad 14 = 10 + 4.$$
$$8 = 8. \qquad 14 = 14.$$

Hence, $x = 7$ and $y = -2$.

13. Solve for x and y:

$$x - y = -2. \quad (1)$$
$$2x - 3y = -1. \quad (2)$$

Solve the first equation for x in terms of y.

$$x = y - 2. \quad (1a)$$

Substitute the value of x from equation (1a) in equation (2).

$$2x - 3y = -1. \quad (2)$$
$$2(y - 2) - 3y = -1.$$
$$2y - 4 - 3y = -1.$$
$$-y = 3.$$
$$y = -3. \quad (3)$$

Substitute the value of y from equation (3) in equation (1a).

$$x = y - 2 \quad (1a)$$
$$= -3 - 2$$
$$= -5.$$

Check:

$$x - y = -2. \quad (1) \qquad 2x - 3y = -1. \quad (2)$$
$$-5 - (-3) = -2. \qquad 2(-5) - 3(-3) = -1.$$
$$-5 + 3 = -2. \qquad -10 + 9 = -1.$$
$$-2 = -2. \qquad -1 = -1.$$

Hence, $x = -5$ and $y = -3$.

14. Solve for x and y:

$$3x - 5y = 4. \quad (1)$$
$$x = y. \quad (2)$$

Substitute the value of x from equation (2) in equation (1).

$$3x - 5y = 4. \quad (1)$$
$$3y - 5y = 4.$$
$$-2y = 4.$$
$$y = -2. \quad (3)$$

Substitute the value of y from equation (3) in equation (2).

$$x = y \quad (2)$$
$$= -2.$$

Check:

$$3x - 5y = 4. \quad (1) \qquad x = y. \quad (2)$$
$$3(-2) - 5(-2) = 4. \qquad -2 = -2.$$
$$-6 + 10 = 4.$$
$$4 = 4.$$

Hence, $x = -2$ and $y = -2$.

Changing the Form of Equations. The methods of elimination cannot be applied directly to many sets of equations. In such cases, changes must be made so that the equations are in the proper form for elimination of an unknown by addition or subtraction. This involves such procedures as clearing of fractions or multiplying equations by appropriate constants.

15. Solve for x and y:

$$3x + 2y = 17. \quad (1)$$
$$2x - y = 9. \quad (2)$$

Before this problem can be worked by addition, the $-y$ must become $-2y$ so the y-terms will be eliminated. This involves the use of the multiplication axiom (p. 27). If both members of the second equation are multiplied by 2, the equality is maintained and the system can be solved by addition.

$$3x + 2y = 17 \quad (1)$$
$$4x - 2y = 18 \quad (2a)$$
$$7x = 35.$$
$$x = 5. \quad (3)$$

Substitute the value of x from equation (3) in equation (1).

$$3x + 2y = 17. \quad (1)$$
$$3(5) + 2y = 17.$$
$$15 + 2y = 17.$$
$$2y = 2.$$
$$y = 1.$$

Check:

$$3x + 2y = 17. \quad (1) \qquad 2x - y = 9. \quad (2)$$
$$3(5) + 2(1) = 17. \qquad 2(5) - 1 = 9.$$
$$15 + 2 = 17. \qquad 10 - 1 = 9.$$
$$17 = 17. \qquad 9 = 9.$$

Hence, $x = 5$ and $y = 1$.

16. Solve for x and y:

$$2x + 3y = 5. \quad (1)$$
$$x + 2y = 3. \quad (2)$$

Multiply both members of equation (2) by 2 and then subtract that new equation from equation (1)

$$2x + 3y = 5 \quad (1)$$
$$2x + 4y = 6 \quad (2a)$$
$$-y = -1.$$
$$y = 1. \quad (3)$$

Substitute the value of y from equation (3) in equation (1).

$$2x + 3y = 5. \quad (1)$$
$$2x + 3(1) = 5.$$
$$2x + 3 = 5.$$
$$2x = 2.$$
$$x = 1.$$

Check:

$$2x + 3y = 5. \quad (1) \qquad x + 2y = 3. \quad (2)$$
$$2(1) + 3(1) = 5. \qquad 1 + 2(1) = 3.$$
$$2 + 3 = 5. \qquad 1 + 2 = 3.$$
$$5 = 5. \qquad 3 = 3.$$

Hence, $x = 1$ and $y = 1$.

17. Solve for x and y:

$$2x - 3y = 13. \quad (1)$$
$$4x + 5y = -7. \quad (2)$$

Multiply both members of equation (1) by 2 and subtract the new equation from equation (2).

$$4x + 5y = -7 \quad (2)$$
$$4x - 6y = 26 \quad (1a)$$
$$11y = -33.$$
$$y = -3. \quad (3)$$

Substitute the value of y from equation (3) in equation (1).

$$2x - 3y = 13. \quad (1)$$
$$2x - 3(-3) = 13.$$
$$2x + 9 = 13.$$
$$2x = 4.$$
$$x = 2.$$

Check:

$$2x - 3y = 13. \quad (1) \qquad 4x + 5y = -7. \quad (2)$$
$$2(2) - 3(-3) = 13. \qquad 4(2) + 5(-3) = -7.$$
$$4 + 9 = 13. \qquad 8 - 15 = -7.$$
$$13 = 13. \qquad -7 = -7.$$

Hence, $x = 2$ and $y = -3$.

18. Solve for x and y:

$$3x + 7y = -17. \quad (1)$$
$$5x - 2y = -1. \quad (2)$$

Multiply equation (1) by 2 and equation (2) by 7.

$$6x + 14y = -34 \quad (1a)$$
$$35x - 14y = -7 \quad (2a)$$
$$41x = -41.$$
$$x = -1. \quad (3)$$

Substitute the value of x from equation (3) in equation (1).

$$3x + 7y = -17. \quad (1)$$
$$3(-1) + 7y = -17.$$
$$-3 + 7y = -17.$$
$$7y = -14.$$
$$y = -2.$$

Check:

$$3x + 7y = -17. \quad (1) \qquad 5x - 2y = -1. \quad (2)$$
$$3(-1) + 7(-2) = -17. \qquad 5(-1) - 2(-2) = -1.$$
$$-3 - 14 = -17. \qquad -5 + 4 = -1.$$
$$-17 = -17. \qquad -1 = -1.$$

Hence, $x = -1$ and $y = -2$.

19. Solve for x and y:

$$5x + 2y = 6. \quad (1)$$
$$3x + 9y = -12. \quad (2)$$

Multiply equation (1) by 3 and equation (2) by 5 and subtract.

$$15x + 6y = 18 \quad (1a)$$
$$15x + 45y = -60 \quad (2a)$$
$$-39y = 78.$$
$$y = -2. \quad (3)$$

Substitute the value of y from equation (3) in equation (2).

$$3x + 9y = -12. \quad (2)$$
$$3x + 9(-2) = -12.$$
$$3x - 18 = -12.$$
$$3x = 6.$$
$$x = 2.$$

Check:

$$5x + 2y = 6. \quad (1) \qquad 3x + 9y = -12. \quad (2)$$
$$5(2) + 2(-2) = 6. \qquad 3(2) + 9(-2) = -12.$$
$$10 - 4 = 6. \qquad 6 - 18 = -12.$$
$$6 = 6. \qquad -12 = -12.$$

Hence, $x = 2$ and $y = -2$.

20. Solve for x and y:

$$\frac{2x}{3} - \frac{y}{2} = -\frac{1}{6}. \quad (1)$$
$$\frac{x}{2} + \frac{2y}{3} = 3. \quad (2)$$

Find a common denominator in both equations and then clear the equations of fractions.

$$\frac{4x}{6} - \frac{3y}{6} = -\frac{1}{6}.$$
$$4x - 3y = -1. \quad (1a)$$
$$\frac{3x}{6} + \frac{4y}{6} = \frac{18}{6}.$$
$$3x + 4y = 18. \quad (2a)$$

Now solve as in the preceding problems. Multiply equation (1a) by 4 and equation (2a) by 3 and add.

$$16x - 12y = -4 \quad (1b)$$
$$9x + 12y = 54 \quad (2b)$$
$$25x = 50.$$
$$x = 2. \quad (3)$$

Substitute the value of x from equation (3) in equation (2a).

$$3x + 4y = 18. \quad (2a)$$
$$3(2) + 4y = 18.$$
$$6 + 4y = 18.$$
$$4y = 12.$$
$$y = 3.$$

Check:

$$\frac{2x}{3} - \frac{y}{2} = -\frac{1}{6}. \qquad \frac{x}{2} + \frac{2y}{3} = 3.$$
$$\frac{2(2)}{3} - \frac{3}{2} = -\frac{1}{6}. \qquad \frac{2}{2} + \frac{2(3)}{3} = 3.$$
$$\frac{8}{6} - \frac{9}{6} = -\frac{1}{6}. \qquad 1 + 2 = 3.$$
$$-\frac{1}{6} = -\frac{1}{6}. \qquad 3 = 3.$$

Hence, $x = 2$ and $y = 3$.

21. Solve for x and y:

$$x + \frac{y}{4} = -\frac{1}{2}. \quad (1)$$
$$\frac{x}{3} + \frac{5y}{3} = 3. \quad (2)$$

Find a common denominator in each equation and then clear of fractions in both equations.

$$\frac{4x}{4} + \frac{y}{4} = -\frac{2}{4}.$$
$$4x + y = -2. \quad (1a)$$
$$\frac{x}{3} + \frac{5y}{3} = \frac{9}{3}.$$
$$x + 5y = 9. \quad (2a)$$

Multiply equation (1a) by 5 and from that subtract equation (2a).

$$20x + 5y = -10 \quad (1b)$$
$$x + 5y = 9 \quad (2a)$$
$$19x = -19$$
$$x = -1. \quad (3)$$

Substitute the value of x from equation (3) in equation (1a).

$$4x + y = -2. \quad (1a)$$
$$4(-1) + y = -2.$$
$$-4 + y = -2.$$
$$y = 2.$$

Check:

$$x + \frac{y}{4} = -\frac{1}{2}. \quad (1) \qquad \frac{x}{3} + \frac{5y}{3} = 3. \quad (2)$$
$$(-1) + \frac{2}{4} = -\frac{1}{2}. \qquad \frac{-1}{3} + \frac{5(2)}{3} = 3.$$
$$-\frac{4}{4} + \frac{2}{4} = -\frac{2}{4}. \qquad -\frac{1}{3} + \frac{10}{3} = \frac{9}{3}.$$
$$-\frac{2}{4} = -\frac{2}{4}. \qquad \frac{9}{3} = \frac{9}{3}.$$

Hence, $x = -1$ and $y = 2$.

22. Solve for x and y:

$$\frac{1}{x+y} - \frac{1}{x-y} = -\frac{2}{x^2 - y^2}. \quad (1)$$
$$\frac{2}{x-y} + \frac{3}{x+y} = \frac{9}{x^2 - y^2}. \quad (2)$$

Find a common denominator in each equation and clear both equations of fractions.

$$\frac{1(x-y)}{(x+y)(x-y)} - \frac{1(x+y)}{(x+y)(x-y)} = \frac{-2}{(x+y)(x-y)}.$$
$$x - y - x - y = -2.$$
$$-2y = -2.$$
$$y = 1. \quad (1a)$$
$$\frac{2(x+y)}{(x+y)(x-y)} + \frac{3(x-y)}{(x+y)(x-y)} = \frac{9}{(x+y)(x-y)}.$$
$$2x + 2y + 3x - 3y = 9.$$
$$5x - y = 9. \quad (2a)$$

Substitute the value of y from equation (1a) in equation (2a).

$$5x - y = 9. \quad (2a)$$
$$5x - 1 = 9.$$
$$5x = 10.$$
$$x = 2.$$

Check:

$$\frac{1}{x+y} - \frac{1}{x-y} = -\frac{2}{x^2-y^2}. \quad (1) \qquad \frac{2}{x-y} + \frac{3}{x+y} = \frac{9}{x^2-y^2}. \quad (2)$$

$$\frac{1}{2+1} - \frac{1}{2-1} = -\frac{2}{4-1}. \qquad \frac{2}{2-1} + \frac{3}{2+1} = \frac{9}{4-1}.$$

$$\tfrac{1}{3} - 1 = -\tfrac{2}{3}. \qquad 2 + \tfrac{3}{3} = \tfrac{9}{3}.$$

$$\tfrac{1}{3} - \tfrac{3}{3} = -\tfrac{2}{3}. \qquad 2 + 1 = 3.$$

$$-\tfrac{2}{3} = -\tfrac{2}{3}. \qquad 3 = 3.$$

Hence, $x = 2$ and $y = 1$.

Three Equations in Three Unknowns. The methods of elimination can also be used to solve a set of three equations involving three unknowns. When all equations contain all variables, two of the equations are used to eliminate one of the unknowns by either addition or subtraction. The remaining equation is used with one of the first two to eliminate the same variable. This leaves two equations involving two variables. Now the procedure is the same as was used in the previous problems.

23. Solve for x, y, and z:

$$x + y + z = 4. \quad (1)$$
$$x - y + z = 0. \quad (2)$$
$$x - y - z = -2. \quad (3)$$

Add equations (1) and (2) to eliminate y.

$$x + y + z = 4 \quad (1)$$
$$x - y + z = 0 \quad (2)$$
$$2x + 2z = 4$$
$$x + z = 2. \quad (4)$$

Add equations (1) and (3) to eliminate y.

$$x + y + z = 4 \quad (1)$$
$$x - y - z = -2 \quad (3)$$
$$2x = 2$$
$$x = 1. \quad (5)$$

In this particular system of equations, z also disappears, but this is not always the case.

Substitute the value of x from equation (5) in equation (4) to find the value of z.

$$x + z = 2. \quad (4)$$
$$1 + z = 2.$$
$$z = 1. \quad (6)$$

Substitute the values of x and z from equations (5) and (6) in equation (1) to find the value of y.

$$x + y + z = 4. \quad (1)$$
$$1 + y + 1 = 4.$$
$$y = 2.$$

Check:

$$x + y + z = 4. \quad (1) \qquad x - y + z = 0. \quad (2)$$
$$1 + 2 + 1 = 4. \qquad 1 - 2 + 1 = 0.$$
$$4 = 4. \qquad 0 = 0.$$
$$x - y - z = -2. \quad (3)$$
$$1 - 2 - 1 = -2.$$
$$-2 = -2.$$

Hence, $x = 1$, $y = 2$, and $z = 1$.

24. Solve for x, y, and z:

$$x + y - 2z = 4. \quad (1)$$
$$2x + y + 2z = 0. \quad (2)$$
$$x - 3y - 4z = -2. \quad (3)$$

Add equations (1) and (2).

$$x + y - 2z = 4 \quad (1)$$
$$2x + y + 2z = 0 \quad (2)$$
$$3x + 2y = 4 \quad (4)$$

Multiply equation (2) by 2 and add it to equation (3).

$$x - 3y - 4z = -2 \quad (3)$$
$$4x + 2y + 4z = 0 \quad (2a)$$
$$5x - y = -2 \quad (5)$$

Consider equations (4) and (5), observing that each is composed of the two variables, x and y.

$$3x + 2y = 4. \quad (4)$$
$$5x - y = -2. \quad (5)$$

Multiply equation (5) by 2 and add the result to equation (4).

$$3x + 2y = 4 \quad (4)$$
$$10x - 2y = -4 \quad (5a)$$
$$13x = 0$$
$$x = 0. \quad (6)$$

Substitute the value of x from equation (6) in equation (5).

$$5x - y = -2. \quad (5)$$
$$5(0) - y = -2.$$
$$0 - y = -2.$$
$$y = 2. \quad (7)$$

Substitute the values of x and y from equations (6) and (7) in equation (2) to find the value of z.

$$2x + y + 2z = 0. \quad (2)$$
$$2(0) + 2 + 2z = 0.$$
$$0 + 2 + 2z = 0.$$
$$2z = -2.$$
$$z = -1.$$

Check:

$$x + y - 2z = 4. \quad (1) \qquad 2x + y + 2z = 0. \quad (2)$$
$$0 + 2 - 2(-1) = 4. \qquad 2(0) + 2 + 2(-1) = 0.$$
$$2 + 2 = 4. \qquad 2 - 2 = 0.$$
$$4 = 4. \qquad 0 = 0.$$
$$x - 3y - 4z = -2. \quad (3)$$
$$0 - 3(2) - 4(-1) = -2.$$
$$0 - 6 + 4 = -2.$$
$$-2 = -2.$$

Hence, $x = 0$, $y = 2$, and $z = -1$.

25. Solve for a, b, and c:

$$2a - b + c = 0. \quad (1)$$

$$3a + 2b + 2c = 10. \quad (2)$$
$$2a + 3b - 4c = 17. \quad (3)$$

Multiply equation (1) by 2 and add it to equation (2).

$$\begin{array}{rcl} 3a + 2b + 2c = 10 & & (2) \\ 4a - 2b + 2c = 0 & & (1a) \\ \hline 7a \quad + 4c = 10 & & (4) \end{array}$$

Multiply equation (1) by 3 and add it to equation (3) to eliminate b.

$$\begin{array}{rcl} 2a + 3b - 4c = 17 & & (3) \\ 6a - 3b + 3c = 0 & & (1b) \\ \hline 8a \quad - c = 17 & & (5) \end{array}$$

Multiply equation (5) by 4 and add it to equation (4).

$$\begin{array}{rcl} 7a + 4c = 10 & & (4) \\ 32a - 4c = 68 & & (5a) \\ \hline 39a \quad = 78 & & \\ a = 2. & & (6) \end{array}$$

Substitute the value of a from equation (6) in equation (4).

$$7a + 4c = 10. \quad (4)$$
$$7(2) + 4c = 10.$$
$$14 + 4c = 10.$$
$$4c = -4.$$
$$c = -1. \quad (7)$$

Substitute the values of a and c from equations (6) and (7) in equation (1).

$$2a - b + c = 0.$$
$$2(2) - b + (-1) = 0.$$
$$4 - b - 1 = 0.$$
$$-b = -3.$$
$$b = 3.$$

Check:

$$2a - b + c = 0. \quad (1) \qquad 3a + 2b + 2c = 10. \quad (2)$$
$$2(2) - 3 + (-1) = 0. \qquad 3(2) + 2(3) + 2(-1) = 10.$$
$$4 - 3 - 1 = 0. \qquad 6 + 6 - 2 = 10.$$
$$0 = 0. \qquad 10 = 10.$$
$$2a + 3b - 4c = 17. \quad (3)$$
$$2(2) + 3(3) - 4(-1) = 17.$$
$$4 + 9 + 4 = 17.$$
$$17 = 17.$$

Hence, $a = 2$, $b = 3$, and $c = -1$.

26. Solve for m, n, and p:

$$m - n = 2. \quad (1)$$
$$n + p = -1. \quad (2)$$
$$m - p = 5. \quad (3)$$

Add equations (1) and (2).

$$\begin{array}{rcl} m - n \quad = 2 & & (1) \\ + n + p = -1 & & (2) \\ \hline m \quad + p = 1 & & (4) \end{array}$$

Add equations (3) and (4).

$$\begin{array}{rcl} m - p = 5 & & (3) \\ m + p = 1 & & (4) \\ \hline 2m \quad = 6 & & \\ m = 3. & & (5) \end{array}$$

Substitute the value of m from equation (5) in equation (1).

$$m - n = 2. \quad (1)$$
$$3 - n = 2.$$
$$-n = -1.$$
$$n = 1. \quad (6)$$

Substitute the value of n from equation (6) in equation (2).

$$n + p = -1.$$
$$1 + p = -1.$$
$$p = -2.$$

Check:

$$m - n = 2. \quad (1) \qquad n + p = -1. \quad (2)$$
$$3 - 1 = 2. \qquad 1 + (-2) = -1.$$
$$2 = 2. \qquad 1 - 2 = -1.$$
$$-1 = -1.$$
$$m - p = 5. \quad (3)$$
$$3 - (-2) = 5.$$
$$3 + 2 = 5.$$
$$5 = 5.$$

Hence, $m = 3$, $n = 1$, and $p = -2$.

27. Solve for x, y, and z:

$$x = y + z + 4. \quad (1)$$
$$x + y = -z. \quad (2)$$
$$x - 3y + 3z = -4. \quad (3)$$

By transposing terms, convert equations (1) and (2) into the forms (1a) and (2a) respectively as shown below.

$$x - y - z = 4 \quad (1a)$$
$$x + y + z = 0 \quad (2a)$$
$$x - 3y + 3z = -4. \quad (3)$$

Add equations (1a) and (2a).

$$\begin{array}{rcl} x - y - z = 4 & & (1a) \\ x + y + z = 0 & & (2a) \\ \hline 2x \quad = 4 & & \\ x = 2. & & (4) \end{array}$$

Multiply equation (2a) by 3 and to that result add equation (3).

$$\begin{array}{rcl} 3x + 3y + 3z = 0 & & (2b) \\ x - 3y + 3z = -4 & & (3) \\ \hline 4x \quad + 6z = -4 & & \\ 2x + 3z = -2. & & (5) \end{array}$$

Substitute the value of x from equation (4) in equation (5).

$$2x + 3z = -2. \qquad (5)$$
$$2(2) + 3z = -2.$$
$$4 + 3z = -2.$$
$$3z = -6.$$
$$z = -2. \qquad (6)$$

Substitute the values of x and z from equations (4) and (6) in equation (2).

$$x + y = -z. \qquad (2)$$
$$2 + y = -(-2).$$
$$2 + y = 2.$$
$$y = 0.$$

Check:

$$x = y + z + 4. \quad (1) \qquad x + y = -z. \quad (2)$$
$$2 = 0 + (-2) + 4. \qquad 2 + 0 = -(-2).$$
$$2 = -2 + 4. \qquad 2 = 2.$$
$$2 = 2.$$

$$x - 3y + 3z = -4. \quad (3)$$
$$2 - 3(0) + 3(-2) = -4.$$
$$2 - 0 - 6 = -4.$$
$$-4 = -4.$$

Hence, $x = 2$, $y = 0$, and $z = -2$.

Verbal Problems in More Than One Unknown. The verbal problems in Chapters 4 and 5 were set up in terms of one unknown. In many problems it is more convenient to use two or three variables to represent the unknown quantities.

28. The length of a rectangle is 4 yards more than its width. The perimeter is 76 yards. What are the dimensions of the rectangle?

Let L = length;
W = width.

$$L - W = 4. \qquad (1)$$
$$2L + 2W = 76. \qquad (2)$$

Multiply equation (1) by 2 and to it add equation (2).

$$2L - 2W = 8 \qquad (1a)$$
$$2L + 2W = 76 \qquad (2)$$
$$4L = 84$$
$$L = 21 \text{ yards}. \qquad (3)$$

Substitute the value of L from equation (3) in equation (1).

$$L - W = 4. \qquad (1)$$
$$21 - W = 4.$$
$$-W = -17.$$
$$W = 17 \text{ yards}.$$

Check:

$$L - W = 4. \quad (1) \qquad 2L + 2W = 76. \quad (2)$$
$$21 - 17 = 4. \qquad 2(21) + 2(17) = 76.$$
$$4 = 4. \qquad 42 + 34 = 76.$$
$$76 = 76.$$

Hence, the rectangle is 21 yards long and 17 yards wide.

29. A school has 650 students enrolled. If there were 25 more boys, there would be twice as many boys as girls. How many of each are enrolled in the school?

Let b = number of boys;
g = number of girls.

$$b + g = 650. \qquad (1)$$
$$b + 25 = 2g. \qquad (2)$$

Convert equation (2) into the form shown below.

$$b - 2g = -25. \qquad (2a)$$

Subtract equation (2*a*) from equation (1).

$$b + g = 650 \qquad (1)$$
$$b - 2g = -25 \qquad (2a)$$
$$3g = 675$$
$$g = 225. \qquad (3)$$

Substitute the value of g from equation (3) in equation (1).

$$b + g = 650. \qquad (1)$$
$$b + 225 = 650.$$
$$b = 650 - 225$$
$$= 425.$$

Check:

$$b + g = 650. \quad (1) \qquad b + 25 = 2g. \quad (2)$$
$$425 + 225 = 650. \qquad 425 + 25 = 2(225).$$
$$650 = 650. \qquad 450 = 450.$$

30. A man made two investments with his \$32,000 inheritance. The first one paid 5% interest and the second, 4½%. If his annual income from the two investments was \$1,500, how much did he have invested in each?

Let x = amount invested at 5%;
y = amount invested at 4½%;

$$x + y = \$32{,}000. \qquad (1)$$

$0.05x$ = income from 5% investment;
$0.045y$ = income from 4½% investment;

$$0.05x + 0.045y = \$1{,}500. \qquad (2)$$

Multiply both members of equation (2) by 1,000 to remove the decimals.

$$50x + 45y = \$1{,}500{,}000. \qquad (2a)$$

Divide both members of equation (2a) by 5 to reduce it to lowest terms.

$$10x + 9y = \$300{,}000. \qquad (2b)$$
$$x + y = \$\ 32{,}000. \qquad (1)$$

Multiply equation (1) by 9 and subtract that value from equation (2b).

$$10x + 9y = \$300{,}000 \quad (2b)$$
$$9x + 9y = \$288{,}000 \quad (1a)$$
$$x = \$\ 12{,}000 \quad (3)$$

Substitute the value of x from equation (3) in equation (1).

$$x + y = \$32{,}000. \quad (1)$$
$$\$12{,}000 + y = \$32{,}000.$$
$$y = \$20{,}000.$$

Check:

$$x + y = \$32{,}000. \quad (1)$$
$$\$12{,}000 + \$20{,}000 = \$32{,}000.$$
$$\$32{,}000 = \$32{,}000.$$

$$0.05x + 0.045y = \$1{,}500. \quad (2)$$
$$0.05(\$12{,}000) + 0.045(\$20{,}000) = \$1{,}500.$$
$$\$600 + \$900 = \$1{,}500.$$
$$\$1{,}500 = \$1{,}500.$$

31. John and Jack each had the same number of envelopes. When John gave Jack one-fourth of his envelopes, Jack had 30 envelopes. How many did each have at the beginning?

Let x = the number Jack had;
y = the number John had.

$$y = x. \quad (1)$$
$$x + \frac{y}{4} = 30. \quad (2)$$
$$4x + y = 120. \quad (2a)$$

Substitute equation (1) in equation (2a)

$$4x + y = 120. \quad (2a)$$
$$4x + x = 120.$$
$$5x = 120.$$
$$x = 24. \quad (3)$$

Substitute the value of x from equation (3) in equation (1).

$$y = x. \quad (1)$$
$$y = 24.$$

Check:

$$y = x. \quad (1)$$
$$24 = 24.$$

$$x + \frac{y}{4} = 30. \quad (2)$$
$$24 + \frac{24}{4} = 30.$$
$$24 + 6 = 30.$$
$$30 = 30.$$

Hence, they had 24 envelopes each at the beginning.

32. The ages of two brothers total 35 years. When the first one was two-thirds of his present age and the second one was three-fourths of his present age, the sum of their ages was 25 years. How old is each at the present time?

Let x = age of first brother;
y = age of second brother.

$$x + y = 35. \quad (1)$$
$$\frac{2x}{3} + \frac{3y}{4} = 25. \quad (2)$$
$$\frac{8x}{12} + \frac{9y}{12} = \frac{300}{12}.$$
$$8x + 9y = 300. \quad (2a)$$

Multiply equation (1) by 9 and from that result subtract equation (2a).

$$9x + 9y = 315 \quad (1a)$$
$$8x + 9y = 300 \quad (2a)$$
$$x = 15 \quad (3)$$

Substitute the value of x from equation (3) in equation (1).

$$x + y = 35. \quad (1)$$
$$15 + y = 35.$$
$$y = 20.$$

Check:

$$x + y = 35. \quad (1)$$
$$15 + 20 = 35.$$
$$35 = 35.$$

$$\frac{2x}{3} + \frac{3y}{4} = 25. \quad (2)$$
$$\frac{2(15)}{3} + \frac{3(20)}{4} = 25.$$
$$10 + 15 = 25.$$
$$25 = 25.$$

Hence, their ages are 15 and 20 years.

33. A grocer sells two brands of coffee, one for 80 cents per pound and the other for 90 cents per pound. From the two brands he mixes 50 pounds together to sell for 88 cents per pound. How many pounds of each must he use for the mixture?

Let x = amount of 80-cent coffee;
y = amount of 90-cent coffee.

$$x + y = 50. \quad (1)$$
$$0.80x + 0.90y = .88(50). \quad (2)$$

Multiply both members of equation (2) by 10.

$$8x + 9y = 440. \quad (2a)$$

Multiply equation (1) by 8 and from that value subtract equation (2a).

$$8x + 8y = 400 \quad (1a)$$
$$8x + 9y = 440 \quad (2a)$$
$$-y = -40$$
$$y = 40 \text{ pounds}. \quad (3)$$

Substitute the value of y from equation (3) in equation (1).

$$x + y = 50. \quad (1)$$
$$x + 40 = 50.$$
$$x = 10 \text{ pounds}.$$

Check:

$$x + y = 50. \quad (1) \qquad 0.80x + 0.90y = 0.88(50). \quad (2)$$
$$10 + 40 = 50. \qquad 0.80(10) + 0.90(40) = 0.88(50).$$
$$50 = 50. \qquad \$8 + \$36 = \$44.$$
$$\$44 = \$44.$$

34. A man has three houses from which he receives a total of \$300 rent per month. If he increases the rent 10% on the first two houses and lowers the rent on the third one by \$17.50, his income remains the same. If he increases the rent on the first one by $66\frac{2}{3}\%$ it will equal the rent on the third one. What is the present monthly income from each house?

Let x = monthly income from first house;
y = monthly income from second house;
z = monthly income from third house.

$$x + y + z = \$300. \quad (1)$$
$$1.10x + 1.10y + z - \$17.50 = \$300. \quad (2)$$
$$x + 0.66\tfrac{2}{3}x = z.$$
$$x + \tfrac{2}{3}x = z.$$
$$\frac{5x}{3} = z. \quad (3)$$

Multiply equation (2) by 10.

$$11x + 11y + 10z - \$175 = \$3,000.$$
$$11x + 11y + 10z = \$3,175. \quad (2a)$$

Multiply equation (1) by 11 and from that value subtract equation (2a).

$$11x + 11y + 11z = \$3,300 \quad (1a)$$
$$11x + 11y + 10z = \$3,175 \quad (2a)$$
$$z = \$\ 125 \quad (4)$$

Substitute the value of z from equation (4) in equation (3).

$$\frac{5x}{3} = z. \quad (3)$$
$$\frac{5x}{3} = \$125.$$
$$5x = \$375.$$
$$x = \$75. \quad (5)$$

Substitute the values of z and x from equations (4) and (5) in equation (1).

$$x + y + z = \$300. \quad (1)$$
$$\$75 + y + \$125 = \$300.$$
$$y = \$100.$$

Check:

$$x + y + z = \$300. \quad (1)$$
$$\$75 + \$100 + \$125 = \$300.$$
$$\$300 = \$300.$$

$$1.10x + 1.10y + z - \$17.50 = \$300. \quad (2)$$
$$1.10(\$75) + 1.10(\$100) + \$125 - \$17.50 = \$300.$$
$$\$82.50 + \$110 + \$125 - \$17.50 = \$300.$$
$$\$300 = \$300.$$

$$\frac{5x}{3} = z. \quad (3)$$
$$\frac{5(\$75)}{3} = \$125.$$
$$\frac{\$375}{3} = \$125.$$
$$\$125 = \$125.$$

35. The sum of three numbers is 17. The first plus three times the second and twice the third equals 25. The sum of twice the second and three times the third is equal to the first. What are the three numbers?

Let x = the first number;
y = the second number;
z = the third number.

$$x + y + z = 17. \quad (1)$$
$$x + 3y + 2z = 25. \quad (2)$$
$$2y + 3z = x. \quad (3)$$
$$x - 2y - 3z = 0. \quad (3a)$$

Subtract equation (1) from equation (2).

$$x + 3y + 2z = 25 \quad (2)$$
$$x + y + z = 17 \quad (1)$$
$$2y + z = 8 \quad (4)$$

Subtract equation (3a) from equation (2).

$$x + 3y + 2z = 25 \quad (2)$$
$$x - 2y - 3z = 0 \quad (3a)$$
$$5y + 5z = 25$$
$$y + z = 5. \quad (5)$$

Subtract equation (5) from equation (4).

$$2y + z = 8 \quad (4)$$
$$y + z = 5 \quad (5)$$
$$y = 3 \quad (6)$$

Substitute the value of y from equation (6) in equation (5).

$$y + z = 5. \quad (5)$$
$$3 + z = 5.$$
$$z = 2. \quad (7)$$

Substitute the values of y and z from equations (6) and (7) in equation (1).

$$x + y + z = 17. \quad (1)$$
$$x + 3 + 2 = 17.$$
$$x = 12.$$

Check:

$$x + y + z = 17. \quad (1) \qquad x + 3y + 2z = 25. \quad (2)$$
$$12 + 3 + 2 = 17. \qquad 12 + 3(3) + 2(2) = 25.$$
$$17 = 17. \qquad 12 + 9 + 4 = 25.$$
$$25 = 25.$$

$$2y + 3z = x. \quad (3)$$
$$2(3) + 3(2) = 12.$$
$$6 + 6 = 12.$$
$$12 = 12.$$

Hence, the numbers are 12, 3, and 2.

36. In a two-digit number, the integer in the tens' place is twice the integer in the units' place. The sum of the digits is six more than the difference of the digits. What is the number?

Let x = the units' digit;
y = the tens' digit.

$$2x = y. \quad (1)$$
$$y + x = y - x + 6. \quad (2)$$
$$2x = 6.$$
$$x = 3. \quad (2a)$$

Substitute the value of x from equation (2a) in equation (1).

$$2x = y. \quad (1)$$
$$2(3) = y.$$
$$6 = y.$$
$$y = 6.$$

Check:

$$2x = y. \quad (1) \qquad y + x = y - x + 6. \quad (2)$$
$$2(3) = 6. \qquad 6 + 3 = 6 - 3 + 6.$$
$$6 = 6. \qquad 9 = 9.$$

Hence, the number is 63.

SOLVING SYSTEMS OF LINEAR EQUATIONS BY DETERMINANTS

Determinants. If a, b, c, and d are any real numbers the expression

$$\begin{vmatrix} a & b \\ c & d \end{vmatrix}$$

is called a **second order determinant** because of the *two* rows and *two* columns.

The four symbols flanked by the two vertical lines are the **elements** of the determinant. The **value** of a second order determinant is obtained by subtracting the product of c and b (read diagonally upward from the lower left hand corner) from the product of a and d (read diagonally downward from the upper left hand corner.)

$$\begin{vmatrix} a & b \\ c & d \end{vmatrix} = ad - cb.$$

A **third order determinant** contains nine elements, arranged in a square array in three columns and three rows, and flanked by vertical lines.

$$\begin{vmatrix} a_1 & b_1 & c_1 \\ a_2 & b_2 & c_2 \\ a_3 & b_3 & c_3 \end{vmatrix}$$

One way of evaluating a third order determinant is to annex to the right of the original determinant its two first columns, and multiply diagonally downward to obtain the product of each group of three. From the sum of those three products, we subtract the sum of the three products obtained by multiplying diagonally upward. The following array demonstrates this type of evaluation by using arrows.

$$\begin{vmatrix} a_1 & b_1 & c_1 \\ a_2 & b_2 & c_2 \\ a_3 & b_3 & c_3 \end{vmatrix} = \begin{vmatrix} a_1 & b_1 & c_1 \\ a_2 & b_2 & c_2 \\ a_3 & b_3 & c_3 \end{vmatrix} \begin{matrix} a_1 & b_1 \\ a_2 & b_2 \\ a_3 & b_3 \end{matrix}$$

$$= (a_1b_2c_3 + b_1c_2a_3 + c_1a_2b_3) - (a_3b_2c_1 + b_3c_2a_1 + c_3a_2b_1).$$

The expression, $(a_1b_2c_3 + b_1c_2a_3 + c_1a_2b_3) - (a_3b_2c_1 + b_3c_2a_1 + c_3a_2b_1)$, is called the **expansion** of the determinant; and each of the products, $a_1b_2c_3$, $b_1c_2a_3$, etc. is said to be a **term** of the expansion. The (numerical) sum of the terms of the expansion is called the **value** of the determinant, and *the value defines the determinant.*

The following two problems illustrate the aforementioned methods for evaluating second and third order determinants.

Evaluate: $\begin{vmatrix} 2 & -3 \\ 1 & 2 \end{vmatrix}$.

$$\begin{vmatrix} 2 & -3 \\ 1 & 2 \end{vmatrix} = (2)(2) - (1)(-3)$$
$$= (4) - (-3)$$
$$= 4 + 3$$
$$= 7.$$

Evaluate: $\begin{vmatrix} 2 & 1 & 2 \\ 3 & 4 & 1 \\ 1 & -2 & 0 \end{vmatrix}$.

$$\begin{vmatrix} 2 & 1 & 2 \\ 3 & 4 & 1 \\ 1 & -2 & 0 \end{vmatrix} = \begin{vmatrix} 2 & 1 & 2 \\ 3 & 4 & 1 \\ 1 & -2 & 0 \end{vmatrix} \begin{matrix} 2 & 1 \\ 3 & 4 \\ 1 & -2 \end{matrix}$$
$$= [(2)(4)(0) + (1)(1)(1) + (2)(3)(-2)] - [(1)(4)(2) + (-2)(1)(2) + (0)(3)(1)]$$
$$= [0 + 1 - 12] - [8 - 4 + 0]$$
$$= -11 - 4$$
$$= -15.$$

1. Evaluate: $\begin{vmatrix} 5 & 3 \\ 2 & 4 \end{vmatrix}$.

$$\begin{vmatrix} 5 & 3 \\ 2 & 4 \end{vmatrix} = (5)(4) - (2)(3)$$
$$= 20 - 6$$
$$= 14.$$

2. Evaluate: $\begin{vmatrix} 1 & 0 \\ -2 & 3 \end{vmatrix}$.

$$\begin{vmatrix} 1 & 0 \\ -2 & 3 \end{vmatrix} = (1)(3) - (-2)(0)$$
$$= 3 - 0$$
$$= 3.$$

3. Evaluate: $\begin{vmatrix} -2 & 3 \\ -1 & -1 \end{vmatrix}$.

$$\begin{vmatrix} -2 & 3 \\ -1 & -1 \end{vmatrix} = (-2)(-1) - (-1)(3) = (2) - (-3) = 2 + 3 = 5.$$

4. Evaluate: $\begin{vmatrix} 2 & 1 & 1 \\ 2 & 3 & 2 \\ 1 & 1 & 3 \end{vmatrix}$.

$$\begin{vmatrix} 2 & 1 & 1 \\ 2 & 3 & 2 \\ 1 & 1 & 3 \end{vmatrix} = \left|\begin{matrix} 2 & 1 & 1 \\ 2 & 3 & 2 \\ 1 & 1 & 3 \end{matrix}\right| \begin{matrix} 2 & 1 \\ 2 & 3 \\ 1 & 1 \end{matrix}$$

$$= [(2)(3)(3) + (1)(2)(1) + (1)(2)(1)] - [(1)(3)(1) + (1)(2)(2) + (3)(2)(1)]$$
$$= [18 + 2 + 2] - [3 + 4 + 6]$$
$$= 22 - 13$$
$$= 9.$$

5. Evaluate: $\begin{vmatrix} 2 & 0 & 3 \\ -1 & 6 & 1 \\ 5 & 2 & -2 \end{vmatrix}$.

$$\begin{vmatrix} 2 & 0 & 3 \\ -1 & 6 & 1 \\ 5 & 2 & -2 \end{vmatrix} = \left|\begin{matrix} 2 & 0 & 3 \\ -1 & 6 & 1 \\ 5 & 2 & -2 \end{matrix}\right| \begin{matrix} 2 & 0 \\ -1 & 6 \\ 5 & 2 \end{matrix}$$

$$= [(2)(6)(-2) + (0)(1)(5) + (3)(-1)(2)] - [(5)(6)(3) + (2)(1)(2) + (-2)(-1)(0)]$$
$$= [-24 + 0 - 6] - [90 + 4 + 0]$$
$$= -30 - 94$$
$$= -124.$$

Properties of Determinants. The following seven properties will be useful in evaluating determinants.

Property 1. If every element of a row (or column) is zero, the value of the determinant is zero.

Evaluate: $\begin{vmatrix} 1 & 4 & 5 \\ 0 & 0 & 0 \\ 3 & -3 & 2 \end{vmatrix}$.

$$\begin{vmatrix} 1 & 4 & 5 \\ 0 & 0 & 0 \\ 3 & -3 & 2 \end{vmatrix} = \left|\begin{matrix} 1 & 4 & 5 \\ 0 & 0 & 0 \\ 3 & -3 & 2 \end{matrix}\right| \begin{matrix} 1 & 4 \\ 0 & 0 \\ 3 & -3 \end{matrix}$$

It can be ascertained by observation that every product diagonally downward and diagonally upward is zero. Hence, the value of the determinant is zero.

Property 2. If two rows (or two columns) of a determinant are identical the value of the determinant is zero.

Evaluate: $\begin{vmatrix} 1 & 1 & 2 \\ 3 & 3 & -1 \\ -2 & -2 & -2 \end{vmatrix}$.

Row 1 and row 2 are identical.

$$\begin{vmatrix} 1 & 1 & 2 \\ 3 & 3 & -1 \\ -2 & -2 & -2 \end{vmatrix} = \left|\begin{matrix} 1 & 1 & 2 \\ 3 & 3 & -1 \\ -2 & -2 & -2 \end{matrix}\right| \begin{matrix} 1 & 1 \\ 3 & 3 \\ -2 & -2 \end{matrix}$$

$$= [(1)(3)(-2) + (1)(-1)(-2) + (2)(3)(-2)] - [(-2)(3)(2) + (-2)(-1)(1) + (-2)(3)(1)]$$
$$= [-6 + 2 - 12] - [-12 + 2 - 6]$$
$$= [-16] - [-16]$$
$$= -16 + 16$$
$$= 0.$$

Property 3. If the corresponding elements in two rows (or two columns) of a determinant are proportional the value of the determinant is zero.

Evaluate: $\begin{vmatrix} 1 & 2 & 3 \\ 2 & 4 & 6 \\ -1 & 3 & -2 \end{vmatrix}$.

The corresponding elements in row 1 and row 2 are proportional. That is, the first element of the second row is k times the first element of the first row, the second element of the second row is k times the second element of the first row, and the third element of the second row is k times the third element of the first row.

$$\begin{vmatrix} 1 & 2 & 3 \\ 2 & 4 & 6 \\ -1 & 3 & -2 \end{vmatrix} = \left|\begin{matrix} 1 & 2 & 3 \\ 2 & 4 & 6 \\ -1 & 3 & -2 \end{matrix}\right| \begin{matrix} 1 & 2 \\ 2 & 4 \\ -1 & 3 \end{matrix}$$

$$= [(1)(4)(-2) + (2)(6)(-1) + (3)(2)(3)] - [(-1)(4)(3) + (3)(6)(1) + (-2)(2)(2)]$$
$$= [-8 - 12 + 18] - [-12 + 18 - 8]$$
$$= [-2] - [-2]$$
$$= -2 + 2 = 0.$$

Property 4. If any two rows (or two columns) of a determinant are interchanged the value of the determinant is the negative of the original determinant.

Evaluate: $\begin{vmatrix} a & c \\ b & d \end{vmatrix}$.

$$\begin{vmatrix} a & c \\ b & d \end{vmatrix} = ad - bc.$$

Interchange the rows of the original determinant and evaluate.

$$\begin{vmatrix} b & d \\ a & c \end{vmatrix} = bc - ad = -(ad - bc).$$

Property 5. If every element of a row (or column) is multiplied by some constant C the value of the determinant is multiplied by C.

Evaluate: $\begin{vmatrix} 3 & 2 \\ 4 & 5 \end{vmatrix}$.

$$\begin{vmatrix} 3 & 2 \\ 4 & 5 \end{vmatrix} = 15 - 8$$
$$= 7.$$

Multiply each element in column 2 by 3 and evaluate.

$$\begin{vmatrix} 3 & 6 \\ 4 & 15 \end{vmatrix} = 45 - 24$$
$$= 21.$$

Observe that the second answer is three times the first.

Property 6. The value of a determinant is not altered if corresponding rows and columns are interchanged.

Evaluate: $\begin{vmatrix} 1 & 5 & 2 \\ 3 & 8 & 3 \\ 2 & 1 & 6 \end{vmatrix}$.

$$\begin{vmatrix} 1 & 5 & 2 \\ 3 & 8 & 3 \\ 2 & 1 & 6 \end{vmatrix} = \left|\begin{array}{ccc|cc} 1 & 5 & 2 & 1 & 5 \\ 3 & 8 & 3 & 3 & 8 \\ 2 & 1 & 6 & 2 & 1 \end{array}\right.$$
$$= [(1)(8)(6) + (5)(3)(2) + (2)(3)(1)] - [(2)(8)(2) + (1)(3)(1) + (6)(3)(5)]$$
$$= [48 + 30 + 6] - [32 + 3 + 90]$$
$$= 84 - 125$$
$$= -41.$$

Interchange all columns with their corresponding rows and all rows with their corresponding columns. That is, interchange the 1st row with the 1st column, the 2nd row with the 2nd column, and the 3rd row with the 3rd column.

$$\begin{vmatrix} 1 & 3 & 2 \\ 5 & 8 & 1 \\ 2 & 3 & 6 \end{vmatrix} = \left|\begin{array}{ccc|cc} 1 & 3 & 2 & 1 & 3 \\ 5 & 8 & 1 & 5 & 8 \\ 2 & 3 & 6 & 2 & 3 \end{array}\right.$$
$$= [(1)(8)(6) + (3)(1)(2) + (2)(5)(3)] - [(2)(8)(2) + (3)(1)(1) + (6)(5)(3)]$$
$$= [48 + 6 + 30] - [32 + 3 + 90]$$
$$= 84 - 125$$
$$= -41.$$

Property 7. If C is any number, the value of the determinant is unchanged if to each element in any row (or any column) is added C times the corresponding element of any other row (or column).

Evaluate: $\begin{vmatrix} 1 & 3 & -1 \\ 2 & 4 & 2 \\ -1 & -2 & 1 \end{vmatrix}$.

$$\begin{vmatrix} 1 & 3 & -1 \\ 2 & 4 & 2 \\ -1 & -2 & 1 \end{vmatrix} = \left|\begin{array}{ccc|cc} 1 & 3 & -1 & 1 & 3 \\ 2 & 4 & 2 & 2 & 4 \\ -1 & -2 & 1 & -1 & -2 \end{array}\right.$$
$$= [(1)(4)(1) + (3)(2)(-1) + (-1)(2)(-2)] - [(-1)(4)(-1) + (-2)(2)(1) + (1)(2)(3)]$$
$$= [4 - 6 + 4] - [4 - 4 + 6]$$
$$= 2 - 6$$
$$= -4.$$

Now apply Property 7 to the original determinant by multiplying each element of row 3 by 1 and adding the result to each corresponding element in row 1. We get

$$\begin{vmatrix} 1+(-1) & 3+(-2) & -1+1 \\ 2 & 4 & 2 \\ - & -2 & 1 \end{vmatrix}$$
$$= \begin{vmatrix} 0 & 1 & 0 \\ 2 & 4 & 2 \\ -1 & -2 & 1 \end{vmatrix}$$
$$= \left|\begin{array}{ccc|cc} 0 & 1 & 0 & 0 & 1 \\ 2 & 4 & 2 & 2 & 4 \\ -1 & -2 & 1 & -1 & -2 \end{array}\right.$$
$$= [(0)(4)(1) + (1)(2)(-1) + (0)(2)(-2)] - [(-1)(4)(0) + (-2)(2)(0) + (1)(2)(1)]$$
$$= [0-2+0]-[0+0+2]$$
$$= -2 - 2$$
$$= -4.$$

If the value of C in the above problem is carefully chosen, it is often possible to obtain zero elements in the determinant. This can reduce considerably the amount of work involved in evaluating the determinant. At this time it should be observed that when one of the elements to be used to obtain a term in the expansion of a determinant is zero, that term will be equal to zero and need not even be written in the expansion. In the above problem, for example, all that need be written in the expansion is

$$[(1)(2)(-1)] - [(1)(2)(1)]$$
$$= -2 - 2$$
$$= -4.$$

In the next topic to be discussed, the use of Property 7 will reduce substantially the amount of work required to evaluate determinants.

Evaluation by Use of Cofactors. Any determinant whose order is greater than that of a second order determinant may be evaluated by a method utilizing **cofactors** (the minor with the appropriate sign preceding it).

The **minor** of any element in a determinant is the determinant formed when all elements in the same row and column as the given element are eliminated. If a given determinant is in the form

$$\begin{vmatrix} a_1 & b_1 & c_1 \\ a_2 & b_2 & c_2 \\ a_3 & b_3 & c_3 \end{vmatrix}$$

each element has a minor. The minor of a_1, for example, is the determinant $\begin{vmatrix} b_2 & c_2 \\ b_3 & c_3 \end{vmatrix}$. The minor of a determinant is usually designated by the capital letter corresponding to the given element, using with it the same subscript as is found with the element. Thus, from the above given determinant, $B_3 = \begin{vmatrix} a_1 & c_1 \\ a_2 & c_2 \end{vmatrix}$, $B_2 = \begin{vmatrix} a_1 & c_1 \\ a_3 & c_3 \end{vmatrix}$, $C_2 = \begin{vmatrix} a_1 & b_1 \\ a_3 & b_3 \end{vmatrix}$, and so on.

The cofactor of an element is the value of the minor of that element with the appropriate sign preceding the value. To determine whether such a value is positive or negative we locate the element under discussion and determine the row and column in which it is situated. Then we add the number of the row (3 if in the third row) and the number of the column (2 if the second column). If that sum is **odd** (as with B_3 and C_2) the cofactor is negative; if that sum is **even** (as with B_2) the cofactor is positive.

The value of a determinant may be found by using cofactors according to the following procedure:

1. Select any row (or column) of elements.
2. Multiply each element in that row (or column) by its cofactor. Do not neglect to affix the appropriate positive or negative sign to the minor.
3. The value of the determinant is the algebraic sum of all of the products obtained by multiplying the elements in the selected row (or column) by their respective cofactors.

To minimize the labor involved in evaluating a determinant, change the elements in the chosen row (or column) to zeros by applying Property 7. In converting the elements to zero a caution must be heeded: The elements from two rows (or columns) may not be used with each other to change *both* to zero.

In some cases, it may be impractical to convert all except one element of a row (or column) to zero because fractional elements will result in other rows or columns; and the labor saved will be far less than that added as a result of having to work with fractional elements.

Evaluate by using cofactors: $\begin{vmatrix} 2 & 5 & 6 \\ 1 & -2 & -2 \\ 3 & 3 & 4 \end{vmatrix}$.

Select column 1. Then multiply each element of column 1 by its cofactor. We get

$$\begin{vmatrix} 2 & 5 & 6 \\ 1 & -2 & -2 \\ 3 & 3 & 4 \end{vmatrix} = +(2)\begin{vmatrix} -2 & -2 \\ 3 & 4 \end{vmatrix} - (1)\begin{vmatrix} 5 & 6 \\ 3 & 4 \end{vmatrix} + (3)\begin{vmatrix} 5 & 6 \\ -2 & -2 \end{vmatrix}$$

$$= 2[(-2)(4) - (3)(-2)] - 1[(5)(4) - (3)(6)] + 3[(5)(-2) - (-2)(6)]$$

$$= 2[-8 + 6] - 1[20 - 18] + 3[-10 + 12]$$

$$= 2[-2] - 1[2] + 3[2]$$

$$= -4 - 2 + 6 = 0.$$

Now evaluate the same determinant in a different way. Begin by converting the 2 and 3 in the first column to zeros. This will be accomplished through the application of Property 7. First, multiply each element in row 2 by -2 and add each product to the corresponding element in row 1.

$$\begin{vmatrix} 2 & 5 & 6 \\ 1 & -2 & -2 \\ 3 & 3 & 4 \end{vmatrix} = \begin{vmatrix} 2+(-2) & 5+4 & 6+4 \\ 1 & -2 & -2 \\ 3 & 3 & 4 \end{vmatrix} = \begin{vmatrix} 0 & 9 & 10 \\ 1 & -2 & -2 \\ 3 & 3 & 4 \end{vmatrix}.$$

Now multiply each element in row by 2 by -3 and add the products to the corresponding elements in row 3.

$$\begin{vmatrix} 0 & 9 & 10 \\ 1 & -2 & -2 \\ 3 & 3 & 4 \end{vmatrix} = \begin{vmatrix} 0 & 9 & 10 \\ 1 & -2 & -2 \\ 3+(-3) & 3+6 & 4+6 \end{vmatrix} = \begin{vmatrix} 0 & 9 & 10 \\ 1 & -2 & -2 \\ 0 & 9 & 10 \end{vmatrix}.$$

The two zeros will produce products of zero so our only task is to use the element 1 in the first column. We affix a negative sign to the cofactor. The product of the element and its cofactor is

$$-(1)\begin{vmatrix} 9 & 10 \\ 9 & 10 \end{vmatrix}$$

$$= -(1)[(9)(10) - (9)(10)]$$

$$= -(1)[90 - 90] = -1[0] = 0.$$

Note that the value of the determinant in the last step could have been ascertained without the use of the cofactor. Since the first and third rows are identical (both being 0 9 10), Property 2 tells us that the value of the determinant is zero.

6. Evaluate: $\begin{vmatrix} 1 & -1 & -2 \\ 3 & 0 & -1 \\ 2 & 3 & 2 \end{vmatrix}$.

Work with column 2. If the 3 in column 2 is converted to zero by multiplying each element in row 1 by 3 and adding the products to the corresponding elements in row 3, we will have only one product to find. Thus,

$$\begin{vmatrix} 1 & -1 & -2 \\ 3 & 0 & -1 \\ 2 & 3 & 2 \end{vmatrix} = \begin{vmatrix} 1 & -1 & -2 \\ 3 & 0 & -1 \\ 5 & 0 & -4 \end{vmatrix}$$

$$= -(-1)\begin{vmatrix} 3 & -1 \\ 5 & -4 \end{vmatrix}$$

$$= 1[(3)(-4) - (5)(-1)]$$

$$= 1[-12 + 5]$$

$$= 1[-7] = -7.$$

7. Evaluate: $\begin{vmatrix} 2 & 1 & 1 \\ 3 & -1 & 2 \\ 5 & 3 & -2 \end{vmatrix}$.

Select column 3 with which to work, and convert the elements 2 and -2 in that column to zeros. To accomplish this, multiply every element in row 1 by -2 and add the products to the corresponding elements in row 2. Then multiply each element in row 2 by 1 and add the products to the corresponding elements in row 3.

$$\begin{vmatrix} 2 & 1 & 1 \\ 3 & -1 & 2 \\ 5 & 3 & -2 \end{vmatrix} = \begin{vmatrix} 2 & 1 & 1 \\ -1 & -3 & 0 \\ 8 & 2 & 0 \end{vmatrix}$$

$$= +(1)\begin{vmatrix} -1 & -3 \\ 8 & 2 \end{vmatrix}$$

$$= 1[(-1)(2) - (8)(-3)]$$

$$= 1[22] = 22.$$

8. Evaluate: $\begin{vmatrix} 2 & 3 & 5 \\ 4 & -1 & 2 \\ -1 & 2 & -3 \end{vmatrix}$.

Select column 2, and convert the elements 3 and 2 in that column to zeros. To accomplish this, multiply each element in row 2 by 3 and add the products to the corresponding elements in row 1. Then multiply each element in row 2 by 2 and add the products to the corresponding elements in row 3.

$$\begin{vmatrix} 2 & 3 & 5 \\ 4 & -1 & 2 \\ -1 & 2 & -3 \end{vmatrix} = \begin{vmatrix} 14 & 0 & 11 \\ 4 & -1 & 2 \\ 7 & 0 & 1 \end{vmatrix}$$

$$= +(-1)\begin{vmatrix} 14 & 11 \\ 7 & 1 \end{vmatrix}$$

$$= -1[14 - 77]$$

$$= -1[-63] = 63.$$

9. Evaluate: $\begin{vmatrix} 2 & 2 & 3 \\ -2 & 3 & 5 \\ 3 & -5 & 2 \end{vmatrix}$

In order to keep the elements in integral form only one zero can be set up. Select column 1 and convert the -2 to zero by multiplying each element in row 1 by 1 and adding the products to the corresponding elements in row 2.

$$\begin{vmatrix} 2 & 2 & 3 \\ -2 & 3 & 5 \\ 3 & -5 & 2 \end{vmatrix} = \begin{vmatrix} 2 & 2 & 3 \\ 0 & 5 & 8 \\ 3 & -5 & 2 \end{vmatrix}$$

$$= +(2)\begin{vmatrix} 5 & 8 \\ -5 & 2 \end{vmatrix} + (3)\begin{vmatrix} 2 & 3 \\ 5 & 8 \end{vmatrix}$$

$$= 2[(5)(2) - (-5)(8)] + 3[(2)(8) - (5)(3)]$$

$$= 2[10 + 40] + 3[16 - 15]$$

$$= 2[50] + 3[1]$$

$$= 100 + 3 = 103.$$

10. Evaluate: $\begin{vmatrix} 3 & -1 & 2 \\ -1 & 1 & 4 \\ 2 & 4 & 3 \end{vmatrix}$

Select column 1, and convert the elements 3 and 2 in that column to zeros. To accomplish this, multiply all elements in row 2 by 3 and add the products to the corresponding elements in row 1. Then multiply each element in row 2 by 2 and add the products to the corresponding elements in row 3.

$$\begin{vmatrix} 3 & -1 & 2 \\ -1 & 1 & 4 \\ 2 & 4 & 3 \end{vmatrix} = \begin{vmatrix} 0 & 2 & 14 \\ -1 & -1 & 4 \\ 0 & 6 & 11 \end{vmatrix}$$

$$= -(-1)\begin{vmatrix} 2 & 14 \\ 6 & 11 \end{vmatrix}$$

$$= +1[(2)(11) - (6)(14)]$$

$$= 1[22 - 84]$$

$$= 1[-62] = -62.$$

Cramer's Rule. The equations

$$a_1x + b_1y = c_1 \quad (1)$$

$$a_2x + b_2y = c_2 \quad (2)$$

represent a general system of equations in the two variables x and y. The a_1 and a_2 represent numerical coefficients of the variable x, and b_1 and b_2 represent numerical coefficients of y; c_1 and c_2 represent constants. To solve the above system of equations for x we eliminate y by multiplying Equation (1) by b_2 and Equation (2) by b_1 and then subtracting. Thus

$$a_1b_2x + b_1b_2y = b_2c_1 \quad (1a)$$

$$a_2b_1x + b_1b_2y = b_1c_2 \quad (2a)$$

$$a_1b_2x - a_2b_1x = b_2c_1 - b_1c_2$$

$$x(a_1b_2 - a_2b_1) = b_2c_1 - b_1c_2$$

$$x = \frac{b_2c_1 - b_1c_2}{a_1b_2 - a_2b_1} \quad \text{if } a_1b_2 - a_2b_1 \neq 0.$$

If, in a similar manner, we eliminate the terms involving x, we obtain the value of y.

$$y = \frac{a_1c_2 - a_2c_1}{a_1b_2 - a_2b_1} \quad \text{if } a_1b_2 - a_2b_1 \neq 0.$$

Now observe that the denominators for x and y are identical, and could be obtained by evaluating the determinant

$$\begin{vmatrix} a_1 & b_1 \\ a_2 & b_2 \end{vmatrix} = a_1b_2 - a_2b_1.$$

Thus, when solving a system of equations by the use of determinants, the denominators are found by using the coefficients of x and y in both equations as the elements of that determinant. This denominator is sometimes called **delta,** and is represented by the Greek symbol Δ (read "delta").

The numerator for x can be found by the use of determinants if the constants are used as replacements for the coefficients of the x-terms in Δ. Likewise, the numerator for y is obtained by replacing the coefficients of y in Δ with the constants. This method of solving a system of linear equations is known as Cramer's Rule. As long as Δ is not equal to 0, a solution can be found for each variable in the system by using this rule. If Δ is equal to zero and one or more of the numerators is not equal to zero, then the system is inconsistent. If Δ is equal to zero and all of the numerators are also equal to zero, then the system may be either inconsistent or dependent.

Solve the following system of equations by the use of determinants.

$$2x - y = 5.$$
$$x + y = 1.$$

In order to avoid error and confusion, it is sometimes desireable to begin the application of Cramer's Rule by setting up a table of the numbers that are to be used as elements of the determinants. This table should have three columns: x-coefficients (a), y-coefficients (b), and constants (c).

(a)	(b)	(c)
2	−1	5
1	1	1

$$x = \frac{\begin{vmatrix} 5 & -1 \\ 1 & 1 \end{vmatrix}}{\begin{vmatrix} 2 & -1 \\ 1 & 1 \end{vmatrix}} = \frac{(5)(1) - (1)(-1)}{(2)(1) - (1)(-1)} = \frac{5+1}{2+1} = \frac{6}{3} = 2.$$

$$y = \frac{\begin{vmatrix} 2 & 5 \\ 1 & 1 \end{vmatrix}}{\Delta} = \frac{(2)(1) - (1)(5)}{3} = \frac{2-5}{3} = \frac{-3}{3} = -1.$$

Check:

$$2x - y = 5.$$
$$2(2) - (-1) = 5.$$
$$5 = 5.$$

$$x + y = 1.$$
$$2 + (-1) = 1.$$
$$1 = 1.$$

Hence, $x = 2$ and $y = -1$.

11. Solve the following system of equations for x and y. Use determinants.

$$x + 2y = 4.$$
$$3x + 4y = 6.$$

(a)	(b)	(c)
1	2	4
3	4	6

$$x = \frac{\begin{vmatrix} 4 & 2 \\ 6 & 4 \end{vmatrix}}{\begin{vmatrix} 1 & 2 \\ 3 & 4 \end{vmatrix}} = \frac{16-12}{4-6} = \frac{4}{-2} = -2.$$

$$y = \frac{\begin{vmatrix} 1 & 4 \\ 3 & 6 \end{vmatrix}}{\Delta} = \frac{6-12}{-2} = \frac{-6}{-2} = 3.$$

Check:

$$x + 2y = 4.$$
$$-2 + 2(3) = 4.$$
$$-2 + 6 = 4$$
$$4 = 4.$$
$$3x + 4y = 6.$$
$$3(-2) + 4(3) = 6.$$
$$-6 + 12 = 6.$$
$$6 = 6.$$

Hence, $x = -2$ and $y = 3$.

12. Solve the following system of equations for x and y by using determinants.

$$3x - y = 3.$$
$$x + 2y = 8.$$

(a)	(b)	(c)
3	−1	3
1	2	8

$$x = \frac{\begin{vmatrix} 3 & -1 \\ 8 & 2 \end{vmatrix}}{\begin{vmatrix} 3 & -1 \\ 1 & 2 \end{vmatrix}} = \frac{6+8}{6+1} = \frac{14}{7} = 2.$$

$$y = \frac{\begin{vmatrix} 3 & 3 \\ 1 & 8 \end{vmatrix}}{\Delta} = \frac{24-3}{7} \quad \frac{21}{7} = 3.$$

Check:

$$3x - y = 3.$$
$$3(2) - 3 = 3.$$
$$6 - 3 = 3.$$
$$3 = 3.$$
$$x + 2y = 8.$$
$$2 + 2(3) = 8.$$
$$2 + 6 = 8.$$
$$8 = 8.$$

Hence, $x = 2$ and $y = 3$.

13. Solve the following system of equations for x and y by using determinants.

$$2x - 3y = 4.$$
$$x - 5y = 9.$$

(a)	(b)	(c)
2	−3	4
1	−5	9

$$x = \frac{\begin{vmatrix} 4 & -3 \\ 9 & -5 \end{vmatrix}}{\begin{vmatrix} 2 & -3 \\ 1 & -5 \end{vmatrix}} = \frac{-20 + 27}{-10 + 3} = \frac{7}{-7} = -1.$$

$$y = \frac{\begin{vmatrix} 2 & 4 \\ 1 & 9 \end{vmatrix}}{\Delta} = \frac{18 - 4}{-7} = \frac{14}{-7} = -2.$$

Check:

$$2x - 3y = 4.$$
$$2(-1) - 3(-2) = 4.$$
$$-2 + 6 = 4.$$
$$4 = 4.$$
$$x - 5y = 9.$$
$$-1 - 5(-2) = 9.$$
$$-1 + 10 = 9.$$
$$9 = 9.$$

Hence, $x = -1$ and $y = 2$.

14. By using determinants, solve the following system of equations for x and y.

$$x + y = 1.$$
$$3x + 2y = 6.$$

(a)	(b)	(c)
1	1	1
3	2	6

$$x = \frac{\begin{vmatrix} 1 & 1 \\ 6 & 2 \end{vmatrix}}{\begin{vmatrix} 1 & 1 \\ 3 & 2 \end{vmatrix}} = \frac{2 - 6}{2 - 3} = \frac{-4}{-1} = 4.$$

$$y = \frac{\begin{vmatrix} 1 & 1 \\ 3 & 6 \end{vmatrix}}{\Delta} = \frac{6 - 3}{-1} = \frac{3}{-1} = -3.$$

Check:

$$x + y = 1.$$
$$4 + (-3) = 1.$$
$$4 - 3 = 1.$$
$$1 = 1.$$
$$3x + 2y = 6.$$
$$3(4) + 2(-3) = 6.$$
$$12 - 6 = 6.$$
$$6 = 6.$$

Hence, $x = 4$ and $y = -3$.

15. Solve the following system of equations by using determinants.

$$2x + 3y = -10.$$
$$-2x + y = 2.$$

(a)	(b)	(c)
2	3	−10
−2	1	2

$$x = \frac{\begin{vmatrix} -10 & 3 \\ 2 & 1 \end{vmatrix}}{\begin{vmatrix} 2 & 3 \\ -2 & 1 \end{vmatrix}} = \frac{-10 - 6}{2 + 6} = \frac{-16}{8} = -2.$$

$$y = \frac{\begin{vmatrix} 2 & -10 \\ -2 & 2 \end{vmatrix}}{\Delta} = \frac{4 - 20}{8} = \frac{-16}{8} = -2.$$

Check:

$$2x + 3y = -10. \qquad -2x + y = 2.$$
$$2(-2) + 3(-2) = -10. \qquad -2(-2) + (-2) = 2.$$
$$-10 = -10. \qquad 2 = 2.$$

Systems of Linear Equations in Three Unknowns. The equations

$$a_1x + b_1y + c_1z = d_1$$
$$a_2x + b_2y + c_2z = d_2$$
$$a_3x + b_3y + c_3z = d_3$$

represents a general system of three equations in x, y, and z, where a, b, and c, with their subscripts, are the coefficients of x, y, and z, respectively, and each d is a constant.

By a solution similar to that used for a system of equations in two variables, the above system in three variables may be solved for x, y, and z. To do this, we set up and evaluate third order determinants in the same way as we previously set up and evaluated second order determinants. In general, then, the solution of the above system is

(a)	(b)	(c)	(d)
a_1	b_1	c_1	d_1
a_2	b_2	c_2	d_2
a_3	b_3	c_3	d_3

$$x = \frac{\begin{vmatrix} d_1 & b_1 & c_1 \\ d_2 & b_2 & c_2 \\ d_3 & b_3 & c_3 \end{vmatrix}}{\begin{vmatrix} a_1 & b_1 & c_1 \\ a_2 & b_2 & c_2 \\ a_3 & b_3 & c_3 \end{vmatrix}}; \; y = \frac{\begin{vmatrix} a_1 & d_1 & c_1 \\ a_2 & d_2 & c_2 \\ a_3 & d_3 & c_3 \end{vmatrix}}{\Delta}; \; z = \frac{\begin{vmatrix} a_1 & b_1 & d_1 \\ a_2 & b_2 & d_2 \\ a_3 & b_3 & d_3 \end{vmatrix}}{\Delta}.$$

A system of four equations in four variables or a system of five equations in five variables may be solved by the use of determinants in a similar manner. Cofactors may be used in all such solutions. At any time that one or more of the variables are missing in an equation of a system, its coefficient is zero; and, therefore, zero must be included in the appropriate place in the determinant.

Again, the conversion of the row or column that is selected into as many zero elements as possible will reduce substantially the amount of work necessary to complete the solution.

Solve by determinants:

$$x + y + z = 2$$
$$x - y - z = 0$$
$$2x - y + 3z = 9.$$

(a)	(b)	(c)	(d)
1	1	1	2
1	−1	−1	0
2	−1	3	9

$$x = \frac{\begin{vmatrix} 2 & 1 & 1 \\ 0 & -1 & -1 \\ 9 & -1 & 3 \end{vmatrix}}{\begin{vmatrix} 1 & 1 & 1 \\ 1 & -1 & -1 \\ 2 & -1 & 3 \end{vmatrix}}. \qquad y = \frac{\begin{vmatrix} 1 & 2 & 1 \\ 1 & 0 & -1 \\ 2 & 9 & 3 \end{vmatrix}}{\Delta}. \qquad z = \frac{\begin{vmatrix} 1 & 1 & 2 \\ 1 & -1 & 0 \\ 2 & -1 & 9 \end{vmatrix}}{\Delta}.$$

In the numerator multiply row 2 by 1 and add the products to the corresponding elements in row 1. In the denominator multiply row 2 by 1 and add the products to the corresponding elements in row 1.

$$x = \frac{\begin{vmatrix} 2 & 0 & 0 \\ 0 & -1 & -1 \\ 9 & -1 & 3 \end{vmatrix}}{\begin{vmatrix} 2 & 0 & 0 \\ 1 & -1 & -1 \\ 2 & -1 & 3 \end{vmatrix}}$$

$$= \frac{+(2)\begin{vmatrix} -1 & -1 \\ -1 & 3 \end{vmatrix}}{+(2)\begin{vmatrix} -1 & -1 \\ -1 & 3 \end{vmatrix}}$$

$$= \frac{2[-3 - 1]}{2[-3 - 1]} = \frac{-8}{-8}$$

$$= 1.$$

In the numerator multiply column 1 by 1 and add the products to the corresponding elements in column 3.

$$y = \frac{\begin{vmatrix} 1 & 2 & 2 \\ 1 & 0 & 0 \\ 2 & 9 & 5 \end{vmatrix}}{-8}$$

$$= \frac{-(1)\begin{vmatrix} 2 & 2 \\ 9 & 5 \end{vmatrix}}{-8}$$

$$= \frac{-1[10 - 18]}{-8}$$

$$= \frac{-1[-8]}{-8}$$

$$= -1[1]$$

$$= -1.$$

Multiply column 1 by 1 and add the products to the corresponding elements in column 2.

$$z = \frac{\begin{vmatrix} 1 & 2 & 2 \\ 1 & 0 & 0 \\ 2 & 1 & 9 \end{vmatrix}}{-8}$$

$$= \frac{-(1)\begin{vmatrix} 2 & 2 \\ 1 & 9 \end{vmatrix}}{-8}$$

$$= \frac{-1[18 - 2]}{-8}$$

$$= \frac{-1[16]}{-8}$$

$$= -1[-2]$$

$$= 2.$$

Check:

$$x + y + z = 2.$$
$$1 - 1 + 2 = 2.$$
$$2 = 2.$$

$$x - y - z = 0.$$
$$1 - (-1) - (2) = 0.$$
$$1 + 1 - 2 = 0.$$
$$0 = 0.$$

$$2x - y + 3z = 9.$$
$$2(1) - (-1) + 3(2) = 9.$$
$$2 + 1 + 6 = 9.$$
$$9 = 9.$$

Hence, $x = 1$, $y = -1$, and $z = 2$.

16. Solve by determinants:

$$x - y + z = -3$$
$$2x + y - 3z = 5$$
$$x + y - z = 3.$$

(a)	(b)	(c)	(d)
1	−1	1	−3
2	1	−3	5
1	1	−1	3

$$x = \frac{\begin{vmatrix} -3 & -1 & 1 \\ 5 & 1 & -3 \\ 3 & 1 & -1 \end{vmatrix}}{\begin{vmatrix} 1 & -1 & 1 \\ 2 & 1 & -3 \\ 1 & 1 & -1 \end{vmatrix}}.$$

In the numerator, the corresponding elements of rows 1 and 3 are proportional; for, if each element of row 1 is multiplied by the constant -1, the products will be the elements of row 3. By Property 3, therefore, the value of the numerator is zero. Moreover, by the third property of zero (p. 2), if the numerator of a fraction is zero, the value of the fraction is zero if the denominator is not zero. Let us, then, evaluate the denominator. If it is not zero, then the value of x will be zero. Select column 3 and multiply row 1 by 3 and add the products to the corresponding elements in row 2. Then multiply row 1 by 1 and add the products to the corresponding elements in row 3. Then

$$\Delta = \begin{vmatrix} 1 & -1 & 1 \\ 5 & -2 & 0 \\ 2 & 0 & 0 \end{vmatrix}$$

$$= +(1)\begin{vmatrix} 5 & -2 \\ 2 & 0 \end{vmatrix}$$

$$= 4.$$

Thus, $x = 0$ and $\Delta = 4$.

$$y = \frac{\begin{vmatrix} 1 & -3 & 1 \\ 2 & 5 & -3 \\ 1 & 3 & -1 \end{vmatrix}}{\Delta}.$$

In the numerator, select column 3 with which to work. Then multiply row 1 by 3 and add the products to the corresponding elements in row 2. Then multiply row 1 by 1 and add the products to the corresponding elements in row 3.

$$y = \frac{\begin{vmatrix} 1 & -3 & 1 \\ 5 & -4 & 0 \\ 2 & 0 & 0 \end{vmatrix}}{4}$$

$$= \frac{+(1)\begin{vmatrix} 5 & -4 \\ 2 & 0 \end{vmatrix}}{4}$$

$$= \frac{1[0 + 8]}{4}$$

$$= 2.$$

$$z = \frac{\begin{vmatrix} 1 & -1 & -3 \\ 2 & 1 & 5 \\ 1 & 1 & 3 \end{vmatrix}}{\Delta}$$

Select row 1 in the numerator. Multiply column 1 by 1 and add the products to the corresponding elements in column 2. Then multiply column 1 by 3 and add the products to the corresponding elements in column 3.

$$z = \frac{\begin{vmatrix} 1 & 0 & 0 \\ 2 & 3 & 11 \\ 1 & 2 & 6 \end{vmatrix}}{4}$$

$$= \frac{+(1)\begin{vmatrix} 3 & 11 \\ 2 & 6 \end{vmatrix}}{4}$$

$$= \frac{1[18 - 22]}{4}$$

$$= \frac{1[-4]}{4} = -1.$$

Check:

$$x - y + z = -3. \qquad 2x + y - 3z = 5. \qquad x + y - z = 3.$$
$$0 - 2 - 1 = -3. \qquad 2(0) + 1(2) - 3(-1) = 5. \qquad 0 + 2 + 1 = 3.$$
$$-3 = -3. \qquad 0 + 2 + 3 = 5. \qquad 3 = 3.$$
$$5 = 5.$$

Hence, $x = 0$, $y = 2$, and $z = -1$.

17. Solve by determinants:

$$x + 4y + z = 1.$$
$$2x + 2y + 7z = 4.$$
$$2x + 4y - 3z = -5.$$

(a)	(b)	(c)	(d)
1	4	1	1
2	2	7	4
2	4	−3	−5

$$x = \frac{\begin{vmatrix} 1 & 4 & 1 \\ 4 & 2 & 7 \\ -5 & 4 & -3 \end{vmatrix}}{\begin{vmatrix} 1 & 4 & 1 \\ 2 & 2 & 7 \\ 2 & 4 & -3 \end{vmatrix}}.$$

In the numerator, select column 2 with which to work. Multiply row 2 by -2 and add the products to the corresponding elements in row 1. Then multiply row 2 by -2 and add the products to the corresponding elements in row 3. In the denominator, select column 1 with which to work. Multiply row 1 by -2 and add the products to the corresponding elements in row 2. Then multiply row 1 by -2 and add the products to the corresponding elements in row 3.

$$x = \frac{\begin{vmatrix} -7 & 0 & -13 \\ 4 & 2 & 7 \\ -13 & 0 & -17 \end{vmatrix}}{\begin{vmatrix} 1 & 4 & 1 \\ 0 & -6 & 5 \\ 0 & -4 & -5 \end{vmatrix}}$$

$$= \frac{+(2)\begin{vmatrix} -7 & -13 \\ -13 & -17 \end{vmatrix}}{+(1)\begin{vmatrix} -6 & 5 \\ -4 & -5 \end{vmatrix}}$$

$$= \frac{+(2)[119 - 169]}{+(1)[30 + 20]}$$

$$= \frac{2[-50]}{1[50]}$$

$$= -2.$$

$$y = \frac{\begin{vmatrix} 1 & 1 & 1 \\ 2 & 4 & 7 \\ 2 & -5 & -3 \end{vmatrix}}{\Delta}.$$

In the numerator, select row 1 with which to work. Multiply column 1 by -1 and add the products to the corresponding elements in column 2. Then multiply column 1 by -1 and add the products to the corresponding elements in column 3.

$$y = \frac{\begin{vmatrix} 1 & 0 & 0 \\ 2 & 2 & 5 \\ 2 & -7 & -5 \end{vmatrix}}{50}$$

$$= \frac{+(1)\begin{vmatrix} 2 & 5 \\ -7 & -5 \end{vmatrix}}{50}$$

$$= \frac{1[-10 + 35]}{50}$$

$$= \frac{1[25]}{50}$$

$$= \frac{1}{2}.$$

$$z = \frac{\begin{vmatrix} 1 & 4 & 1 \\ 2 & 2 & 4 \\ 2 & 4 & -5 \end{vmatrix}}{\Delta}.$$

For the numerator, select row 1 with which to work. Multiply column 3 by -1 and add the products to the corresponding elements in column 1. Then multiply column 3 by -4 and add the products to the corresponding elements in column 2.

$$z = \frac{\begin{vmatrix} 0 & 0 & 1 \\ -2 & -14 & 4 \\ 7 & 24 & -5 \end{vmatrix}}{50}$$

$$= \frac{+(1)\begin{vmatrix} -2 & -14 \\ 7 & 24 \end{vmatrix}}{50}$$

$$= \frac{1[-48 + 98]}{50}$$

$$= \frac{1[50]}{50}$$

$$= 1.$$

Check:

$$x + 4y + z = 1. \qquad 2x + 2y + 7z = 4.$$
$$-2 + 4(\tfrac{1}{2}) + (1) = 1. \qquad 2(-2) + 2(\tfrac{1}{2}) + 7(1) = 4.$$
$$-2 + 2 + 1 = 1. \qquad -4 + 1 + 7 = 4.$$
$$1 = 1. \qquad 4 = 4.$$

$$2x + 4y - 3z = -5.$$
$$2(-2) + 4(\tfrac{1}{2}) - 3(1) = -5.$$
$$-4 + 2 - 3 = -5.$$
$$-5 = -5.$$

Hence, $x = -2$, $y = \frac{1}{2}$, and $z = 1$.

18. Solve by use of determinants:

$$x + y - 3z = 2$$
$$3x - y - 2z = 1$$
$$2x + 2y + 3z = 1.$$

(a)	(b)	(c)	(d)
1	1	−3	2
3	−1	−2	1
2	2	3	1

$$x = \frac{\begin{vmatrix} 2 & 1 & -3 \\ 1 & -1 & -2 \\ 1 & 2 & 3 \end{vmatrix}}{\begin{vmatrix} 1 & 1 & -3 \\ 3 & -1 & -2 \\ 2 & 2 & 3 \end{vmatrix}}.$$

In the numerator select column 2 with which to work. Multiply row 2 by 1 and add the products to the corresponding elements in row 1. Then multiply row 2 by 2 and add the products to the corresponding elements in row 3. In the denominator, select column 1 with which to work. Multiply row 1 by −3 and add the products to the corresponding elements in row 2. Then multiply row 1 by −2 and add the products to the corresponding elements in row 3.

$$x = \frac{\begin{vmatrix} 3 & 0 & -5 \\ 2 & -1 & -2 \\ 3 & 0 & -1 \end{vmatrix}}{\begin{vmatrix} 1 & 1 & -3 \\ 0 & -4 & 7 \\ 0 & 0 & 9 \end{vmatrix}}$$

$$= \frac{+(-1)\begin{vmatrix} 3 & -5 \\ 3 & -1 \end{vmatrix}}{+(1)\begin{vmatrix} -4 & 7 \\ 0 & 9 \end{vmatrix}}$$

$$= \frac{-1[-3 + 15]}{1[-36 - 0]}$$

$$= \frac{-12}{-36}$$

$$= \frac{1}{3}.$$

$$y = \frac{\begin{vmatrix} 1 & 2 & -3 \\ 3 & 1 & -2 \\ 2 & 1 & 3 \end{vmatrix}}{\Delta}.$$

In the numerator select row 1 with which to work. Multiply column 1 by −2 and add the products to the corresponding elements in column 2. Then multiply column 1 by 3 and add the products to the corresponding elements in column 3.

$$y = \frac{\begin{vmatrix} 1 & 0 & 0 \\ 3 & -5 & 7 \\ 2 & -3 & 9 \end{vmatrix}}{-36}$$

$$= \frac{+(1)\begin{vmatrix} -5 & 7 \\ -3 & 9 \end{vmatrix}}{-36}$$

$$= \frac{1[-45 + 21]}{-36}$$

$$= \frac{-24}{-36}$$

$$= \frac{2}{3}.$$

$$z = \frac{\begin{vmatrix} 1 & 1 & 2 \\ 3 & -1 & 1 \\ 2 & 2 & 1 \end{vmatrix}}{\Delta}.$$

Select column 2 with which to work. Multiply column 1 by −1 and add the products to the corresponding elements of column 2.

$$z = \frac{\begin{vmatrix} 1 & 0 & 2 \\ 3 & -4 & 1 \\ 2 & 0 & 1 \end{vmatrix}}{-36}$$

$$= \frac{+(-4)\begin{vmatrix} 1 & 2 \\ 2 & 1 \end{vmatrix}}{-36}$$

$$= \frac{-4[1 - 4]}{-36}$$

$$= \frac{-4[-3]}{-36}$$

$$= \frac{12}{-36}$$

$$= -\frac{1}{3}.$$

Check:

$$x + y - 3z = 2. \qquad 3x - y - 2z = 1.$$
$$\tfrac{1}{3} + \tfrac{2}{3} - 3(-\tfrac{1}{3}) = 2. \qquad 3(\tfrac{1}{3}) - (\tfrac{2}{3}) - 2(-\tfrac{1}{3}) = 1.$$
$$\tfrac{1}{3} + \tfrac{2}{3} + \tfrac{3}{3} = 2. \qquad \tfrac{3}{3} - \tfrac{2}{3} + \tfrac{2}{3} = 1.$$
$$2 = 2. \qquad 1 = 1.$$

$$2x + 2y + 3z = 1.$$
$$2(\tfrac{1}{3}) + 2(\tfrac{2}{3}) + 3(-\tfrac{1}{3}) = 1.$$
$$\tfrac{2}{3} + \tfrac{4}{3} - \tfrac{3}{3} = 1.$$
$$1 = 1.$$

Hence, $x = \frac{1}{3}$, $y = \frac{2}{3}$, and $z = -\frac{1}{3}$.

19. Solve the following system of equations by determinants.

$$w + x + y + z = 5.$$
$$w - x + 2y - z = 3.$$
$$2w - 3x + y - 2z = 1.$$
$$w - x - y - z = -3.$$

(a)	(b)	(c)	(d)	(e)
1	1	1	1	5
1	−1	2	−1	3
2	−3	1	−2	1
1	−1	−1	−1	−3

In this case, a, b, c, and d are the coefficients of w, x, y, and z, respectively. The constant here is e.

$$w = \frac{\begin{vmatrix} 5 & 1 & 1 & 1 \\ 3 & -1 & 2 & -1 \\ 1 & -3 & 1 & -2 \\ -3 & -1 & -1 & -1 \end{vmatrix}}{\begin{vmatrix} 1 & 1 & 1 & 1 \\ 1 & -1 & 2 & -1 \\ 2 & -3 & 1 & -2 \\ 1 & -1 & -1 & -1 \end{vmatrix}}.$$

In this case, both the numerator and denominator can be treated in the same way. In each, select column 4 with which to work. Then multiply row 1 by 1 and add the products to the corresponding elements of row 2. Then multiply row 1 by 2 and add the products to the corresponding elements in row 3. Finally, multiply row 1 by 1 and add the products to the corresponding elements in row 4.

$$w = \frac{\begin{vmatrix} 5 & 1 & 1 & 1 \\ 8 & 0 & 3 & 0 \\ 11 & -1 & 3 & 0 \\ 2 & 0 & 0 & 0 \end{vmatrix}}{\begin{vmatrix} 1 & 1 & 1 & 1 \\ 2 & 0 & 3 & 0 \\ 4 & -1 & 3 & 0 \\ 2 & 0 & 0 & 0 \end{vmatrix}} = \frac{-(1)\begin{vmatrix} 8 & 0 & 3 \\ 11 & -1 & 3 \\ 2 & 0 & 0 \end{vmatrix}}{-(1)\begin{vmatrix} 2 & 0 & 3 \\ 4 & -1 & 3 \\ 2 & 0 & 0 \end{vmatrix}}.$$

Now look at row 3 in both the numerator and the denominator. Since there are two zeroes in each, we can evaluate this fraction by using the minor of the 2. Each product will have three factors: the $-(1)$ above, the $(+2)$ required for the new cofactor, and the minor.

$$w = \frac{-(1)[+(2)]\begin{vmatrix} 0 & 3 \\ -1 & 3 \end{vmatrix}}{-(1)[+(2)]\begin{vmatrix} 0 & 3 \\ -1 & 3 \end{vmatrix}}$$
$$= \frac{-2[(0)(3) - (-1)(3)]}{-2[(0)(3) - (-1)(3)]}$$
$$= \frac{-2[3]}{-2[3]}$$
$$= \frac{-6}{-6}$$
$$= 1.$$

$$x = \frac{\begin{vmatrix} 1 & 5 & 1 & 1 \\ 1 & 3 & 2 & -1 \\ 2 & 1 & 1 & -2 \\ 1 & -3 & -1 & -1 \end{vmatrix}}{\Delta}.$$

In the numerator, select column 3 with which to work. Multiply row 1 by -2 and add the products to the corresponding elements in row 2. Then multiply row 1 by -1 and add the products to the corresponding elements in row 3. Finally, multiply row 1 by 1 and add to the corresponding elements in row 4.

$$x = \frac{\begin{vmatrix} 1 & 5 & 1 & 1 \\ -1 & -7 & 0 & -3 \\ 1 & -4 & 0 & -3 \\ 2 & 2 & 0 & 0 \end{vmatrix}}{-6}$$
$$= \frac{+(1)\begin{vmatrix} -1 & -7 & -3 \\ 1 & -4 & -3 \\ 2 & 2 & 0 \end{vmatrix}}{-6}.$$

Now select row 3 with which to work. Multiply column 1 by -1 and add the products to the corresponding elements in column 2.

$$x = \frac{+(1)\begin{vmatrix} -1 & -6 & -3 \\ 1 & -5 & -3 \\ 2 & 0 & 0 \end{vmatrix}}{-6}$$
$$= \frac{+(1)[+(2)]\begin{vmatrix} -6 & -3 \\ -5 & -3 \end{vmatrix}}{-6}$$
$$= \frac{(+1)(+2)[18 - 15]}{-6}$$
$$= \frac{2[3]}{-6}$$
$$= \frac{6}{-6}$$
$$= -1.$$

$$y = \frac{\begin{vmatrix} 1 & 1 & 5 & 1 \\ 1 & -1 & 3 & -1 \\ 2 & -3 & 1 & -2 \\ 1 & -1 & -3 & -1 \end{vmatrix}}{\Delta}.$$

Select column 4 with which to work. Multiply row 1 by 1 and add the products to the corresponding elements in row 2. Then multiply row 1 by 2 and add the products to the corresponding elements in row 3. Finally, multiply row 1 by 1 and add the products to the corresponding elements in row 4.

$$y = \frac{\begin{vmatrix} 1 & 1 & 5 & 1 \\ 2 & 0 & 8 & 0 \\ 4 & -1 & 11 & 0 \\ 2 & 0 & 2 & 0 \end{vmatrix}}{-6} = \frac{-(1)\begin{vmatrix} 2 & 0 & 8 \\ 4 & -1 & 11 \\ 2 & 0 & 2 \end{vmatrix}}{-6}.$$

Take the minor of the -1 in row 2 column 2 and we have

$$y = \frac{-(1)[+(-1)]\begin{vmatrix} 2 & 8 \\ 2 & 2 \end{vmatrix}}{-6} = \frac{(-1)(-1)[(2)(2)-(2)(8)]}{-6} = \frac{1[4-16]}{-6} = \frac{-12}{-6} = 2.$$

Since we have values for w, x and y we can now substitute these values in one of the original equations and find z. Let us use the first equation. Then

$$w + x + y + z = 5.$$
$$1 - 1 + 2 + z = 5.$$
$$2 + z = 5.$$
$$z = 3.$$

Check:

$$w + x + y + z = 5. \quad 1 - 1 + 2 + 3 = 5. \quad 5 = 5.$$

$$w - x + 2y - z = 3. \quad 1 + 1 + 4 - 3 = 3. \quad 3 = 3.$$

$$2w - 3x + y - 2z = 1. \quad 2(1) - 3(-1) + 2 - 2(3) = 1. \quad 2 + 3 + 2 - 6 = 1. \quad 1 = 1.$$

$$w - x - y - z = -3. \quad 1 - (-1) - 2 - 3 = -3. \quad 1 + 1 - 2 - 3 = -3. \quad -3 = -3.$$

Hence, $w = 1$, $x = -1$, $y = 2$, and $z = 3$.

20. Using determinants until sufficient values are obtained to complete the solution by substitution, solve the following system of equations.

$$w + x + y + z = 4.$$
$$2w - 3x - y = 7.$$
$$w + x = 2.$$
$$x - y + z = -3.$$

(a)	(b)	(c)	(d)	(e)
1	1	1	1	4
2	−3	−1	0	7
1	1	0	0	2
0	1	−1	1	−3

Let us solve for w first, and then substitute that value into the third equation in order to find the value of x. Then the values of w and x may be substituted into the second equation to find the value of y. Finally, the values of x and y may be used in the last equation to find the value of z. Hence, it is necessary to evaluate only one determinant—the one which will yield the value of w. (Had we wished, we could have evaluated the determinant for x first and solved for w by substituting the value of x in the third equation.)

$$w = \frac{\begin{vmatrix} 4 & 1 & 1 & 1 \\ 7 & -3 & -1 & 0 \\ 2 & 1 & 0 & 0 \\ -3 & 1 & -1 & 1 \end{vmatrix}}{\begin{vmatrix} 1 & 1 & 1 & 1 \\ 2 & -3 & -1 & 0 \\ 1 & 1 & 0 & 0 \\ 0 & 1 & -1 & 1 \end{vmatrix}}.$$

In both the numerator and denominator, select column 4 with which to work. In both cases, multiply row 1 by -1 and add the products to the corresponding elements in row 4.

$$w = \frac{\begin{vmatrix} 4 & 1 & 1 & 1 \\ 7 & -3 & -1 & 0 \\ 2 & 1 & 0 & 0 \\ -7 & 0 & -2 & 0 \end{vmatrix}}{\begin{vmatrix} 1 & 1 & 1 & 1 \\ 2 & -3 & -1 & 0 \\ 1 & 1 & 0 & 0 \\ -1 & 0 & -2 & 0 \end{vmatrix}} = \frac{-(1)\begin{vmatrix} 7 & -3 & -1 \\ 2 & 1 & 0 \\ -7 & 0 & -2 \end{vmatrix}}{-(1)\begin{vmatrix} 2 & -3 & -1 \\ 1 & 1 & 0 \\ -1 & 0 & -2 \end{vmatrix}}.$$

In both the numerator and denominator, select column 2 with which to work. In both cases, multiply row 2 by 3 and add the products to the corresponding elements in row 1.

$$w = \frac{-(1)\begin{vmatrix} 13 & 0 & -1 \\ 2 & 1 & 0 \\ -7 & 0 & -2 \end{vmatrix}}{-(1)\begin{vmatrix} 5 & 0 & -1 \\ 1 & 1 & 0 \\ -1 & 0 & -2 \end{vmatrix}}.$$

In both numerator and denominator, take the element in row 2 and column 2 and its minor to complete the solution.

$$w = \frac{-(1)[+(1)]\begin{vmatrix} 13 & -1 \\ -7 & -2 \end{vmatrix}}{-(1)[+(1)]\begin{vmatrix} 5 & -1 \\ -1 & -2 \end{vmatrix}} = \frac{-1[(13)(-2)-(-7)(-1)]}{-1[(5)(-2)-(-1)(-1)]}$$

$$= \frac{-1[-26-7]}{-1[-10-1]}$$
$$= \frac{-1[-33]}{-1[-11]}$$
$$= \frac{33}{11}$$
$$= 3.$$

Substitute this value of w in the third equation to obtain the value of x.

$$w + x = 2.$$
$$3 + x = 2.$$
$$x = -1.$$

Substitute the values found, $w = 3$ and $x = -1$, in the second equation to obtain the value of y.

$$2w - 3x - y = 7.$$
$$2(3) - 3(-1) - y = 7.$$
$$6 + 3 - y = 7.$$
$$9 - y = 7.$$
$$-y = -2.$$
$$y = 2.$$

To find the value of z, substitute -1 for x and 2 for y in the last equation.

$$x - y + z = -3.$$
$$-1 - (2) + z = -3.$$
$$-3 + z = -3.$$
$$z = 0.$$

Check:

$$w + x + y + z = 4.$$
$$3 - 1 + 2 + 0 = 4.$$
$$4 = 4.$$

$$2w - 3x - y = 7.$$
$$2(3) - 3(-1) - (2) = 7.$$
$$6 + 3 - 2 = 7.$$
$$7 = 7.$$

$$w + x = 2.$$
$$3 - 1 = 2.$$
$$2 = 2.$$

$$x - y + z = -3.$$
$$-1 - 2 + 0 = -3.$$
$$-3 = -3.$$

Hence, $w = 3$, $x = -1$, $y = 2$, and $z = 0$.

SUPPLEMENTARY PROBLEMS

Solve the following systems of equations by elimination.

1. $x + y = 2.$
$2x - 3y = 9.$

2. $2x - y = 1.$
$3x + 2y = 12.$

3. $4x - 3y = 2.$
$-x + 2y = -3.$

4. $2x - y = 1.$
$x + 2y = 13.$

5. $\frac{3x}{4} + \frac{5y}{3} = 8.$
$\frac{x}{3} - \frac{y}{5} = \frac{11}{15}.$

6. $\frac{5x}{2} + \frac{3y}{4} = \frac{7}{8}.$
$\frac{3x}{4} + \frac{y}{3} = \frac{5}{24}.$

7. $\frac{x+1}{3} - \frac{y-1}{2} = 0.$
$\frac{2x-1}{2} + \frac{y+2}{4} = \frac{11}{4}$

8. $\frac{3x+2}{3} - \frac{3y-3}{2} = \frac{7}{2}$
$x - y = 1.$

9. $x + y + z = 3.$
$x - y - z = -1.$
$x - y + z = 9.$

10. $2x + 3y - z = 0.$
$x + 2y + 3z = -3.$
$2x + y - 4z = 1.$

11. $3x - 2y + z = 0.$
$4x - 2y + 5z = 2.$
$3x - y - 2z = 3.$

12. $x + y + z = 0.$
$2x - y + 3z = 1.$
$3x - 2y - 2z = 15.$

13. $x + y + z = 2\frac{2}{3}.$
$3x - 4y + 3z = 4\frac{1}{2}.$
$3x + 2y + 2z = 6.$

14. $x + 3y + z = 2.$
$x + y - z = 0.$
$2x + 6y + 5z = 5.$

Solve the following systems of equations by determinants.

15. $3x + 2y = 5.$
$2x - 3y = 12.$

16. $x + 2y = 5.$
$2x - y = -5.$

17. $3x + 2y + 3z = 1.$
$x - 2y + 2z = 3.$
$2x - y - z = 3.$

18. $2x + 2y + z = 3.$
$x - 2y + 2z = 12.$
$3x + 2y - z = -1.$

19. $a + 2c - d = 4.$
$2a - c + 2d = 3.$
$-2a + 2c - d = -2.$
$b + c + d = 0.$

20. $w + x + y + z = 3.$
$w - x + z = 6.$
$2w - 2y + 3z = 15.$
$w + x + y = 0.$

Chapter 7

The Formula

Solving by Formula. A formula is a set of symbols expressing a mathematical relationship or stating some mathematical or scientific principle. In work with geometric figures, the formula is used to determine the value of any one part of the figure when the values of the other parts are given. By use of the formula, we are able to determine the value of one part of a rule or scientific principle when the other parts are given.

> What is the perimeter of a square whose side is 5 inches?
>
> $$P = 4s$$
> $$= 4(5)$$
> $$= 20 \text{ inches.}$$

1. What is the area of a rectangular city lot that is 60 feet wide and 120 feet long?

   The formula for the area of a rectangle is $A = lw$.

   $$A = lw$$
   $$= (120)(60)$$
   $$= 7{,}200 \text{ square feet.}$$

2. What is the perimeter of a rectangular field 12 yards wide and 15 yards long?

   The formula for the perimeter of a rectangle is $P = 2l + 2w$.

   $$P = 2l + 2w$$
   $$= 2(15) + 2(12)$$
   $$= 30 + 24$$
   $$= 54 \text{ yards.}$$

3. What is the circumference of a circle whose radius is 4 inches?

   The formula for the circumference of a circle is $C = 2\pi r$ or $C = \pi d$. (Use $\pi = 3.14$.)

   $$C = 2\pi r$$
   $$= 2(3.14)(4)$$
   $$= 25.1 \text{ inches.}$$

4. What is the circumference of a circle whose diameter is 8 feet?

   $$C = \pi d$$
   $$= (3.14)(8)$$
   $$= 25.1 \text{ feet.}$$

   The answer can be found by using the formula $C = 2\pi r$. Remember that the diameter is twice the radius.

   $$C = 2\pi r$$
   $$= 2(3.14)(4)$$
   $$= 25.1 \text{ feet.}$$

5. What is the area of a circle whose radius is 2 inches?

   The formula for the area of a circle is $A = \pi r^2$.

   $$A = \pi r^2$$
   $$= (3.14)(2)^2$$
   $$= (3.14)(4)$$
   $$= 12.56 \text{ square inches.}$$

6. Find the perimeter of a rectangle whose sides are 36 inches and 5 feet.

   Both dimensions must be in terms of the same unit of measure. Therefore, we must change the 36 inches to 3 feet (1) or the 5 feet to 60 inches (2).

   $$P = 2l + 2w \quad (1)$$
   $$= 2(3) + 2(5)$$
   $$= 6 + 10$$
   $$= 16 \text{ feet.}$$

   $$P = 2l + 2w \quad (2)$$
   $$= 2(36) + 2(60)$$
   $$= 72 + 120$$
   $$= 192 \text{ inches.}$$

7. Find the perimeter of a triangle whose sides are 8 inches, 2 feet, and $2\frac{1}{3}$ feet.

   The formula for the perimeter of a triangle is $P = a + b + c$, where a, b, and c are the sides of the triangle.

   $$P = a + b + c$$
   $$= 8 + 24 + 28$$
   $$= 60 \text{ inches.}$$

   The perimeter, in terms of feet, would be

   $$P = a + b + c$$
   $$= \tfrac{2}{3} + 2 + 2\tfrac{1}{3}$$
   $$= 5 \text{ feet.}$$

8. What is the area of a triangle whose base is 6 inches and whose altitude is 4 inches?

The formula for the area of a triangle is $A = \frac{1}{2}bh$.

$$\begin{aligned} A &= \tfrac{1}{2}bh \\ &= \tfrac{1}{2}(6)(4) \\ &= 12 \text{ square inches.} \end{aligned}$$

9. What is the volume of a cube whose edge is 4 inches?

The formula for the volume of a cube is $V = e^3$.

$$\begin{aligned} V &= e^3 \\ &= (4)^3 \\ &= (4)(4)(4) \\ &= 64 \text{ cubic inches.} \end{aligned}$$

10. What is the volume of a cone whose base has a radius of 3 inches and whose height is 7 inches?

The formula for the volume of a cone is $V = \frac{1}{3}\pi r^2h$. It will be more convenient to use the common fractional approximation of the value of π ($\frac{22}{7}$) in this problem since a multiple of seven is involved.

$$\begin{aligned} V &= \tfrac{1}{3}\pi r^2h \\ &= \tfrac{1}{3}(\tfrac{22}{7})(3)^2(7) \\ &= \tfrac{1}{3}(\tfrac{22}{7})(9)(7) \\ &= 66 \text{ cubic inches.} \end{aligned}$$

11. What is the volume of a block of wood 2 feet wide, 6 feet long, and one foot high?

The block of wood represents a rectangular solid. The formula for the volume of a rectangular solid is $V = lwh$.

$$\begin{aligned} V &= lwh \\ &= (6)(2)(1) \\ &= 12 \text{ cubic feet.} \end{aligned}$$

12. What is the volume of a block of wood 6 inches wide, 12 feet long, and 4 inches high?

Again, all dimensions must be stated in terms of the same unit of measure.

$$\begin{aligned} V &= lwh & (1) \\ &= (12)(\tfrac{1}{2})(\tfrac{1}{3}) \\ &= 2 \text{ cubic feet.} \end{aligned}$$

$$\begin{aligned} V &= lwh & (2) \\ &= (144)(6)(4) \\ &= 3{,}456 \text{ cubic inches.} \end{aligned}$$

13. Find the volume of a right circular cylinder whose radius is 5 inches and whose height is 14 inches.

The formula for the volume of a right circular cylinder is $V = \pi r^2h$.

$$\begin{aligned} V &= \pi r^2h \\ &= (\tfrac{22}{7})(5)^2(14) \\ &= (\tfrac{22}{7})(25)(14) \\ &= 1{,}100 \text{ cubic inches.} \end{aligned}$$

14. What is the interest on \$1,000 loaned at 7% per year for 3 years?

The formula for the amount of interest due is $I = Prt$, where P is the principal or amount loaned, r is the rate, and t is the time in years.

$$\begin{aligned} I &= Prt \\ &= (\$1{,}000)(0.07)(3) \\ &= \$210.00. \end{aligned}$$

15. What is the area of a trapezoid whose bases are 15 inches and 7 inches respectively and whose altitude is 5 inches?

The formula for the area of a trapezoid is $A = \frac{(b + b')h}{2}$, where b and b' are the bases and h is the altitude or height.

$$\begin{aligned} A &= \frac{(b + b')h}{2} \\ &= \frac{(15 + 7)5}{2} \\ &= \frac{(22)5}{2} \\ &= \tfrac{110}{2} \\ &= 55 \text{ square inches.} \end{aligned}$$

Changing the Subject. The left member of a formula can be thought of as representing the subject of that formula. The variable with which we are concerned frequently does not occupy the subject position in the formula. In such cases, the solution is usually simpler if the subject of the formula is changed.

Solve the formula $P = 4s$ for s.

$$\begin{aligned} P &= 4s. \\ 4s &= P. \\ s &= \frac{P}{4}. \end{aligned}$$

16. Solve the formula $I = Prt$ for P.

$$\begin{aligned} I &= Prt. \\ Prt &= I. \end{aligned}$$

Divide both members of the equation by rt.

$$P = \frac{I}{rt}.$$

17. Solve the formula $I = Prt$ for r.

$$\begin{aligned} I &= Prt. \\ Prt &= I. \end{aligned}$$

Divide both members of the equation by Pt.

$$r = \frac{I}{Pt}.$$

18. Solve the formula $A = lw$ for w.

$$A = lw.$$
$$lw = A.$$

Divide both members of the equation by l.

$$w = \frac{A}{l}.$$

19. Solve the formula $P = 2l + 2w$ for w.

$$P = 2l + 2w.$$
$$-2w = -P + 2l.$$

Divide both members of the equation by -2.

$$w = \frac{P - 2l}{2}.$$

20. Solve the formula $A = \frac{1}{2}bh$ for h.

$$A = \tfrac{1}{2}bh$$
$$= \frac{bh}{2}.$$

Clear of fractions and we have

$$bh = 2A.$$

Divide both members of the equation by b.

$$h = \frac{2A}{b}.$$

21. Solve the formula $A = \pi r^2$ for r^2.

$$A = \pi r^2.$$
$$\pi r^2 = A.$$
$$r^2 = \frac{A}{\pi}.$$

22. Solve the formula $A = \pi r^2$ for r.

$$A = \pi r^2.$$
$$\pi r^2 = A.$$
$$r^2 = \frac{A}{\pi}.$$
$$r = \pm\sqrt{\frac{A}{\pi}}.$$

23. Solve the formula $A = P + Prt$ for r.

$$A = P + Prt.$$
$$-Prt = -A + P.$$

Divide both members of the equation by $-Pt$.

$$r = \frac{A - P}{Pt}.$$

24. Solve the formula $A = P + Prt$ for P.

$$A = P + Prt.$$
$$P + Prt = A.$$

Factor P out of the left member of the equation.

$$P(1 + rt) = A.$$

Divide both members of the equation by $(1 + rt)$.

$$P = \frac{A}{1 + rt}.$$

25. Solve the formula $A = \frac{(b + b')h}{2}$ for h.

$$A = \frac{(b + b')h}{2}.$$

Clear of fractions and we have

$$(b + b')h = 2A.$$

Divide both members of the equation by $(b + b')$.

$$h = \frac{2A}{b + b'}.$$

26. Solve the formula $A = \frac{(b + b')h}{2}$ for b'.

$$A = \frac{(b + b')h}{2}.$$

Clear of fractions and we have

$$(b + b')h = 2A.$$
$$bh + b'h = 2A.$$
$$b'h = 2A - bh.$$

Divide both members of the equation by h.

$$b' = \frac{2A - bh}{h}.$$

27. Solve the formula $x = ab + ac + d$ for a.

$$x = ab + ac + d.$$
$$-ab - ac = -x + d.$$

Factor $-a$ out of each term in the left member of the equation.

$$-a(b + c) = -x + d.$$

Divide both members of the equation by $-(b + c)$.

$$a = \frac{x - d}{b + c}.$$

28. Solve the formula $F = \frac{9}{5}C + 32$ for C.

This is the formula for converting centigrade to Fahrenheit temperature.

$$F = \tfrac{9}{5}C + 32.$$

Clear of fractions and we have

$$5F = 9C + 160.$$
$$-9C = -5F + 160.$$

Divide both members of the equation by -9.

$$C = \frac{5F - 160}{9}$$
$$= \tfrac{5}{9}(F - 32).$$

This is the formula for converting Fahrenheit to centigrade temperature.

29. Solve the formula $S = \frac{n}{2}(a + l)$ for n.

$$S = \frac{n}{2}(a + l).$$

Clear of fractions and we have

$$2S = n(a + l).$$
$$n(a + l) = 2S.$$

Divide both members of the equation by $(a + l)$.

$$n = \frac{2S}{a + l}.$$

30. Solve the formula $S = \frac{n}{2}(a + l)$ for a.

$$S = \frac{n}{2}(a + l);$$

or

$$S = \frac{n(a + l)}{2}.$$

Clear of fractions and we get

$$n(a + l) = 2S.$$

Divide both members of the equation by n.

$$a + l = \frac{2S}{n}.$$
$$a = \frac{2S}{n} - l;$$

or

$$a = \frac{2S - nl}{n}.$$

31. Solve the formula $F = m \cdot \frac{v^2}{R}$ for m.

This is a formulation of Newton's second law of circular motion.

$$m \cdot \frac{v^2}{R} = F.$$

Clear of fractions and we have

$$mv^2 = FR.$$
$$m = \frac{FR}{v^2}.$$

32. Solve the formula $\frac{1}{a} = \frac{1}{x} + \frac{1}{y}$ for y.

$$\frac{1}{a} = \frac{1}{x} + \frac{1}{y}.$$

Clear of fractions and we get

$$\frac{xy}{axy} = \frac{ay}{axy} + \frac{ax}{axy}.$$
$$xy = ay + ax.$$
$$xy - ay = ax.$$

Factor out the y in the left member of the equation.

$$y(x - a) = ax.$$
$$y = \frac{ax}{x - a}.$$

33. Solve the formula $y = \frac{V - v}{m}$ for v.

$$y = \frac{V - v}{m}.$$

Clear of fractions and we have

$$ym = V - v.$$
$$v = V - ym.$$

34. Solve the formula $y = mx + b$ for m.

$$y = mx + b.$$
$$-mx = -y + b.$$
$$m = \frac{y - b}{x}.$$

35. Solve the formula $x = \frac{(V_1 + V_2)h}{V_3}$ for V_2.

$$x = \frac{(V_1 + V_2)h}{V_3}.$$

Clear of fractions and we get

$$V_3x = (V_1 + V_2)h.$$

Remove the parentheses and we have

$$V_3x = V_1h + V_2h.$$
$$-V_2h = -V_3x + V_1h.$$
$$V_2 = \frac{V_3x - V_1h}{h}.$$

36. Solve the formula $2P = \frac{5(a + b + c)}{3}$ for a.

$$2P = \frac{5(a + b + c)}{3}.$$

Clear of fractions and we have

$$6P = 5a + 5b + 5c.$$
$$-5a = -6P + 5b + 5c.$$
$$a = \frac{6P - 5b - 5c}{5}.$$

SUPPLEMENTARY PROBLEMS

In each of the following formulas, solve for the indicated letter.

1. $I = Prt$, for t.
2. $y = mx + b$, for b.
3. $V = \frac{4}{3}\pi r^3 h$, for r^3.
4. $A = \frac{h(b + b')}{2}$, for b.
5. $\frac{1}{a} + \frac{1}{b} = \frac{1}{c}$, for b.
6. $V = lwh$, for w.
7. $C = 2\pi r$, for r.
8. $V = \pi r^2 h$, for h.
9. $\frac{1}{a} + \frac{1}{b} = \frac{1}{c}$, for c.
10. $V = \pi r^2 h$, for r^2.
11. $S = \frac{n}{2}(a + l)$, for l.

12. $y = mx + b$, for x.

13. $P = 2l + 2w$, for l.

14. $V = \frac{4}{3}\pi r^3$, for r.

15. $A = P + Prt$, for t.

16. $F = m \cdot \frac{v^2}{R}$, for v.

Solve the following problems.

17. Find the volume of a sphere whose radius is 8 inches. The formula for the volume of a sphere is $V = \frac{4}{3}\pi r^3$. Use 3.14 for the approximate value of π.

18. A Fahrenheit thermometer registers a temperature of 50°. What would be the corresponding reading on the Centigrade thermometer?

19. A cylindrical container holds 1,100 cubic inches of liquid. If the radius is 5 inches, how deep is the container? Use the formula $V = \pi r^2 h$ and let $\pi = \frac{22}{7}$.

20. A man borrowed $4,000 for 9 months. If he pays the lender $4,210 when the note is due, what annual rate of interest did he pay?

21. The length of a rectangle is $\frac{4}{3}$ its width. If the perimeter is 196 feet what are the dimensions of the rectangle?

Chapter 8

Exponents and Radicals

Definitions. A symbol placed at the upper right-hand corner of an expression to indicate how many times the expression is to be used as a factor is known as an **exponent.** In the expression x^3, the 3 is the **exponent** and the x is the **base.** This expression means that x is to be multiplied by itself three times. Thus, $x^3 = (x)(x)(x)$. In the expression $(2)^4$, the 2 is used as a factor four times. Thus, $2^4 = 2 \cdot 2 \cdot 2 \cdot 2 = 16$.

Laws of Exponents.

1. $x^n = (x)(x)(x) \cdots$ to n factors.

$$\boxed{\begin{aligned} 2^3 &= 2 \cdot 2 \cdot 2 \\ &= 8. \end{aligned}}$$

2. $x^m \cdot x^n = x^{m+n}$.

$$\boxed{x^4 \cdot x^3 = x^7.}$$

3. $\dfrac{x^m}{x^n} = x^{m-n}$ if $m > n$.

$$\boxed{\frac{x^8}{x^5} = x^3.}$$

4. $\dfrac{x^m}{x^n} = \dfrac{1}{x^{n-m}}$ if $n > m$.

$$\boxed{\frac{x^3}{x^5} = \frac{1}{x^2}.}$$

5. $(x^m)^n = x^{mn}$.

$$\boxed{(x^2)^3 = x^6.}$$

6. $(xy)^m = x^m y^m$.

$$\boxed{\begin{aligned} (3ab)^2 &= 3^2a^2b^2 \\ &= 9a^2b^2. \end{aligned}}$$

7. $\left(\dfrac{x}{y}\right)^m = \dfrac{x^m}{y^m}$.

$$\boxed{\begin{aligned} \left(\frac{a}{3}\right)^2 &= \frac{a^2}{3^2} \\ &= \frac{a^2}{9}. \end{aligned}}$$

8. $x^0 = 1$.

$$\boxed{\begin{aligned} x^0 &= x^{m-m} \\ &= \frac{x^m}{x^m} \\ &= 1. \end{aligned}}$$

1. Simplify $x^2 \cdot x^3$.

$$x^2 \cdot x^3 = x^5.$$

2. Simplify $2x^3 \cdot 3x^2$.

$$2x^3 \cdot 3x^2 = 6x^5.$$

3. Simplify $\dfrac{x^6}{x^2}$.

$$\frac{x^6}{x^2} = x^4.$$

4. Simplify $\dfrac{8x^7}{2x^2}$.

$$\frac{8x^7}{2x^2} = 4x^5.$$

5. Simplify $\dfrac{12x}{3x^3}$.

$$\frac{12x}{3x^3} = \frac{4}{x^2}.$$

6. Simplify $\dfrac{2^8}{2^2}$.

$$\begin{aligned} \frac{2^8}{2^2} &= 2^6 \\ &= 64. \end{aligned}$$

7. Simplify $3^2 \cdot 3^3$.

$$\begin{aligned} 3^2 \cdot 3^3 &= 3^5 \\ &= 243. \end{aligned}$$

8. Simplify 7^0.

$$7^0 = 1.$$

9. Simplify $7x^0$.

$$\begin{aligned} 7x^0 &= (7)(1) \\ &= 7. \end{aligned}$$

10. Simplify $(a^2b)^3$.

$$(a^2b)^3 = a^6b^3.$$

11. Simplify $\left(\frac{x^2}{y^3}\right)^2$.

$$\left(\frac{x^2}{y^3}\right)^2 = \frac{x^4}{y^6}.$$

12. Simplify $\frac{12x^4y^2}{2xy^5}$.

$$\frac{12x^4y^2}{2xy^5} = \frac{6x^3}{y^3}.$$

Negative Exponents. The laws of exponents as set forth at the beginning of this chapter apply to negative as well as positive exponents as long as the laws of signed numbers are also observed. Law 4 can be replaced by the definition of a negative exponent below.

4a. $x^{-m} = \frac{1}{x^m}$.

$$\frac{x^3}{x^5} = x^{-2} \quad \text{(by Law 3)}$$
$$= \frac{1}{x^2}. \quad \text{(by definition above)}$$

13. Simplify $(x^2)^{-2}$.

$$(x^2)^{-2} = x^{-4} = \frac{1}{x^4}.$$

14. Simplify $x^{-2} \cdot x^3$.

$$x^{-2} \cdot x^3 = x^{-2+3} = x.$$

15. Simplify $(3a^{-2})^3$.

$$(3a^{-2})^3 = 3^3a^{-6} = \frac{27}{a^6}.$$

16. Simplify $\frac{x^{2n}}{x^{-n}}$.

$$\frac{x^{2n}}{x^{-n}} = x^{2n} \cdot x^n = x^{2n+n} = x^{3n}.$$

17. Simplify $\frac{a^{-3x}}{a^{-5x}}$.

$$\frac{a^{-3x}}{a^{-5x}} = a^{-3x} \cdot a^{5x} = a^{-3x+5x} = a^{2x}.$$

18. Simplify $(y^{-3})^{-3}$.

$$(y^{-3})^{-3} = y^9.$$

19. Simplify $\left(\frac{x^{-2}}{y^3}\right)^{-2}$.

$$\left(\frac{x^{-2}}{y^3}\right)^{-2} = \frac{x^4}{y^{-6}} = x^4y^6.$$

20. Simplify $\frac{x}{y^{-1}} + \frac{2x^2}{y^{-2}}$.

$$\frac{x}{y^{-1}} + \frac{2x^2}{y^{-2}} = xy + 2x^2y^2.$$

21. Simplify $7(x^{-1})^0$.

$$7(x^{-1})^0 = 7(1) = 7.$$

22. Simplify $(7x^{-1})^0$.

$$(7x^{-1})^0 = 1.$$

23. Simplify $(x^{-1} + y^{-1})^2$.

$$(x^{-1} + y^{-1})^2 = \left(\frac{1}{x} + \frac{1}{y}\right)^2 = \frac{1}{x^2} + \frac{2}{xy} + \frac{1}{y^2}.$$

This problem may be solved without changing to positive exponents before using the rule for squaring a binomial. This is shown in the solution below.

$$(x^{-1} + y^{-1})^2 = x^{-2} + 2x^{-1}y^{-1} + y^{-2} = \frac{1}{x^2} + \frac{2}{xy} + \frac{1}{y^2}.$$

24. Simplify $\frac{x^{-1} + y^{-1}}{z^{-1}}$.

$$\frac{x^{-1} + y^{-1}}{z^{-1}} = \frac{x^{-1}}{z^{-1}} + \frac{y^{-1}}{z^{-1}} = \frac{z}{x} + \frac{z}{y} = \frac{zy + xz}{xy}.$$

25. Simplify $\frac{x^{-1} + y^{-1}}{x^{-2}y^{-2}}$.

$$\frac{x^{-1} + y^{-1}}{x^{-2}y^{-2}} = \frac{x^{-1}}{x^{-2}y^{-2}} + \frac{y^{-1}}{x^{-2}y^{-2}} = x^{-1}x^2y^2 + x^2y^2y^{-1} = xy^2 + x^2y.$$

Fractional Exponents. The expression $a^{\frac{1}{n}}$ is defined as the nth root of a. Hence, $a^{\frac{1}{n}} = \sqrt[n]{a}$, $a^{\frac{2}{n}} = \sqrt[n]{a^2}$, $27^{\frac{1}{3}} = \sqrt[3]{27} = 3$, etc. It can be seen from these examples that the numerator of a fractional exponent indicates the power and the denominator indicates the root.

The laws of exponents as set forth at the beginning of this chapter hold true for fractional exponents as well as for integral exponents.

26. Simplify $a^{\frac{1}{3}} \cdot a^{\frac{2}{3}}$.

$$a^{\frac{1}{3}} \cdot a^{\frac{2}{3}} = a^{\frac{3}{3}} = a.$$

27. Simplify $x^{-\frac{1}{3}} \cdot x^{\frac{2}{3}}$.

$$x^{-\frac{1}{3}} \cdot x^{\frac{2}{3}} = x^{-\frac{1}{3}+\frac{2}{3}} = x^{\frac{1}{3}}.$$

28. Simplify $\dfrac{x^{\frac{2}{3}}}{x^{\frac{1}{3}}}$.

$$\frac{x^{\frac{2}{3}}}{x^{\frac{1}{3}}} = x^{\frac{2}{3}-\frac{1}{3}} = x^{\frac{1}{3}}.$$

29. Simplify $m^2 \cdot m^{\frac{1}{4}}$.

$$m^2 \cdot m^{\frac{1}{4}} = m^{\frac{8}{4}+\frac{1}{4}} = m^{\frac{9}{4}}.$$

30. Simplify $\dfrac{b^3}{b^{\frac{2}{3}}}$.

$$\frac{b^3}{b^{\frac{2}{3}}} = b^{3-\frac{2}{3}} = b^{\frac{9}{3}-\frac{2}{3}} = b^{\frac{7}{3}}.$$

31. Simplify $(x^{-\frac{1}{3}})^3$.

$$(x^{-\frac{1}{3}})^3 = x^{-\frac{3}{3}} = x^{-1} = \frac{1}{x}.$$

32. Simplify $(ab^{-\frac{1}{3}})^2$.

$$(ab^{-\frac{1}{3}})^2 = a^2b^{-\frac{2}{3}} = \frac{a^2}{b^{\frac{2}{3}}}.$$

33. Simplify $\dfrac{12x^{\frac{1}{3}}y}{4xy^{\frac{1}{2}}}$.

$$\frac{12x^{\frac{1}{3}}y}{4xy^{\frac{1}{2}}} = \frac{3y^{\frac{1}{2}}}{x^{\frac{2}{3}}}.$$

34. Simplify $(25x^{-2}y^{-4})^{\frac{1}{2}}$.

$$(25x^{-2}y^{-4})^{\frac{1}{2}} = \sqrt{25}x^{-\frac{2}{2}}y^{-\frac{4}{2}} = 5x^{-1}y^{-2} = \frac{5}{xy^2}.$$

35. Simplify $(27a^{-2}b^{-3})^{\frac{2}{3}}$.

$$(27a^{-2}b^{-3})^{\frac{2}{3}} = 27^{\frac{2}{3}}a^{-\frac{4}{3}}b^{-\frac{6}{3}} = \sqrt[3]{27^2}a^{-\frac{4}{3}}b^{-2} = \frac{(3)^2}{a^{\frac{4}{3}}b^2} = \frac{9}{a^{\frac{4}{3}}b^2}.$$

36. Simplify $\left(\dfrac{8^{-\frac{1}{3}}x^{-2}}{2y^{-\frac{1}{4}}}\right)^2$.

$$\left(\frac{8^{-\frac{1}{3}}x^{-2}}{2y^{-\frac{1}{4}}}\right)^2 = \left(\frac{y^{\frac{1}{4}}}{(2)(8^{\frac{1}{3}})x^2}\right)^2 = \left(\frac{y^{\frac{1}{4}}}{(2)(2)x^2}\right)^2 = \left(\frac{y^{\frac{1}{4}}}{4x^2}\right)^2 = \frac{y^{\frac{2}{4}}}{16x^4} = \frac{y^{\frac{1}{2}}}{16x^4}.$$

37. Simplify $(64x^{-1}y^{\frac{3}{2}})^{-\frac{2}{3}}$.

$$(64x^{-1}y^{\frac{3}{2}})^{-\frac{2}{3}} = (64)^{-\frac{2}{3}}x^{\frac{2}{3}}y^{-\frac{6}{6}} = \frac{x^{\frac{2}{3}}}{(64)^{\frac{2}{3}}y} = \frac{x^{\frac{2}{3}}}{\sqrt[3]{(64)^2}y} = \frac{x^{\frac{2}{3}}}{(4)^2y} = \frac{x^{\frac{2}{3}}}{16y}.$$

38. Simplify $\dfrac{(32bc^{-1})^{-\frac{2}{5}}}{a^{-2}}$.

$$\frac{(32bc^{-1})^{-\frac{2}{5}}}{a^{-2}} = \frac{(32)^{-\frac{2}{5}}b^{-\frac{2}{5}}c^{\frac{2}{5}}}{a^{-2}} = \frac{a^2c^{\frac{2}{5}}}{(32)^{\frac{2}{5}}b^{\frac{2}{5}}} = \frac{a^2c^{\frac{2}{5}}}{b^{\frac{2}{5}}\sqrt[5]{(32)^2}} = \frac{a^2c^{\frac{2}{5}}}{(2)^2b^{\frac{2}{5}}} = \frac{a^2c^{\frac{2}{5}}}{4b^{\frac{2}{5}}}.$$

39. Simplify $\left(\dfrac{a^{\frac{4}{3}}b^{-1}c}{16}\right)^{-\frac{1}{4}}$.

$$\left(\frac{a^{\frac{4}{3}}b^{-1}c}{16}\right)^{-\frac{1}{4}} = \frac{a^{-\frac{4}{12}}b^{\frac{1}{4}}c^{-\frac{1}{4}}}{(16)^{-\frac{1}{4}}} = \frac{(16)^{\frac{1}{4}}b^{\frac{1}{4}}}{a^{\frac{4}{12}}c^{\frac{1}{4}}} = \frac{b^{\frac{1}{4}}\sqrt[4]{16}}{a^{\frac{1}{3}}c^{\frac{1}{4}}} = \frac{2b^{\frac{1}{4}}}{a^{\frac{1}{3}}c^{\frac{1}{4}}}.$$

40. Simplify $x^{\frac{1}{2}}(x^{\frac{1}{2}} + x^4 - x^6)$.

$$x^{\frac{1}{2}}(x^{\frac{1}{2}} + x^4 - x^6) = x^{\frac{1}{2}+\frac{1}{2}} + x^{\frac{1}{2}+\frac{8}{2}} - x^{\frac{1}{2}+\frac{12}{2}}$$
$$= x + x^{\frac{9}{2}} - x^{\frac{13}{2}}.$$

41. Simplify $a^{\frac{2}{3}}(a - a^{\frac{2}{3}} + a^3)$.

$$a^{\frac{2}{3}}(a - a^{\frac{2}{3}} + a^3) = a^{\frac{2}{3}+\frac{3}{3}} - a^{\frac{2}{3}+\frac{2}{3}} + a^{\frac{2}{3}+\frac{9}{3}}$$
$$= a^{\frac{5}{3}} - a^{\frac{4}{3}} + a^{\frac{11}{3}}.$$

42. Simplify $m^{-1}(m - m^{-\frac{1}{2}} + m^{-\frac{1}{3}})$.

$$m^{-1}(m - m^{-\frac{1}{2}} + m^{-\frac{1}{3}}) = m^{1-1} - m^{-1-\frac{1}{2}} + m^{-1-\frac{1}{3}}$$
$$= m^0 - m^{-\frac{3}{2}} + m^{-\frac{4}{3}}$$
$$= 1 - \frac{1}{m^{\frac{3}{2}}} + \frac{1}{m^{\frac{4}{3}}}.$$

Conversion of Radicals. It is often convenient to change a radical to an equivalent expression having a fractional exponent. Convert the following radicals into expressions with fractional exponents and simplify.

43. $\sqrt{x^2}$.

$$\sqrt{x^2} = x^{\frac{2}{2}}$$
$$= x.$$

44. $\sqrt[3]{x^6}$.

$$\sqrt[3]{x^6} = x^{\frac{6}{3}}$$
$$= x^2.$$

45. $\sqrt{a^2b^4c^8}$.

$$\sqrt{a^2b^4c^8} = a^{\frac{2}{2}}b^{\frac{4}{2}}c^{\frac{8}{2}}$$
$$= ab^2c^4.$$

46. $\sqrt[4]{16ab^2}$.

$$\sqrt[4]{16ab^2} = (16)^{\frac{1}{4}}a^{\frac{1}{4}}b^{\frac{2}{4}}$$
$$= 2a^{\frac{1}{4}}b^{\frac{1}{2}}.$$

47. $\sqrt[4]{625a^2b^8}$.

$$\sqrt[4]{625a^2b^8} = \sqrt[4]{(25)^2a^2b^8}$$
$$= (25)^{\frac{2}{4}}a^{\frac{2}{4}}b^{\frac{8}{4}}$$
$$= 25^{\frac{1}{2}}a^{\frac{1}{2}}b^2$$
$$= 5a^{\frac{1}{2}}b^2.$$

48. $\sqrt[3]{\dfrac{54x^4y^{-6}}{2x^{-2}}}$.

$$\sqrt[3]{\frac{54x^4y^{-6}}{2x^{-2}}} = \sqrt[3]{\frac{27x^6}{y^6}}$$
$$= \frac{27^{\frac{1}{3}}x^{\frac{6}{3}}}{y^{\frac{6}{3}}}$$
$$= \frac{3x^2}{y^2}.$$

Reduce the following expressions to the same order and simplify.

49. $\sqrt{x^2} \cdot \sqrt[3]{x^2}$.

$$\sqrt{x^2} \cdot \sqrt[3]{x^2} = x^{\frac{2}{2}} \cdot x^{\frac{2}{3}}$$
$$= x^{\frac{6}{6}} \cdot x^{\frac{4}{6}}$$
$$= x^{\frac{10}{6}}$$
$$= x^{\frac{5}{3}}$$
$$= \sqrt[3]{x^5}.$$

50. $\sqrt[3]{x^2} \cdot \sqrt{x^2} \cdot \sqrt[4]{x^2}$.

$$\sqrt[3]{x^2} \cdot \sqrt{x^2} \cdot \sqrt[4]{x^2} = x^{\frac{2}{3}} \cdot x^{\frac{2}{2}} \cdot x^{\frac{1}{2}}$$
$$= x^{\frac{4}{6}} \cdot x^{\frac{6}{6}} \cdot x^{\frac{3}{6}}$$
$$= x^{\frac{13}{6}}$$
$$= \sqrt[6]{x^{13}}.$$

51. $\dfrac{\sqrt{2}}{\sqrt{8}}$.

$$\frac{\sqrt{2}}{\sqrt{8}} = \frac{\sqrt{2}}{\sqrt{2^3}}$$
$$= \frac{2^{\frac{1}{2}}}{2^{\frac{3}{2}}}$$
$$= \frac{1}{2^{\frac{2}{2}}}$$
$$= \tfrac{1}{2}.$$

52. $\sqrt[4]{a} \cdot \sqrt{a} \cdot \sqrt{a^3} \cdot \sqrt[4]{a^3}$.

$$\sqrt[4]{a} \cdot \sqrt{a} \cdot \sqrt{a^3} \cdot \sqrt[4]{a^3} = a^{\frac{1}{4}} \cdot a^{\frac{1}{2}} \cdot a^{\frac{3}{2}} \cdot a^{\frac{3}{4}}$$
$$= a^{\frac{1}{4}} \cdot a^{\frac{2}{4}} \cdot a^{\frac{6}{4}} \cdot a^{\frac{3}{4}}$$
$$= a^{\frac{12}{4}}$$
$$= a^3.$$

53. $\sqrt{32a} \cdot \sqrt{8a^3}$.

$$\sqrt{32a} \cdot \sqrt{8a^3} = \sqrt{2^5a} \cdot \sqrt{2^3a^3}$$
$$= 2^{\frac{5}{2}}a^{\frac{1}{2}} \cdot 2^{\frac{3}{2}}a^{\frac{3}{2}}$$
$$= 2^{\frac{8}{2}}a^{\frac{4}{2}}$$
$$= 2^4a^2$$
$$= 16a^2.$$

Rules for Simplification of Radicals.

1. Each factor under a radical may be placed under a separate radical. The reverse is not always true (see p. 81, Problem 17).

$$\boxed{\begin{aligned}\sqrt{16x} &= \sqrt{16} \cdot \sqrt{x} \\ &= 4\sqrt{x}.\end{aligned}}$$

2. If a fraction is under a radical, the numerator and denominator may be placed under separate radicals. The reverse is true.

$$\sqrt{\frac{x}{4}} = \frac{\sqrt{x}}{\sqrt{4}} = \frac{\sqrt{x}}{2}.$$

$$\frac{\sqrt{32}}{\sqrt{2}} = \sqrt{\frac{32}{2}} = \sqrt{16} = 4.$$

3. To be added or subtracted, radicals must have the same index *and* the same radicand. The **index** indicates what root is to be extracted; the **radicand** is the expression under the radical.

$$4\sqrt{2} + 3\sqrt{2} = 7\sqrt{2}.$$
$$5\sqrt[3]{2} - 2\sqrt[3]{2} = 3\sqrt[3]{2}.$$

4. Only radicals of the same **order** (radicals which have the same index) may be multiplied or divided by one another.

$$(2\sqrt[3]{5})(\sqrt[3]{5}) = 2\sqrt[3]{25}.$$
$$\frac{\sqrt[3]{a^2}}{\sqrt[3]{a}} = \sqrt[3]{\frac{a^2}{a}} = \sqrt[3]{a}.$$

54. Simplify $\sqrt{32}$.

$$\sqrt{32} = \sqrt{(16)(2)} = 4\sqrt{2}.$$

55. Simplify $\sqrt{98}$.

$$\sqrt{98} = \sqrt{(49)(2)} = 7\sqrt{2}.$$

56. Simplify $\sqrt[3]{54}$.

$$\sqrt[3]{54} = \sqrt[3]{(27)(2)} = 3\sqrt[3]{2}.$$

57. Simplify $\sqrt[3]{128}$.

$$\sqrt[3]{128} = \sqrt[3]{(64)(2)} = 4\sqrt[3]{2}.$$

58. Simplify $\sqrt{256}$.

$$\sqrt{256} = \sqrt{(16)(16)} = \sqrt{16}\cdot\sqrt{16} = 4\cdot 4 = 16.$$

59. Simplify $\sqrt{8} + \sqrt{32}$.

$$\sqrt{8} + \sqrt{32} = \sqrt{(4)(2)} + \sqrt{(16)(2)} = 2\sqrt{2} + 4\sqrt{2} = 6\sqrt{2}.$$

60. Simplify $3\sqrt{5} + \sqrt{125} - \sqrt{80}$.

$$3\sqrt{5} + \sqrt{125} - \sqrt{80} = 3\sqrt{5} + \sqrt{(25)(5)} - \sqrt{(16)(5)} = 3\sqrt{5} + 5\sqrt{5} - 4\sqrt{5} = 4\sqrt{5}.$$

61. Simplify $\sqrt[3]{54} + 3\sqrt[3]{2} - \sqrt[3]{128}$.

$$\sqrt[3]{54} + 3\sqrt[3]{2} - \sqrt[3]{128} = \sqrt[3]{(27)(2)} + 3\sqrt[3]{2} - \sqrt[3]{(64)(2)} = 3\sqrt[3]{2} + 3\sqrt[3]{2} - 4\sqrt[3]{2} = 2\sqrt[3]{2}.$$

62. Simplify $\frac{\sqrt[3]{256}}{\sqrt[3]{32}}$.

$$\frac{\sqrt[3]{256}}{\sqrt[3]{32}} = \frac{\sqrt[3]{(64)(4)}}{\sqrt[3]{(8)(4)}} = \frac{4\sqrt[3]{4}}{2\sqrt[3]{4}} = 2.$$

This problem may also be solved in the following manner:

$$\frac{\sqrt[3]{256}}{\sqrt[3]{32}} = \sqrt[3]{\frac{256}{32}} = \sqrt[3]{8} = 2.$$

63. Simplify $\sqrt[3]{-8} + \sqrt[3]{-27} + \sqrt[3]{-64}$.

$$\sqrt[3]{-8} + \sqrt[3]{-27} + \sqrt[3]{-64} = -2 + (-3) + (-4) = -2 - 3 - 4 = -9.$$

64. Simplify $3\sqrt{75} + 2\sqrt{48} - 5\sqrt{27}$.

$$3\sqrt{75} + 2\sqrt{48} - 5\sqrt{27}$$
$$= 3\sqrt{(25)(3)} + 2\sqrt{(16)(3)} - 5\sqrt{(9)(3)}$$
$$= 3(5\sqrt{3}) + 2(4\sqrt{3}) - 5(3\sqrt{3})$$
$$= 15\sqrt{3} + 8\sqrt{3} - 15\sqrt{3}$$
$$= 8\sqrt{3}.$$

65. Simplify $\sqrt{72} + 3\sqrt{98} - 2\sqrt{32}$.

$$\sqrt{72} + 3\sqrt{98} - 2\sqrt{32}$$
$$= \sqrt{(36)(2)} + 3\sqrt{(49)(2)} - 2\sqrt{(16)(2)}$$
$$= 6\sqrt{2} + 3(7\sqrt{2}) - 2(4\sqrt{2})$$
$$= 6\sqrt{2} + 21\sqrt{2} - 8\sqrt{2}$$
$$= 19\sqrt{2}.$$

66. Simplify $3\sqrt[3]{54} - 5\sqrt[3]{16} + \sqrt[3]{2}$.

$$
\begin{aligned}
&3\sqrt[3]{54} - 5\sqrt[3]{16} + \sqrt[3]{2} \\
&= 3\sqrt[3]{(27)(2)} - 5\sqrt[3]{(8)(2)} + \sqrt[3]{2} \\
&= 3(3\sqrt[3]{2}) - 5(2\sqrt[3]{2}) + (\sqrt[3]{2}) \\
&= 9\sqrt[3]{2} - 10\sqrt[3]{2} + \sqrt[3]{2} \\
&= 0.
\end{aligned}
$$

Rationalizing the Denominator. Converting an expression which has a radical in the denominator into one whose denominator is a rational expression is known as rationalizing the denominator. When the denominator is a single term under a radical, multiply both the numerator and denominator by a radical such that the radicand in the denominator of the product is a perfect square. This is the same as multiplying by one, and does not change the value of the original fraction.

Rationalize the denominators of the following expressions and simplify the results if necessary.

67. $\dfrac{3}{\sqrt{2}}$.

$$
\begin{aligned}
\frac{3}{\sqrt{2}} &= \frac{3}{\sqrt{2}} \cdot \frac{\sqrt{2}}{\sqrt{2}} \\
&= \frac{3\sqrt{2}}{2}.
\end{aligned}
$$

68. $\dfrac{4}{\sqrt{8}}$.

$$
\begin{aligned}
\frac{4}{\sqrt{8}} &= \frac{4}{\sqrt{8}} \cdot \frac{\sqrt{2}}{\sqrt{2}} \\
&= \frac{4\sqrt{2}}{\sqrt{16}} \\
&= \frac{4\sqrt{2}}{4} \\
&= \sqrt{2}.
\end{aligned}
$$

69. $\dfrac{2x}{\sqrt{3}}$.

$$
\begin{aligned}
\frac{2x}{\sqrt{3}} &= \frac{2x}{\sqrt{3}} \cdot \frac{\sqrt{3}}{\sqrt{3}} \\
&= \frac{2x\sqrt{3}}{3}.
\end{aligned}
$$

70. $\dfrac{x+3}{\sqrt{5}}$.

$$
\begin{aligned}
\frac{x+3}{\sqrt{5}} &= \frac{x+3}{\sqrt{5}} \cdot \frac{\sqrt{5}}{\sqrt{5}} \\
&= \frac{x\sqrt{5} + 3\sqrt{5}}{5}.
\end{aligned}
$$

71. $\dfrac{2\sqrt{3}}{\sqrt{6}}$.

$$
\begin{aligned}
\frac{2\sqrt{3}}{\sqrt{6}} &= \frac{2\sqrt{3}}{\sqrt{6}} \cdot \frac{\sqrt{6}}{\sqrt{6}} \\
&= \frac{2\sqrt{3} \cdot \sqrt{3 \cdot 2}}{6} \\
&= \frac{2(3\sqrt{2})}{6} \\
&= \sqrt{2}.
\end{aligned}
$$

72. $\dfrac{\sqrt{3} - \sqrt{2}}{\sqrt{2}}$.

$$
\begin{aligned}
\frac{\sqrt{3} - \sqrt{2}}{\sqrt{2}} &= \frac{\sqrt{3} - \sqrt{2}}{\sqrt{2}} \cdot \frac{\sqrt{2}}{\sqrt{2}} \\
&= \frac{\sqrt{6} - 2}{2}.
\end{aligned}
$$

73. $\dfrac{\sqrt{18} - \sqrt{32}}{\sqrt{2}}$.

$$
\begin{aligned}
\frac{\sqrt{18} - \sqrt{32}}{\sqrt{2}} &= \frac{\sqrt{18} - \sqrt{32}}{\sqrt{2}} \cdot \frac{\sqrt{2}}{\sqrt{2}} \\
&= \frac{\sqrt{36} - \sqrt{64}}{2} \\
&= \frac{6 - 8}{2} \\
&= -1.
\end{aligned}
$$

74. $\dfrac{\sqrt[3]{16}}{\sqrt[3]{2}}$.

$$
\begin{aligned}
\frac{\sqrt[3]{16}}{\sqrt[3]{2}} &= \frac{\sqrt[3]{16}}{\sqrt[3]{2}} \cdot \frac{\sqrt[3]{4}}{\sqrt[3]{4}} \\
&= \frac{\sqrt[3]{64}}{\sqrt[3]{8}} \\
&= \tfrac{4}{2} \\
&= 2.
\end{aligned}
$$

This problem could also be solved as shown below.

$$
\begin{aligned}
\frac{\sqrt[3]{16}}{\sqrt[3]{2}} &= \sqrt[3]{\frac{16}{2}} \\
&= \sqrt[3]{8} \\
&= 2.
\end{aligned}
$$

75. $\dfrac{2a - 3\sqrt{2}}{\sqrt[3]{2}}$.

$$
\begin{aligned}
\frac{2a - 3\sqrt{2}}{\sqrt[3]{2}} &= \frac{2a - 3\sqrt{2}}{\sqrt[3]{2}} \cdot \frac{\sqrt[3]{4}}{\sqrt[3]{4}} \\
&= \frac{2a\sqrt[3]{4} - (3\sqrt{2} \cdot \sqrt[3]{4})}{\sqrt[3]{8}}
\end{aligned}
$$

$$= \frac{2a\sqrt[3]{4} - (3)(2)^{\frac{1}{2}}(4)^{\frac{1}{3}}}{2}$$

$$= \frac{2a\sqrt[3]{4} - (3)(2)^{\frac{3}{6}}(4)^{\frac{2}{6}}}{2}$$

$$= \frac{2a\sqrt[3]{4} - (3\sqrt[6]{2^3} \cdot \sqrt[6]{4^2})}{2}$$

$$= \frac{2a\sqrt[3]{4} - (3\sqrt[6]{8} \cdot \sqrt[6]{16})}{2}$$

$$= \frac{2a\sqrt[3]{4} - 3\sqrt[6]{128}}{2}$$

$$= \frac{2a\sqrt[3]{4} - 3\sqrt[6]{(64)(2)}}{2}$$

$$= \frac{2a\sqrt[3]{4} - 3(2\sqrt[6]{2})}{2}$$

$$= a\sqrt[3]{4} - 3\sqrt[6]{2}.$$

Irrational Binomial Denominators. If the denominator is a binomial with either one or both terms an irrational expression, it must be multiplied by its **conjugate.** The conjugate of any binomial is the same pair of terms, used in the same order, but with the opposite sign between them. Illustrations of binomials and their conjugates are: $(x + y)$ and $(x - y)$; $(a - b)$ and $(a + b)$; $(2 - \sqrt{3})$ and $(2 + \sqrt{3})$; $(\sqrt{6} - \sqrt{5})$ and $(\sqrt{6} + \sqrt{5})$.

76. $\dfrac{3}{\sqrt{2} - \sqrt{3}}$.

$$\frac{3}{\sqrt{2} - \sqrt{3}} = \frac{3}{\sqrt{2} - \sqrt{3}} \cdot \frac{\sqrt{2} + \sqrt{3}}{\sqrt{2} + \sqrt{3}}$$

$$= \frac{3\sqrt{2} + 3\sqrt{3}}{2 - 3}$$

$$= \frac{3\sqrt{2} + 3\sqrt{3}}{-1}$$

$$= -3\sqrt{2} - 3\sqrt{3}.$$

77. $\dfrac{2x}{\sqrt{3} - 1}$.

$$\frac{2x}{\sqrt{3} - 1} = \frac{2x}{\sqrt{3} - 1} \cdot \frac{\sqrt{3} + 1}{\sqrt{3} + 1}$$

$$= \frac{2x\sqrt{3} + 2x}{3 - 1}$$

$$= \frac{2x\sqrt{3} + 2x}{2}$$

$$= x\sqrt{3} + x.$$

78. $\dfrac{\sqrt{3} + 1}{\sqrt{3} - 1}$.

$$\frac{\sqrt{3} + 1}{\sqrt{3} - 1} = \frac{\sqrt{3} + 1}{\sqrt{3} - 1} \cdot \frac{\sqrt{3} + 1}{\sqrt{3} + 1}$$

$$= \frac{3 + 2\sqrt{3} + 1}{3 - 1}$$

$$= \frac{4 + 2\sqrt{3}}{2}$$

$$= 2 + \sqrt{3}.$$

79. $\dfrac{8}{1 - \sqrt{2}}$.

$$\frac{8}{1 - \sqrt{2}} = \frac{8}{1 - \sqrt{2}} \cdot \frac{1 + \sqrt{2}}{1 + \sqrt{2}}$$

$$= \frac{8 + 8\sqrt{2}}{1 - 2}$$

$$= \frac{8 + 8\sqrt{2}}{-1}$$

$$= -8 - 8\sqrt{2}.$$

80. $\dfrac{2\sqrt{3} - \sqrt{2}}{\sqrt{2} + \sqrt{3}}$.

$$\frac{2\sqrt{3} - \sqrt{2}}{\sqrt{2} + \sqrt{3}} = \frac{2\sqrt{3} - \sqrt{2}}{\sqrt{2} + \sqrt{3}} \cdot \frac{\sqrt{2} - \sqrt{3}}{\sqrt{2} - \sqrt{3}}$$

$$= \frac{2\sqrt{6} - 2(3) - 2 + \sqrt{6}}{2 - 3}$$

$$= \frac{3\sqrt{6} - 8}{-1}$$

$$= 8 - 3\sqrt{6}.$$

81. $\dfrac{\sqrt{5} - 3\sqrt{3}}{\sqrt{5} + \sqrt{3}}$.

$$\frac{\sqrt{5} - 3\sqrt{3}}{\sqrt{5} + \sqrt{3}} = \frac{\sqrt{5} - 3\sqrt{3}}{\sqrt{5} + \sqrt{3}} \cdot \frac{\sqrt{5} - \sqrt{3}}{\sqrt{5} - \sqrt{3}}$$

$$= \frac{5 - \sqrt{15} - 3\sqrt{15} + 3(3)}{5 - 3}$$

$$= \frac{14 - 4\sqrt{15}}{2}$$

$$= 7 - 2\sqrt{15}.$$

82. $\dfrac{4\sqrt{3} - 5\sqrt{2}}{\sqrt{5} - \sqrt{2}}$.

$$\frac{4\sqrt{3} - 5\sqrt{2}}{\sqrt{5} - \sqrt{2}} = \frac{4\sqrt{3} - 5\sqrt{2}}{\sqrt{5} - \sqrt{2}} \cdot \frac{\sqrt{5} + \sqrt{2}}{\sqrt{5} + \sqrt{2}}$$

$$= \frac{4\sqrt{15} + 4\sqrt{6} - 5\sqrt{10} - 5(2)}{5 - 2}$$

$$= \frac{4\sqrt{15} + 4\sqrt{6} - 5\sqrt{10} - 10}{3}.$$

83. $\dfrac{a\sqrt{3}-\sqrt{2}}{a\sqrt{2}-b}$.

$$\frac{a\sqrt{3}-\sqrt{2}}{a\sqrt{2}-b} = \frac{a\sqrt{3}-\sqrt{2}}{a\sqrt{2}-b}\cdot\frac{a\sqrt{2}+b}{a\sqrt{2}+b} = \frac{a^2\sqrt{6}+ab\sqrt{3}-2a-b\sqrt{2}}{2a^2-b^2}.$$

Radicals under Radicals. Convert the following expressions into a form that contains not more than one radical.

84. $\sqrt{\sqrt{32}}$.

$$\begin{aligned}\sqrt{\sqrt{32}} &= (\sqrt{32})^{\frac{1}{2}}\\ &= [(32)^{\frac{1}{2}}]^{\frac{1}{2}}\\ &= 32^{\frac{1}{4}}\\ &= \sqrt[4]{32}\\ &= \sqrt[4]{(16)(2)}\\ &= 2\sqrt[4]{2}.\end{aligned}$$

85. $\sqrt{\sqrt{b^{12}}}$.

$$\begin{aligned}\sqrt{\sqrt{b^{12}}} &= (\sqrt{b^{12}})^{\frac{1}{2}}\\ &= (b^{\frac{12}{2}})^{\frac{1}{2}}\\ &= b^{\frac{12}{4}}\\ &= b^3.\end{aligned}$$

86. $\sqrt[3]{\sqrt{256}}$.

$$\begin{aligned}\sqrt[3]{\sqrt{256}} &= (\sqrt{256})^{\frac{1}{3}}\\ &= [(256)^{\frac{1}{2}}]^{\frac{1}{3}}\\ &= (256)^{\frac{1}{6}}\\ &= \sqrt[6]{256}\\ &= \sqrt[6]{(64)(4)}\\ &= 2\sqrt[6]{4}.\end{aligned}$$

87. $\sqrt[4]{\sqrt[3]{64}}$.

$$\begin{aligned}\sqrt[4]{\sqrt[3]{64}} &= \sqrt[4]{\sqrt[3]{4^3}}\\ &= \sqrt[4]{4^{\frac{3}{3}}}\\ &= \sqrt[4]{4}.\end{aligned}$$

88. $\sqrt[4]{\sqrt{x^5y^3}}$.

$$\begin{aligned}\sqrt[4]{\sqrt{x^5y^3}} &= \sqrt[4]{(x^5y^3)^{\frac{1}{2}}}\\ &= [(x^5y^3)^{\frac{1}{2}}]^{\frac{1}{4}}\\ &= (x^5y^3)^{\frac{1}{8}}\\ &= \sqrt[8]{x^5y^3}.\end{aligned}$$

89. $\sqrt{\sqrt[3]{x^{15}y^{12}}}$.

$$\begin{aligned}\sqrt{\sqrt[3]{x^{15}y^{12}}} &= \sqrt{x^{\frac{15}{3}}y^{\frac{12}{3}}}\\ &= \sqrt{x^5y^4}\\ &= (x^5y^4)^{\frac{1}{2}}\\ &= x^{\frac{5}{2}}y^{\frac{4}{2}}\\ &= x^{\frac{4}{2}}\cdot x^{\frac{1}{2}}\cdot y^2\\ &= x^2y^2\sqrt{x}.\end{aligned}$$

SUPPLEMENTARY PROBLEMS

Simplify each of the following expressions.

1. $x^{-5}\cdot x^2$
2. $x^3\cdot\dfrac{1}{x^2}$
3. $(x^4)^{-3}$
4. $3x^{-1}\cdot 4x^{-3}$
5. $x^{-1}y\cdot x^2y^{-3}$
6. $\dfrac{15a^{-1}b^3}{3a^{-3}b}$
7. $\dfrac{(2m)^{-1}}{x^{-2}}$
8. $\left(\dfrac{m^3}{n^{-2}}\right)^{-2}$
9. $\dfrac{(a-1)^{-3}}{(a-1)^{-5}}$
10. $\dfrac{b^{-1}+3}{b}$
11. $\dfrac{c^{-1}+d^{-1}}{\frac{1}{d}}$
12. $3x^0y^{-1}$
13. $m^{\frac{2}{3}}\cdot m^{-\frac{1}{3}}$
14. $(d^{\frac{1}{2}}x^{-\frac{1}{2}})^{-\frac{1}{2}}$
15. $(y^{-\frac{1}{2}}x^{\frac{1}{2}})^2$
16. $\sqrt{128}$
17. $\sqrt[3]{-72}$
18. $\sqrt{192}$
19. $\sqrt{\dfrac{125x^4y^7}{5xy^2}}$
20. $3x^{\frac{1}{2}}\cdot 2x^{\frac{1}{3}}$
21. $(2x)^0\cdot 7x^2$
22. $3\sqrt{128}+2\sqrt{50}-4\sqrt{32}$
23. $\sqrt{125}+\sqrt{20}-\sqrt{80}$
24. $\sqrt{3}\cdot\sqrt{15}\cdot\sqrt{5}$
25. $\sqrt[3]{2}\cdot\sqrt{6}$
26. $\sqrt[3]{6}\cdot\sqrt[3]{4}\cdot\sqrt[3]{10}$
27. $\sqrt[3]{16}\cdot\sqrt{2}$
28. $\dfrac{2\sqrt{12}}{\sqrt{3}}$
29. $\dfrac{5\sqrt{7}}{\sqrt{12}}$
30. $\dfrac{1-\sqrt{5}}{\sqrt{3}}$
31. $\dfrac{\sqrt{7}}{\sqrt{3}+\sqrt{5}}$
32. $\dfrac{2\sqrt{7}-\sqrt{3}}{\sqrt{3}-\sqrt{7}}$
33. $\sqrt[3]{\sqrt{64a^7b^8}}$
34. $\sqrt{\sqrt[3]{729x^2y^5}}$
35. $\sqrt[3]{\sqrt{a^4b^6}}$

Chapter 9

Imaginary and Complex Numbers

Square Roots of Negative Numbers. When working with radicals, we are sometimes faced with a negative number under the radical symbol. If we consider the meaning of the radical symbol, we soon discover that the indicated operation, taking the square root of a negative number, is impossible within the realm of the real number system. The square root of any number is some number which when multiplied by itself gives the original number. Any number, whether positive or negative, when multiplied by itself gives a positive result. Hence, the square root of a negative number is called **imaginary.** The symbol i represents $\sqrt{-1}$ and its use enables us to work with imaginary numbers.

Powers of i. Let us consider some of the integral powers of i. By definition, $i = \sqrt{-1}$ and $i^2 = (\sqrt{-1})^2 = -1$. Further powers of i are developed and tabulated as follows.

$i = i.$	$i^5 = (i)(i^4) = (i)(1) = i.$	$i^9 = (i)(i^4)^2 = (i)(1) = i.$
$i^2 = -1.$	$i^6 = (i^2)(i^4) = (-1)(1) = -1.$	$i^{10} = (i^2)(i^4)^2 = (-1)(1) = -1.$
$i^3 = (i)(i^2) = (i)(-1) = -i.$	$i^7 = (i^3)(i^4) = (-i)(1) = -i.$	$i^{11} = (i^3)(i^4)^2 = (-i)(1) = -i.$
$i^4 = (i^2)(i^2) = (-1)(-1) = 1.$	$i^8 = (i^4)^2 = (1)^2 = 1.$	$i^{12} = (i^4)^3 = (1)^3 = 1.$

It can be observed from the above table that the integral powers of i repeat themselves at intervals of 4. This means that any integral power of i can be reduced to i, -1, $-i$, or 1.

Using the information above, evaluate the following expressions.

1. i^{16}.

$$\begin{aligned} i^{16} &= (i^4)^4 \\ &= (1)^4 \\ &= 1. \end{aligned}$$

2. i^{21}.

$$\begin{aligned} i^{21} &= (i)(i^4)^5 \\ &= (i)(1) \\ &= i. \end{aligned}$$

3. i^{19}.

$$\begin{aligned} i^{19} &= (i^3)(i^4)^4 \\ &= (-i)(1) \\ &= -i. \end{aligned}$$

4. i^{27}.

$$\begin{aligned} i^{27} &= (i^3)(i^4)^6 \\ &= (-i)(1) \\ &= -i. \end{aligned}$$

5. i^{40}.

$$\begin{aligned} i^{40} &= (i^4)^{10} \\ &= (1)^{10} \\ &= 1. \end{aligned}$$

Simplification. Unlike their positive counterparts, *negative* numbers must be removed from under the radical symbol before any valid mathematical operations can be performed with them. The square root of a negative number in the form $\sqrt{-ax}$ must be changed into its equivalent form $i\sqrt{ax}$ in which it appears as a multiple of i.

Simplify the following expressions.

6. $\sqrt{-9}$.

$$\begin{aligned} \sqrt{-9} &= \sqrt{(9)(-1)} \\ &= 3i. \end{aligned}$$

7. $\sqrt{-25}$.

$$\begin{aligned} \sqrt{-25} &= \sqrt{(25)(-1)} \\ &= 5i. \end{aligned}$$

8. $\sqrt{-12}$.

$$\begin{aligned} \sqrt{-12} &= \sqrt{(4)(3)(-1)} \\ &= 2i\sqrt{3}. \end{aligned}$$

9. $\sqrt{-32}$.

$$\begin{aligned}\sqrt{-32} &= \sqrt{(16)(2)(-1)}\\ &= 4i\sqrt{2}.\end{aligned}$$

10. $\sqrt{-48x^3y}$.

$$\begin{aligned}\sqrt{-48x^3y} &= \sqrt{(16x^2)(3xy)(-1)}\\ &= 4xi\sqrt{3xy}.\end{aligned}$$

11. $\sqrt{-2x^4y^3}$.

$$\begin{aligned}\sqrt{-2x^4y^3} &= \sqrt{(x^4y^2)(2y)(-1)}\\ &= x^2yi\sqrt{2y}.\end{aligned}$$

12. $\sqrt{-108ab^2}$.

$$\begin{aligned}\sqrt{-108ab^2} &= \sqrt{(36b^2)(3a)(-1)}\\ &= 6bi\sqrt{3a}.\end{aligned}$$

13. $\sqrt{-6}+\sqrt{-24}$.

$$\begin{aligned}\sqrt{-6}+\sqrt{-24} &= \sqrt{(6)(-1)}+\sqrt{(4)(6)(-1)}\\ &= i\sqrt{6}+2i\sqrt{6}\\ &= 3i\sqrt{6}.\end{aligned}$$

14. $\sqrt{-72}+\sqrt{-98}-\sqrt{-32}$.

$$\begin{aligned}&\sqrt{-72}+\sqrt{-98}-\sqrt{-32}\\ &= \sqrt{(36)(2)(-1)}+\sqrt{(49)(2)(-1)}-\sqrt{(16)(2)(-1)}\\ &= 6i\sqrt{2}+7i\sqrt{2}-4i\sqrt{2}\\ &= 9i\sqrt{2}.\end{aligned}$$

15. $\sqrt{-27}-\sqrt{-48}-\sqrt{-192}$.

$$\begin{aligned}&\sqrt{-27}-\sqrt{-48}-\sqrt{-192}\\ &= \sqrt{(9)(3)(-1)}-\sqrt{(16)(3)(-1)}-\sqrt{(64)(3)(-1)}\\ &= 3i\sqrt{3}-4i\sqrt{3}-8i\sqrt{3}\\ &= -9i\sqrt{3}.\end{aligned}$$

16. $\sqrt{-7x^2}+\sqrt{-28y^2}$.

$$\begin{aligned}\sqrt{-7x^2}+\sqrt{-28y^2} &= \sqrt{(x^2)(7)(-1)}+\sqrt{(4y^2)(7)(-1)}\\ &= xi\sqrt{7}+2yi\sqrt{7}\\ &= (x+2y)i\sqrt{7}.\end{aligned}$$

17. $\sqrt{-4}\cdot\sqrt{-9}$.

$$\begin{aligned}\sqrt{-4}\cdot\sqrt{-9} &= (2i)(3i)\\ &= 6i^2\\ &= 6(-1)\\ &= -6.\end{aligned}$$

Putting the two radicals under one symbol *before* changing them into multiples of i would result in the following *incorrect* solution.

$$\begin{aligned}\sqrt{-4}\cdot\sqrt{-9} &= \sqrt{(-4)(-9)}\\ &= \sqrt{+36}\\ &= +6.\end{aligned}$$

This is why the reverse of Rule 1 on page 75 does not hold.

18. $\sqrt{-25}\cdot\sqrt{-16}$.

$$\begin{aligned}\sqrt{-25}\cdot\sqrt{-16} &= (5i)(4i)\\ &= 20i^2\\ &= 20(-1)\\ &= -20.\end{aligned}$$

19. $\sqrt{-3}\cdot\sqrt{-5}$.

$$\begin{aligned}\sqrt{-3}\cdot\sqrt{-5} &= i\sqrt{3}\cdot i\sqrt{5}\\ &= i^2\sqrt{15}\\ &= (-1)\sqrt{15}\\ &= -\sqrt{15}.\end{aligned}$$

20. $\sqrt{-2}\cdot\sqrt{-4}$.

$$\begin{aligned}\sqrt{-2}\cdot\sqrt{-4} &= i\sqrt{2}\cdot 2i\\ &= 2i^2\sqrt{2}\\ &= 2(-1)\sqrt{2}\\ &= -2\sqrt{2}.\end{aligned}$$

21. $(\sqrt{-9})^3$.

$$\begin{aligned}(\sqrt{-9})^3 &= (3i)^3\\ &= 27i^3\\ &= 27i(i^2)\\ &= 27i(-1)\\ &= -27i.\end{aligned}$$

22. $(2\sqrt{-3})^3$.

$$\begin{aligned}(2\sqrt{-3})^3 &= (2i\sqrt{3})^3\\ &= 8i^3\sqrt{27}\\ &= 8i^3\sqrt{(9)(3)}\\ &= 8i^3(3)\sqrt{3}\\ &= 24i^3\sqrt{3}\\ &= 24i(i^2)\sqrt{3}\\ &= -24i\sqrt{3}.\end{aligned}$$

23. $\dfrac{\sqrt{-9}}{\sqrt{-4}}$.

$$\begin{aligned}\frac{\sqrt{-9}}{\sqrt{-4}} &= \frac{3i}{2i}\\ &= \tfrac{3}{2}.\end{aligned}$$

24. $\dfrac{\sqrt{-32}}{\sqrt{2}}$.

$$\begin{aligned}\frac{\sqrt{-32}}{\sqrt{2}} &= \frac{\sqrt{(16)(2)(-1)}}{\sqrt{2}}\\ &= \frac{4i\sqrt{2}}{\sqrt{2}}\\ &= 4i.\end{aligned}$$

25. $\frac{8}{\sqrt{-4}}$.

$$\frac{8}{\sqrt{-4}} = \frac{8}{2i}$$
$$= \frac{4}{i}.$$

Rationalize the denominator by multiplying both the numerator and denominator by i.

$$= \frac{4i}{i^2}$$
$$= \frac{4i}{-1}$$
$$= -4i.$$

26. $\frac{\sqrt{-18}}{\sqrt{-2}}$.

$$\frac{\sqrt{-18}}{\sqrt{-2}} = \frac{\sqrt{(9)(2)(-1)}}{\sqrt{(2)(-1)}}$$
$$= \frac{3i\sqrt{2}}{i\sqrt{2}}$$
$$= 3.$$

27. $\frac{x\sqrt{-12x^2}}{\sqrt{-9x^4}}$.

$$\frac{x\sqrt{-12x^2}}{\sqrt{-9x^4}} = \frac{x\sqrt{(4x^2)(3)(-1)}}{\sqrt{(9x^4)(-1)}}$$
$$= \frac{x(2xi)\sqrt{3}}{3x^2i}$$
$$= \frac{2x^2i\sqrt{3}}{3x^2i}$$
$$= \frac{2\sqrt{3}}{3}.$$

Complex Numbers. The sum of a real number and an imaginary number is known as a **complex number.** The expression $a + bi$ is a complex number if a and b are real numbers.

Two complex numbers are equal if and only if their real parts are equal and their imaginary parts are equal. Thus, if $a + bi = c + di$, $a = c$ and $bi = di$. Dividing both members of $bi = di$ by i gives us $b = d$.

Solve for x and y: $x + yi = 4 - 3i$.

The real parts must be equal and the imaginary parts must be equal.

$x = 4$. $\quad yi = -3i$.
$y = -3$.

Check:

$$x + yi = 4 - 3i.$$
$$4 + (-3)i = 4 - 3i.$$
$$4 - 3i = 4 - 3i.$$

Solve the following equations for x and y.

28. $x + 2yi = 3 - i$.

$x = 3$. $\quad 2yi = -i$.
$2y = -1$.
$y = -\frac{1}{2}$.

Check:

$$x + 2yi = 3 - i.$$
$$3 + 2(-\tfrac{1}{2})i = 3 - i.$$
$$3 - i = 3 - i.$$

29. $4x - yi = 8 + 2i$.

$4x = 8$. $\quad -yi = 2i$.
$x = 2$. $\quad -y = 2$.
$y = -2$.

Check:

$$4x - yi = 8 + 2i.$$
$$4(2) - (-2)i = 8 + 2i.$$
$$8 + 2i = 8 + 2i$$

30. $(x + y) + (2x - y)i = 1 + 5i$.

$x + y = 1$. (1) $\quad (2x - y)i = 5i$.
$2x - y = 5$. (2)

Solve equations (1) and (2) by addition to find the value of x.

$x + y = 1$ (1)
$2x - y = 5$ (2)
$3x = 6$
$x = 2$. (3)

Substitute the value of x from equation (3) in equation (1).

$x + y = 1$. (1)
$2 + y = 1$.
$y = -1$.

Check:

$$(x + y) + (2x - y)i = 1 + 5i.$$
$$(2 + [-1]) + (2[2] - [-1])i = 1 + 5i.$$
$$(2 - 1) + (4 + 1)i = 1 + 5i.$$
$$1 + 5i = 1 + 5i.$$

Hence, $x = 2$ and $y = -1$.

31. $(2x - y) + (x + y)i = 6i$.

In this problem the real part of the right member of the equation is equal to zero.

$2x - y = 0$. (1) $\quad (x + y)i = 6i$.
$x + y = 6$. (2)

$2x - y = 0$ (1)
$x + y = 6$ (2)
$3x = 6$
$x = 2$. (3)

Substitute the value of x from equation (3) in equation (2).

$x + y = 6$. (2)
$2 + y = 6$.
$y = 4$.

Check:

$$(2x - y) + (x + y)i = 6i.$$
$$(2[2] - 4) + (2 + 4)i = 6i.$$
$$(4 - 4) + 6i = 6i.$$
$$6i = 6i.$$

Hence, $x = 2$ and $y = 4$.

32. $3x - 2y = 5i - xi - 1 - 2yi$.

Place all terms with x or y in them in the left member of the equation and arrange them so they can be grouped with the real parts together and the imaginary parts together.

$$3x - 2y + xi + 2yi = -1 + 5i.$$
$$(3x - 2y) + (x + 2y)i = -1 + 5i.$$
$$3x - 2y = -1. \quad (1) \qquad (x + 2y)i = 5i.$$
$$x + 2y = 5. \quad (2)$$

$$\begin{array}{rl} 3x - 2y = -1 & (1) \\ x + 2y = 5 & (2) \\ \hline 4x \quad = 4 & \\ x = 1. & (3) \end{array}$$

Substitute the value of x from equation (3) in equation (2).

$$x + 2y = 5.$$
$$1 + 2y = 5.$$
$$2y = 4.$$
$$y = 2.$$

Check:

$$3x - 2y = 5i - xi - 1 - 2yi.$$
$$3(1) - 2(2) = 5i - (1)i - 1 - 2(2)i.$$
$$3 - 4 = 5i - i - 1 - 4i.$$
$$-1 = -1.$$

Hence, $x = 1$ and $y = 2$.

Addition. Complex numbers are added together or subtracted from one another by combining first the real parts and then the imaginary parts.

33. Simplify $(x + 6i) - (3x - 2i) + (7x - i)$.

$$(x + 6i) - (3x - 2i) + (7x - i)$$
$$= x + 6i - 3x + 2i + 7x - i$$
$$= 5x + 7i.$$

34. Simplify $(4 + i) - (3 + 5i) + (7 + 6i)$.

$$(4 + i) - (3 + 5i) + (7 + 6i)$$
$$= 4 + i - 3 - 5i + 7 + 6i$$
$$= 8 + 2i.$$

35. Simplify $(3x + y + i) - (2x - y + i) + (x - y - 2i)$.

$$(3x + y + i) - (2x - y + i) + (x - y - 2i)$$
$$= 3x + y + i - 2x + y - i + x - y - 2i$$
$$= 2x + y - 2i.$$

Multiplication. Expressions involving i are multiplied in the same way as other algebraic expressions, treating i as an ordinary algebraic symbol. The only exception is that any power of i is replaced by its simplest equivalent.

36. Find the product of $(6 + i)$ and $(5 - i)$.

$$\begin{array}{r} 6 + i \\ 5 - i \\ \hline 30 + 5i \\ -6i - i^2 \\ \hline 30 - \quad i - i^2 \end{array}$$

$$30 - i - i^2 = 30 - i - (-1)$$
$$= 30 - i + 1$$
$$= 31 - i.$$

37. Find the product of $(x + \sqrt{-4})$ and $(x + \sqrt{-9})$.

Convert the factors into forms involving i.

$$x + \sqrt{-4} = x + 2i$$

and

$$x + \sqrt{-9} = x + 3i.$$

$$\begin{array}{r} x + 2i \\ x + 3i \\ \hline x^2 + 2xi \\ + 3xi + 6i^2 \\ \hline x^2 + 5xi + 6i^2 \end{array}$$

$$x^2 + 5xi + 6i^2 = x^2 + 5xi + 6(-1)$$
$$= x^2 + 5xi - 6$$
$$= x^2 - 6 + 5xi.$$

Complex numbers should always be put into the $a + bi$ form.

38. Find the product of $(\sqrt{2} + \sqrt{-2})$ and $(\sqrt{3} - \sqrt{-3})$.

$$\sqrt{2} + \sqrt{-2} = \sqrt{2} + i\sqrt{2}$$

and

$$\sqrt{3} - \sqrt{-3} = \sqrt{3} - i\sqrt{3}.$$

$$\begin{array}{r} \sqrt{2} + i\sqrt{2} \\ \sqrt{3} - i\sqrt{3} \\ \hline \sqrt{6} + i\sqrt{6} \\ - i\sqrt{6} - i^2\sqrt{6} \\ \hline \sqrt{6} \qquad - i^2\sqrt{6} \end{array}$$

$$\sqrt{6} - i^2\sqrt{6} = \sqrt{6} - (-1)\sqrt{6}$$
$$= \sqrt{6} + \sqrt{6}$$
$$= 2\sqrt{6}.$$

Rationalizing Complex Denominators. To rationalize a denominator that is a complex number, multiply both the numerator and denominator of the expression by the conjugate of the denominator. The **conjugate** of a complex number consists of the same values for both the real and imaginary parts as in the original binomial but with the opposite sign between them. Hence, the conjugate of $a + bi$ is $a - bi$, and the conjugate of $6 - 3i$ is $6 + 3i$.

Rationalize the denominator in the following expressions and simplify:

39. $\dfrac{7}{4-i}$.

$$\frac{7}{4-i}=\frac{7}{4-i}\cdot\frac{4+i}{4+i}=\frac{28+7i}{16-i^2}=\frac{28+7i}{16-(-1)}=\frac{28+7i}{16+1}=\frac{28+7i}{17}=\frac{28}{17}+\frac{7i}{17}.$$

40. $\dfrac{x-3i}{x+2i}$.

$$\frac{x-3i}{x+2i}=\frac{x-3i}{x+2i}\cdot\frac{x-2i}{x-2i}=\frac{x^2-5xi+6i^2}{x^2-4i^2}=\frac{x^2-5xi+6(-1)}{x^2-4(-1)}=\frac{x^2-5xi-6}{x^2+4}=\frac{x^2-6}{x^2+4}-\frac{5xi}{x^2+4}.$$

41. $\dfrac{\sqrt{3}+1}{4-3i}$.

$$\frac{\sqrt{3}+1}{4-3i}=\frac{\sqrt{3}+1}{4-3i}\cdot\frac{4+3i}{4+3i}=\frac{4\sqrt{3}+4+3i\sqrt{3}+3i}{16-9i^2}=\frac{4(\sqrt{3}+1)+3i(\sqrt{3}+1)}{16+9}=\frac{4(\sqrt{3}+1)+3i(\sqrt{3}+1)}{25}=\frac{4(\sqrt{3}+1)}{25}+\frac{3i(\sqrt{3}+1)}{25}.$$

42. $\dfrac{4-i}{2+3i}$.

$$\frac{4-i}{2+3i}=\frac{4-i}{2+3i}\cdot\frac{2-3i}{2-3i}=\frac{8-14i+3i^2}{4-9i^2}=\frac{8-14i+3(-1)}{4-9(-1)}=\frac{8-14i-3}{4+9}=\frac{5-14i}{13}=\frac{5}{13}-\frac{14i}{13}.$$

SUPPLEMENTARY PROBLEMS

Simplify each of the following expressions.

1. i^{13}
2. i^{-1}
3. $i^2 \cdot i^3$
4. $1-i^{-2}$
5. $\sqrt{-27}$
6. $\sqrt{-64}$
7. $\sqrt{-3a^3b}$
8. $\sqrt{-5}\cdot\sqrt{-12}$
9. $(a\sqrt{-3})(\sqrt{-3b^2})$
10. $\dfrac{\sqrt{-10}}{\sqrt{-2}}$
11. $\dfrac{\sqrt{-6}}{\sqrt{-3}}$
12. $\sqrt{-3}\cdot\sqrt{-15}$

Find the values of x and y which satisfy each of the following equations.

13. $x+yi=2+i$
14. $2x-2=3i-yi$
15. $x-i=3yi+2$
16. $2x+y+xi-yi=3+3i$

Perform the indicated operations and simplify each of the following expressions.

17. $(2x+i)-(3x-i)+(4x-i)$
18. $(x-i)+(2x-3i)-(2x-5i)$
19. $(x-3+i)+(x-yi+i)-(3i-2)$
20. $(x+i)(x-2i)$
21. $(2x-yi)(2x+yi)$
22. $(x+yi-1)(x+yi+1)$
23. $\dfrac{2-\sqrt{-1}}{3-\sqrt{-3}}$
24. $\dfrac{4+i}{1-3i}$
25. $\dfrac{1-3\sqrt{-2}}{1+3\sqrt{-2}}$

Chapter 10

Quadratic Equations

SOLVING QUADRATIC EQUATIONS

Definitions. An equation that contains the second power of the variable and no power greater than the second is known as a **quadratic equation.**

The **general** quadratic equation is written in the form

$$ax^2 + bx + c = 0$$

in which a, b, and c can be any constants, except that a cannot be zero. (When a is zero, the equation reduces to linear form.)

Literal quadratic equations contain at least one letter other than the one for which we are solving. The variable for which we are solving must be in quadratic form. Examples of such equations are: $mx^2 + 5mx + 6m = 0$; $px^2 + 5qx - 2r^2 = 0$; and $3x^2 - 13xy - 10y^2 = 0$.

Complete quadratic equations include the first and second powers of the variable as well as a constant. This means that neither a, nor b, nor c is equal to zero. The following are complete quadratic equations: $x^2 - 3x + 6 = 0$; $2x^2 + x - 8 = 0$; and $x^2 - 7x + 9 = 0$.

Incomplete quadratic equations are those in which c is equal to zero and a and b are not. The following are incomplete quadratic equations: $x^2 - 3x = 0$; $x^2 + x = 0$; and $3x^2 = 4x$.

If b equals zero and a and c do not, the equation is called a **pure** quadratic. The following are examples of the pure quadratic equation: $x^2 - 3 = 0$; $x^2 = 36$; and $3x^2 - 48 = 0$.

All quadratic equations have two roots (they may be identical).

Solving Pure Quadratics. To solve a pure quadratic equation, we place the constant in the right member of the equation and extract the square root of both members, keeping in mind that both the positive and negative values of the square root are to be used.

$$x^2 - 81 = 0.$$
$$x^2 = 81.$$
$$x = \pm 9.$$

Check:

$x^2 - 81 = 0.$	$x^2 - 81 = 0.$
$(9)^2 - 81 = 0.$	$(-9)^2 - 81 = 0.$
$81 - 81 = 0.$	$81 - 81 = 0.$
$0 = 0.$	$0 = 0.$

1. Solve for x: $x^2 = 144$. $x = \pm 12$.

Check:

$x^2 = 144.$	$x^2 = 144.$
$(12)^2 = 144.$	$(-12)^2 = 144.$
$144 = 144.$	$144 = 144.$

2. Solve for x: $x^2 - 36 = 0$. $x^2 = 36$. $x = \pm 6$.

Check:

$x^2 - 36 = 0.$	$x^2 - 36 = 0.$
$(6)^2 - 36 = 0.$	$(-6)^2 - 36 = 0.$
$36 - 36 = 0.$	$36 - 36 = 0.$
$0 = 0.$	$0 = 0.$

3. Solve for x: $3x^2 - 75 = 0$. $3x^2 = 75$. $x^2 = 25$. $x = \pm 5$.

Check:

$3x^2 - 75 = 0.$	$3x^2 - 75 = 0.$
$3(5)^2 - 75 = 0.$	$3(-5)^2 - 75 = 0.$
$3(25) - 75 = 0.$	$3(25) - 75 = 0.$
$75 - 75 = 0.$	$75 - 75 = 0.$
$0 = 0.$	$0 = 0.$

Solving Incomplete Quadratics. To solve an incomplete quadratic equation, arrange the equation in standard form with zero as the right member. By simple factoring (see page 14) break the left member into two linear factors. Since this product is equal to zero, the equation will be a true mathematical statement if either or both factors are equal to zero. That is, $ab = 0$ is true if a is equal to zero, or if b is equal to zero, or if both a and b are equal to zero. Therefore, in order to obtain roots which will satisfy the equation, we may set each factor equal to zero, and solve for the required unknown in each of the two resulting equations.

$$x^2 = 2x.$$
$$x^2 - 2x = 0.$$
$$x(x - 2) = 0.$$
$$x = 0. \qquad x - 2 = 0.$$
$$x = 2.$$

Check:

$x^2 = 2x.$	$x^2 = 2x.$
$(0)^2 = 2(0).$	$(2)^2 = 2(2).$
$0 = 0.$	$4 = 4.$

4. Solve for x: $x^2 - 3x = 0$.

$$x(x-3) = 0.$$

$$x = 0. \qquad x - 3 = 0.$$
$$x = 3.$$

Check:

$$x^2 - 3x = 0. \qquad x^2 - 3x = 0.$$
$$(0)^2 - 3(0) = 0. \qquad (3)^2 - 3(3) = 0.$$
$$0 = 0. \qquad 9 - 9 = 0.$$
$$0 = 0.$$

5. Solve for x: $6x^2 + 2x = 0$.

$$2x(3x+1) = 0.$$

$$2x = 0. \qquad 3x + 1 = 0.$$
$$x = 0. \qquad 3x = -1.$$
$$x = -\tfrac{1}{3}.$$

Check:

$$6x^2 + 2x = 0. \qquad 6x^2 + 2x = 0.$$
$$6(0)^2 + 2(0) = 0. \qquad 6(-\tfrac{1}{3})^2 + 2(-\tfrac{1}{3}) = 0.$$
$$0 = 0. \qquad \tfrac{2}{3} - \tfrac{2}{3} = 0.$$
$$0 = 0.$$

Solving Complete Quadratics. Complete quadratic equations may be solved by three principal methods, namely: by **factoring**; by **completing the square**; and by using the **quadratic formula.** Some equations may be solved by more than one of these three methods and others can only be solved by a particular method.

Solving by Factoring. To solve a quadratic equation by the method of factoring, first arrange the equation in standard form with zero as the right member. If the left member can be factored easily (see p. 16), factor it, set each factor in turn equal to zero, and solve for the required unknown in both of the resulting equations.

$$x^2 - 2x - 3 = 0.$$
$$(x-3)(x+1) = 0.$$

$$x - 3 = 0. \qquad x + 1 = 0.$$
$$x = 3. \qquad x = -1.$$

Check:

$$x^2 - 2x - 3 = 0. \qquad x^2 - 2x - 3 = 0.$$
$$(3)^2 - 2(3) - 3 = 0. \qquad (-1)^2 - 2(-1) - 3 = 0.$$
$$9 - 6 - 3 = 0. \qquad 1 + 2 - 3 = 0.$$
$$0 = 0. \qquad 0 = 0.$$

6. Solve for x: $x^2 + 7x + 12 = 0$.

$$(x+4)(x+3) = 0.$$

$$x + 4 = 0. \qquad x + 3 = 0.$$
$$x = -4. \qquad x = -3.$$

Check:

$$x^2 + 7x + 12 = 0. \qquad x^2 + 7x + 12 = 0.$$
$$(-4)^2 + 7(-4) + 12 = 0. \qquad (-3)^2 + 7(-3) + 12 = 0.$$
$$16 - 28 + 12 = 0. \qquad 9 - 21 + 12 = 0.$$
$$0 = 0. \qquad 0 = 0.$$

Therefore, -4 and -3 both satisfy the original equation. No other values of x, when substituted in the equation, will satisfy it.

7. Solve for x: $x^2 - 4x - 21 = 0$.

$$(x-7)(x+3) = 0.$$

$$x - 7 = 0. \qquad x + 3 = 0.$$
$$x = 7. \qquad x = -3.$$

Check:

$$x^2 - 4x - 21 = 0. \qquad x^2 - 4x - 21 = 0.$$
$$(7)^2 - 4(7) - 21 = 0. \qquad (-3)^2 - 4(-3) - 21 = 0.$$
$$49 - 28 - 21 = 0. \qquad 9 + 12 - 21 = 0.$$
$$0 = 0. \qquad 0 = 0.$$

8. Solve for y: $y^2 - 16y + 63 = 0$.

$$(y-9)(y-7) = 0.$$

$$y - 9 = 0. \qquad y - 7 = 0.$$
$$y = 9. \qquad y = 7.$$

Check:

$$y^2 - 16y + 63 = 0. \qquad y^2 - 16y + 63 = 0.$$
$$(9)^2 - 16(9) + 63 = 0. \qquad (7)^2 - 16(7) + 63 = 0.$$
$$81 - 144 + 63 = 0. \qquad 49 - 112 + 63 = 0.$$
$$0 = 0. \qquad 0 = 0.$$

9. Solve for m: $m^2 - 3m - 40 = 0$.

$$(m-8)(m+5) = 0.$$

$$m - 8 = 0. \qquad m + 5 = 0.$$
$$m = 8. \qquad m = -5.$$

Check:

$$m^2 - 3m - 40 = 0. \qquad m^2 - 3m - 40 = 0.$$
$$(8)^2 - 3(8) - 40 = 0. \qquad (-5)^2 - 3(-5) - 40 = 0.$$
$$64 - 24 - 40 = 0. \qquad 25 + 15 - 40 = 0.$$
$$0 = 0. \qquad 0 = 0.$$

10. Solve for x: $x^2 - 10 = 3x$.

$$x^2 - 3x - 10 = 0.$$
$$(x-5)(x+2) = 0.$$

$$x - 5 = 0. \qquad x + 2 = 0.$$
$$x = 5. \qquad x = -2.$$

Check:

$$x^2 - 10 = 3x. \qquad x^2 - 10 = 3x.$$
$$(5)^2 - 10 = 3(5). \qquad (-2)^2 - 10 = 3(-2).$$
$$25 - 10 = 15. \qquad 4 - 10 = -6.$$
$$15 = 15. \qquad -6 = -6.$$

11. Solve for x: $3x^2 - 5x - 2 = 0$.

$$(3x+1)(x-2) = 0.$$

$$3x + 1 = 0. \qquad x - 2 = 0.$$
$$3x = -1. \qquad x = 2.$$
$$x = -\tfrac{1}{3}.$$

Check:

$$3x^2 - 5x - 2 = 0. \qquad 3x^2 - 5x - 2 = 0.$$
$$3(-\tfrac{1}{3})^2 - 5(-\tfrac{1}{3}) - 2 = 0. \qquad 3(2)^2 - 5(2) - 2 = 0.$$
$$3(\tfrac{1}{9}) + \tfrac{5}{3} - 2 = 0. \qquad 3(4) - 10 - 2 = 0.$$
$$\tfrac{1}{3} + \tfrac{5}{3} - 2 = 0. \qquad 12 - 10 - 2 = 0.$$
$$\tfrac{6}{3} - 2 = 0. \qquad 0 = 0.$$
$$2 - 2 = 0.$$
$$0 = 0.$$

12. Solve for x: $8x^2 + 2x - 1 = 0$.

$$(4x - 1)(2x + 1) = 0.$$

$$4x - 1 = 0. \qquad 2x + 1 = 0.$$
$$4x = 1. \qquad 2x = -1.$$
$$x = \tfrac{1}{4}. \qquad x = -\tfrac{1}{2}.$$

Check:

$$8x^2 + 2x - 1 = 0. \qquad 8x^2 + 2x - 1 = 0.$$
$$8(\tfrac{1}{4})^2 + 2(\tfrac{1}{4}) - 1 = 0. \qquad 8(-\tfrac{1}{2})^2 + 2(-\tfrac{1}{2}) - 1 = 0.$$
$$8(\tfrac{1}{16}) + \tfrac{2}{4} - 1 = 0. \qquad 8(\tfrac{1}{4}) - \tfrac{2}{2} - 1 = 0.$$
$$\tfrac{8}{16} + \tfrac{1}{2} - 1 = 0. \qquad \tfrac{8}{4} - 1 - 1 = 0.$$
$$\tfrac{1}{2} + \tfrac{1}{2} - 1 = 0. \qquad 2 - 1 - 1 = 0.$$
$$0 = 0. \qquad 0 = 0.$$

13. Solve for x: $x^2 + 2xy + y^2 = 0$.

$$(x + y)(x + y) = 0.$$
$$x + y = 0. \qquad x + y = 0.$$
$$x = -y. \qquad x = -y.$$

The two roots are identical.

Check:

$$x^2 + 2xy + y^2 = 0.$$
$$(-y)^2 + 2(-y)y + y^2 = 0.$$
$$y^2 - 2y^2 + y^2 = 0.$$
$$0 = 0.$$

14. Solve for x: $2x^2 - 5bx - 3b^2 = 0$.

$$(2x + b)(x - 3b) = 0.$$
$$2x + b = 0. \qquad x - 3b = 0.$$
$$2x = -b. \qquad x = 3b.$$
$$x = -\frac{b}{2}.$$

Check:

$$2x^2 - 5bx - 3b^2 = 0.$$
$$2\left(-\frac{b}{2}\right)^2 - 5b\left(-\frac{b}{2}\right) - 3b^2 = 0.$$
$$2\left(\frac{b^2}{4}\right) + \frac{5b^2}{2} - 3b^2 = 0.$$
$$\frac{b^2}{2} + \frac{5b^2}{2} - 3b^2 = 0.$$
$$+\frac{6b^2}{2} - 3b^2 = 0.$$
$$0 = 0.$$

$$2x^2 - 5bx - 3b^2 = 0.$$
$$2(3b)^2 - 5b(3b) - 3b^2 = 0.$$
$$2(9b^2) - 15b^2 - 3b^2 = 0.$$
$$18b^2 - 15b^2 - 3b^2 = 0.$$
$$0 = 0.$$

15. Solve for a: $3a^2 + 11abc - 4b^2c^2 = 0$.

$$(3a - bc)(a + 4bc) = 0.$$
$$3a - bc = 0. \qquad a + 4bc = 0.$$
$$3a = bc. \qquad a = -4bc.$$
$$a = \frac{bc}{3}.$$

Check:

$$3a^2 + 11abc - 4b^2c^2 = 0.$$
$$3\left(\frac{bc}{3}\right)^2 + 11\left(\frac{bc}{3}\right)bc - 4b^2c^2 = 0.$$
$$\frac{b^2c^2}{3} + \frac{11b^2c^2}{3} - 4b^2c^2 = 0.$$
$$\frac{12b^2c^2}{3} - 4b^2c^2 = 0.$$
$$4b^2c^2 - 4b^2c^2 = 0.$$
$$0 = 0.$$

$$3a^2 + 11abc - 4b^2c^2 = 0.$$
$$3(-4bc)^2 + 11(-4bc)bc - 4b^2c^2 = 0.$$
$$3(16b^2c^2) - 44b^2c^2 - 4b^2c^2 = 0.$$
$$48b^2c^2 - 44b^2c^2 - 4b^2c^2 = 0.$$
$$0 = 0.$$

Solving by Completing the Square. When solving a quadratic equation by completing the square, subtract the constant from both members of the equation. With the two remaining terms in the left member, form a perfect square trinomial. To get the constant term of the perfect square, take one-half the coefficient of the x-term and square it. (This rule can only be applied when the coefficient of the x^2-term is 1.) The value of the third term of the perfect square must be added to both members of the equation to maintain the equality.

$$x^2 - 10x + 16 = 0.$$
$$x^2 - 10x = -16.$$
$$x^2 - 10x + 25 = 25 - 16.$$
$$(x - 5)^2 = 9.$$
$$x - 5 = \pm 3.$$
$$x = 5 \pm 3.$$
$$x = 8. \qquad x = 2.$$

Check:

$$x^2 - 10x + 16 = 0. \qquad x^2 - 10x + 16 = 0.$$
$$(8)^2 - 10(8) + 16 = 0. \qquad (2)^2 - 10(2) + 16 = 0.$$
$$64 - 80 + 16 = 0. \qquad 4 - 20 + 16 = 0.$$
$$0 = 0. \qquad 0 = 0.$$

16. Solve for x: $x^2 + 8x + 12 = 0$.

$$x^2 + 8x = -12.$$
$$x^2 + 8x + 16 = 16 - 12.$$
$$(x + 4)^2 = 4.$$

Take the square root of each member of the equation.

$$x + 4 = \pm 2.$$
$$x = -4 \pm 2.$$
$$x = -6. \qquad x = -2.$$

Check:

$$x^2 + 8x + 12 = 0. \qquad x^2 + 8x + 12 = 0.$$
$$(-6)^2 + 8(-6) + 12 = 0. \qquad (-2)^2 + 8(-2) + 12 = 0.$$
$$36 - 48 + 12 = 0. \qquad 4 - 16 + 12 = 0.$$
$$0 = 0. \qquad 0 = 0.$$

17. Solve for x: $x^2 + 12x - 45 = 0$.

$$x^2 + 12x = 45.$$
$$x^2 + 12x + 36 = 45 + 36.$$
$$(x + 6)^2 = 81.$$
$$x + 6 = \pm 9.$$
$$x = -6 \pm 9.$$
$$x = -15. \qquad x = 3.$$

Check:

$$x^2 + 12x - 45 = 0. \qquad x^2 + 12x - 45 = 0.$$
$$(-15)^2 + 12(-15) - 45 = 0. \qquad (3)^2 + 12(3) - 45 = 0.$$
$$225 - 180 - 45 = 0. \qquad 9 + 36 - 45 = 0.$$
$$0 = 0. \qquad 0 = 0.$$

18. Solve for y: $y^2 + 3y - 88 = 0$.

$$y^2 + 3y = 88.$$
$$y^2 + 3y + (\tfrac{3}{2})^2 = 88 + (\tfrac{3}{2})^2.$$
$$y^2 + 3y + \tfrac{9}{4} = 88 + \tfrac{9}{4}.$$
$$y^2 + 3y + \tfrac{9}{4} = \tfrac{352}{4} + \tfrac{9}{4}.$$
$$(y + \tfrac{3}{2})^2 = \tfrac{361}{4}.$$
$$y + \tfrac{3}{2} = \pm\tfrac{19}{2}.$$
$$y = -\tfrac{3}{2} \pm \tfrac{19}{2}.$$
$$y = 8. \qquad y = -11.$$

Check:

$$y^2 + 3y - 88 = 0. \qquad y^2 + 3y - 88 = 0.$$
$$(8)^2 + 3(8) - 88 = 0. \qquad (-11)^2 + 3(-11) - 88 = 0.$$
$$64 + 24 - 88 = 0. \qquad 121 - 33 - 88 = 0.$$
$$0 = 0. \qquad 0 = 0.$$

19. Solve for a: $a^2 + 14a + 48 = 0$.

$$a^2 + 14a = -48.$$
$$a^2 + 14a + 49 = 49 - 48.$$
$$(a + 7)^2 = 1.$$
$$a + 7 = \pm 1.$$
$$a = -7 \pm 1.$$
$$a = -8. \qquad a = -6.$$

Check:

$$a^2 + 14a + 48 = 0. \qquad a^2 + 14a + 48 = 0.$$
$$(-8)^2 + 14(-8) + 48 = 0. \qquad (-6)^2 + 14(-6) + 48 = 0.$$
$$64 - 112 + 48 = 0. \qquad 36 - 84 + 48 = 0.$$
$$0 = 0. \qquad 0 = 0.$$

20. Solve for b: $b^2 - 6b - 27 = 0$.

$$b^2 - 6b = 27.$$
$$b^2 - 6b + 9 = 27 + 9.$$
$$(b - 3)^2 = 36.$$
$$b - 3 = \pm 6.$$
$$b = 3 \pm 6.$$
$$b = 9. \qquad b = -3.$$

Check:

$$b^2 - 6b - 27 = 0. \qquad b^2 - 6b - 27 = 0.$$
$$(9)^2 - 6(9) - 27 = 0. \qquad (-3)^2 - 6(-3) - 27 = 0.$$
$$81 - 54 - 27 = 0. \qquad 9 + 18 - 27 = 0.$$
$$0 = 0. \qquad 0 = 0.$$

21. Solve for x: $2x^2 + 5x - 3 = 0$.

Divide both members of the equation by 2 so that the coefficient of the x^2-term will be 1. Then the rule for completing the square will apply.

$$x^2 + \tfrac{5}{2}x = \tfrac{3}{2}.$$
$$x^2 + \tfrac{5}{2}x + (\tfrac{5}{4})^2 = \tfrac{3}{2} + (\tfrac{5}{4})^2.$$
$$(x + \tfrac{5}{4})^2 = \tfrac{24}{16} + \tfrac{25}{16}.$$
$$(x + \tfrac{5}{4})^2 = \tfrac{49}{16}.$$
$$x + \tfrac{5}{4} = \pm\tfrac{7}{4}.$$
$$x = -\tfrac{5}{4} \pm \tfrac{7}{4}.$$
$$x = -\tfrac{12}{4} = -3. \qquad x = \tfrac{2}{4} = \tfrac{1}{2}.$$

Check:

$$2x^2 + 5x - 3 = 0. \qquad 2x^2 + 5x - 3 = 0.$$
$$2(-3)^2 + 5(-3) - 3 = 0. \qquad 2(\tfrac{1}{2})^2 + 5(\tfrac{1}{2}) - 3 = 0.$$
$$2(9) - 15 - 3 = 0. \qquad 2(\tfrac{1}{4}) + \tfrac{5}{2} - 3 = 0.$$
$$18 - 15 - 3 = 0. \qquad \tfrac{2}{4} + \tfrac{5}{2} - 3 = 0.$$
$$0 = 0. \qquad \tfrac{1}{2} + \tfrac{5}{2} - 3 = 0.$$
$$\tfrac{6}{2} - 3 = 0.$$
$$3 - 3 = 0.$$
$$0 = 0.$$

22. Solve for y: $3y^2 + 4y - 4 = 0$.

$$y^2 + \tfrac{4}{3}y = \tfrac{4}{3}.$$
$$y^2 + \tfrac{4}{3}y + (\tfrac{2}{3})^2 = \tfrac{4}{3} + (\tfrac{2}{3})^2.$$
$$(y + \tfrac{2}{3})^2 = \tfrac{12}{9} + \tfrac{4}{9}.$$
$$(y + \tfrac{2}{3})^2 = \tfrac{16}{9}.$$
$$y + \tfrac{2}{3} = \pm\tfrac{4}{3}.$$
$$y = -\tfrac{2}{3} \pm \tfrac{4}{3}.$$
$$y = \tfrac{2}{3}. \qquad x = -2.$$

Check:

$$3y^2 + 4y - 4 = 0. \qquad 3y^2 + 4y - 4 = 0.$$
$$3(\tfrac{2}{3})^2 + 4(\tfrac{2}{3}) - 4 = 0. \qquad 3(-2)^2 + 4(-2) - 4 = 0.$$
$$3(\tfrac{4}{9}) + \tfrac{8}{3} - 4 = 0. \qquad 3(4) - 8 - 4 = 0.$$
$$\tfrac{4}{3} + \tfrac{8}{3} - 4 = 0. \qquad 12 - 8 - 4 = 0.$$
$$\tfrac{12}{3} - 4 = 0. \qquad 0 = 0.$$
$$4 - 4 = 0.$$
$$0 = 0.$$

23. Solve for m: $4m^2 + 11m - 3 = 0$.

$$4m^2 + 11m = 3.$$
$$m^2 + \frac{11m}{4} = \frac{3}{4}.$$
$$m^2 + \tfrac{11}{4}m + (\tfrac{11}{8})^2 = \tfrac{3}{4} + (\tfrac{11}{8})^2.$$
$$(m + \tfrac{11}{8})^2 = \tfrac{48}{64} + \tfrac{121}{64}.$$
$$(m + \tfrac{11}{8})^2 = \tfrac{169}{64}.$$
$$m + \tfrac{11}{8} = \pm\tfrac{13}{8}.$$
$$m = -\tfrac{11}{8} \pm \tfrac{13}{8}.$$
$$m = -\tfrac{24}{8} = -3. \qquad m = \tfrac{2}{8} = \tfrac{1}{4}.$$

Check:

$$4m^2 + 11m - 3 = 0. \qquad 4m^2 + 11m - 3 = 0.$$
$$4(-3)^2 + 11(-3) - 3 = 0. \qquad 4(\tfrac{1}{4})^2 + 11(\tfrac{1}{4}) - 3 = 0.$$
$$4(9) - 33 - 3 = 0. \qquad 4(\tfrac{1}{16}) + \tfrac{11}{4} - 3 = 0.$$
$$36 - 33 - 3 = 0. \qquad \tfrac{1}{4} + \tfrac{11}{4} - 3 = 0.$$
$$0 = 0. \qquad \tfrac{12}{4} - 3 = 0.$$
$$3 - 3 = 0.$$

24. Solve for a: $6a^2 - 13a + 6 = 0$.

$$6a^2 - 13a = -6.$$
$$a^2 - \tfrac{13}{6}a = -1.$$
$$a^2 - \tfrac{13}{6}a + (\tfrac{13}{12})^2 = -1 + (\tfrac{13}{12})^2.$$
$$(a - \tfrac{13}{12})^2 = -\tfrac{144}{144} + \tfrac{169}{144}.$$
$$(a - \tfrac{13}{12})^2 = \tfrac{25}{144}.$$
$$a - \tfrac{13}{12} = \pm\tfrac{5}{12}.$$
$$a = \tfrac{13}{12} \pm \tfrac{5}{12}.$$
$$a = \tfrac{18}{12} = \tfrac{3}{2}. \qquad a = \tfrac{8}{12} = \tfrac{2}{3}.$$

Check:

$$6a^2 - 13a + 6 = 0. \qquad 6a^2 - 13a + 6 = 0.$$
$$6(\tfrac{3}{2})^2 - 13(\tfrac{3}{2}) + 6 = 0. \qquad 6(\tfrac{2}{3})^2 - 13(\tfrac{2}{3}) + 6 = 0.$$
$$6(\tfrac{9}{4}) - \tfrac{39}{2} + 6 = 0. \qquad 6(\tfrac{4}{9}) - \tfrac{26}{3} + 6 = 0.$$
$$\tfrac{27}{2} - \tfrac{39}{2} + 6 = 0. \qquad \tfrac{8}{3} - \tfrac{26}{3} + 6 = 0.$$
$$-\tfrac{12}{2} + 6 = 0. \qquad -\tfrac{18}{3} + 6 = 0.$$
$$-6 + 6 = 0. \qquad -6 + 6 = 0.$$
$$0 = 0. \qquad 0 = 0.$$

25. Solve for x: $x^2 - 3yx + 2y^2 = 0$.

$$x^2 - 3yx = -2y^2.$$
$$x^2 - 3yx + \left(\frac{3y}{2}\right)^2 = -2y^2 + \left(\frac{3y}{2}\right)^2.$$
$$\left(x - \frac{3y}{2}\right)^2 = -\frac{8y^2}{4} + \frac{9y^2}{4}.$$
$$\left(x - \frac{3y}{2}\right)^2 = \frac{y^2}{4}.$$
$$x - \frac{3y}{2} = \pm \frac{y}{2}.$$
$$x = \frac{3y}{2} \pm \frac{y}{2}.$$
$$x = 2y. \qquad x = y.$$

Check:

$$x^2 - 3yx + 2y^2 = 0. \qquad x^2 - 3yx + 2y^2 = 0.$$
$$(2y)^2 - 3y(2y) + 2y^2 = 0. \qquad (y)^2 - 3y(y) + 2y^2 = 0.$$
$$4y^2 - 6y^2 + 2y^2 = 0. \qquad y^2 - 3y^2 + 2y^2 = 0.$$
$$0 = 0. \qquad 0 = 0.$$

Solving by Quadratic Formula. Many quadratic equations, especially those whose roots are irrational numbers, are not easily solved by factoring. Such equations are usually solved by use of the quadratic formula. This formula is derived from the general quadratic equation as shown below. The principles involved in the method of completing the square are used to derive the formula.

$$ax^2 + bx + c = 0.$$
$$x^2 + \frac{b}{a}x = -\frac{c}{a}.$$
$$x^2 + \frac{b}{a}x + \left(\frac{b}{2a}\right)^2 = \left(\frac{b}{2a}\right)^2 - \frac{c}{a}.$$
$$\left(x + \frac{b}{2a}\right)^2 = \frac{b^2}{4a^2} - \frac{4ac}{4a^2}.$$
$$\left(x + \frac{b}{2a}\right)^2 = \frac{b^2 - 4ac}{4a^2}.$$
$$x + \frac{b}{2a} = \frac{\pm\sqrt{b^2 - 4ac}}{2a}.$$
$$x = -\frac{b}{2a} \pm \frac{\sqrt{b^2 - 4ac}}{2a}$$
$$= \frac{-b \pm \sqrt{b^2 - 4ac}}{2a}.$$

Hence, the roots are

$$\frac{-b + \sqrt{b^2 - 4ac}}{2a} \quad \text{and} \quad \frac{-b - \sqrt{b^2 - 4ac}}{2a}.$$

The equation derived above is called the quadratic formula. The values for a, b, and c from any quadratic equation may be substituted in the quadratic formula and the roots found by completing the indicated operations.

Solve for x: $3x^2 - 7x + 2 = 0$.

$$a = 3;$$
$$b = -7;$$
$$c = 2.$$
$$x = \frac{-b \pm \sqrt{b^2 - 4ac}}{2a}$$
$$= \frac{-(-7) \pm \sqrt{(7)^2 - 4(3)(2)}}{2(3)}$$
$$= \frac{7 \pm \sqrt{49 - 24}}{6}$$
$$= \frac{7 \pm \sqrt{25}}{6}$$
$$= \frac{7 \pm 5}{6}.$$
$$x = 2. \qquad x = \tfrac{1}{3}.$$

Check:

$$3x^2 - 7x + 2 = 0. \qquad 3x^2 - 7x + 2 = 0.$$
$$3(2)^2 - 7(2) + 2 = 0. \qquad 3(\tfrac{1}{3})^2 - 7(\tfrac{1}{3}) + 2 = 0.$$
$$3(4) - 14 + 2 = 0. \qquad 3(\tfrac{1}{9}) - \tfrac{7}{3} + 2 = 0.$$
$$12 - 14 + 2 = 0. \qquad \tfrac{1}{3} - \tfrac{7}{3} + 2 = 0.$$
$$0 = 0. \qquad -\tfrac{6}{3} + 2 = 0.$$
$$-2 + 2 = 0.$$
$$0 = 0.$$

26. Solve for x: $x^2 + 8x + 12 = 0$.

$$a = 1;$$
$$b = 8;$$
$$c = 12.$$
$$x = \frac{-b \pm \sqrt{b^2 - 4ac}}{2a}$$
$$= \frac{-8 \pm \sqrt{(8)^2 - 4(1)(12)}}{2(1)}$$
$$= \frac{-8 \pm \sqrt{64 - 48}}{2}$$
$$= \frac{-8 \pm \sqrt{16}}{2}$$
$$= \frac{-8 \pm 4}{2}$$
$$= -4 \pm 2.$$
$$x = -6. \qquad x = -2.$$

Check:

$$x^2 + 8x + 12 = 0. \qquad x^2 + 8x + 12 = 0.$$
$$(-6)^2 + 8(-6) + 12 = 0. \qquad (-2)^2 + 8(-2) + 12 = 0.$$
$$36 - 48 + 12 = 0. \qquad 4 - 16 + 12 = 0.$$
$$0 = 0. \qquad 0 = 0.$$

27. Solve for x: $2x^2 - 5x - 3 = 0$.

$$a = 2;\quad b = -5;\quad c = -3.$$

$$x = \frac{-b \pm \sqrt{b^2 - 4ac}}{2a}$$
$$= \frac{-(-5) \pm \sqrt{(-5)^2 - 4(2)(-3)}}{2(2)}$$
$$= \frac{5 \pm \sqrt{25 + 24}}{4}$$
$$= \frac{5 \pm \sqrt{49}}{4}$$
$$= \frac{5 \pm 7}{4}.$$
$$x = \tfrac{12}{4} = 3. \qquad x = -\tfrac{2}{4} = -\tfrac{1}{2}.$$

Check:

$$2x^2 - 5x - 3 = 0.$$
$$2(3)^2 - 5(3) - 3 = 0.$$
$$2(9) - 15 - 3 = 0.$$
$$18 - 15 - 3 = 0.$$
$$0 = 0.$$

$$2x^2 - 5x - 3 = 0.$$
$$2(-\tfrac{1}{2})^2 - 5(-\tfrac{1}{2}) - 3 = 0.$$
$$2(\tfrac{1}{4}) + \tfrac{5}{2} - 3 = 0.$$
$$\tfrac{1}{2} + \tfrac{5}{2} - 3 = 0.$$
$$\tfrac{6}{2} - 3 = 0.$$
$$3 - 3 = 0.$$
$$0 = 0.$$

28. Solve for x: $6x^2 - 5x - 6 = 0$.

$$a = 6;\quad b = -5;\quad c = -6.$$

$$x = \frac{-b \pm \sqrt{b^2 - 4ac}}{2a}$$
$$= \frac{-(-5) \pm \sqrt{(-5)^2 - 4(6)(-6)}}{2(6)}$$
$$= \frac{5 \pm \sqrt{25 + 144}}{12}$$
$$= \frac{5 \pm \sqrt{169}}{12}$$
$$= \frac{5 \pm 13}{12}.$$
$$x = \tfrac{18}{12} = \tfrac{3}{2}. \qquad x = -\tfrac{8}{12} = -\tfrac{2}{3}.$$

Check:

$$6x^2 - 5x - 6 = 0.$$
$$6(\tfrac{3}{2})^2 - 5(\tfrac{3}{2}) - 6 = 0.$$
$$6(\tfrac{9}{4}) - \tfrac{15}{2} - 6 = 0.$$
$$\tfrac{27}{2} - \tfrac{15}{2} - 6 = 0.$$
$$\tfrac{12}{2} - 6 = 0.$$
$$6 - 6 = 0.$$
$$0 = 0.$$

$$6x^2 - 5x - 6 = 0.$$
$$6(-\tfrac{2}{3})^2 - 5(-\tfrac{2}{3}) - 6 = 0.$$
$$6(\tfrac{4}{9}) + \tfrac{10}{3} - 6 = 0.$$
$$\tfrac{8}{3} + \tfrac{10}{3} - 6 = 0.$$
$$\tfrac{18}{3} - 6 = 0.$$
$$6 - 6 = 0.$$
$$0 = 0.$$

29. Solve for x: $x^2 + 4x - 3 = 0$.

$$a = 1;\quad b = 4;\quad c = -3.$$

$$x = \frac{-b \pm \sqrt{b^2 - 4ac}}{2a}$$
$$= \frac{-4 \pm \sqrt{(4)^2 - 4(1)(-3)}}{2(1)}$$
$$= \frac{-4 \pm \sqrt{16 + 12}}{2}$$
$$= \frac{-4 \pm \sqrt{28}}{2}$$
$$= \frac{-4 \pm 2\sqrt{7}}{2}.$$
$$x = -2 + \sqrt{7}. \qquad x = -2 - \sqrt{7}.$$

Notice that the roots of this equation are conjugates. In general, whenever a root has the form $a + \sqrt{b}$, where $\sqrt{b}$ is a quadratic surd, the conjugate of $a + \sqrt{b}$ is also a root of the equation.

Check:

$$x^2 + 4x - 3 = 0.$$
$$(-2 + \sqrt{7})^2 + 4(-2 + \sqrt{7}) - 3 = 0.$$
$$11 - 4\sqrt{7} - 8 + 4\sqrt{7} - 3 = 0.$$
$$0 = 0.$$

$$x^2 + 4x - 3 = 0.$$
$$(-2 - \sqrt{7})^2 + 4(-2 - \sqrt{7}) - 3 = 0.$$
$$11 + 4\sqrt{7} - 8 - 4\sqrt{7} - 3 = 0.$$
$$0 = 0.$$

30. Solve for x: $3x^2 - 7x - 2 = 0$.

$$a = 3;\quad b = -7;\quad c = -2.$$

$$x = \frac{-b \pm \sqrt{b^2 - 4ac}}{2a}$$
$$= \frac{-(-7) \pm \sqrt{(-7)^2 - 4(3)(-2)}}{2(3)}$$
$$= \frac{7 \pm \sqrt{49 + 24}}{6}$$
$$= \frac{7 \pm \sqrt{73}}{6}.$$
$$x = \frac{7 + \sqrt{73}}{6}. \qquad x = \frac{7 - \sqrt{73}}{6}.$$

Check:

$$3x^2 - 7x - 2 = 0.$$
$$3\left(\frac{7 + \sqrt{73}}{6}\right)^2 - 7\left(\frac{7 + \sqrt{73}}{6}\right) - 2 = 0.$$
$$3\left(\frac{49 + 14\sqrt{73} + 73}{36}\right) - \frac{49 + 7\sqrt{73}}{6} - 2 = 0.$$
$$\frac{122 + 14\sqrt{73}}{12} - \frac{98 + 14\sqrt{73}}{12} - \frac{24}{12} = 0.$$
$$122 + 14\sqrt{73} - 98 - 14\sqrt{73} - 24 = 0.$$
$$0 = 0.$$

$$3x^2 - 7x - 2 = 0.$$
$$3\left(\frac{7 - \sqrt{73}}{6}\right)^2 - 7\left(\frac{7 - \sqrt{73}}{6}\right) - 2 = 0.$$
$$3\left(\frac{49 - 14\sqrt{73} + 73}{36}\right) - \frac{49 - 7\sqrt{73}}{6} - 2 = 0.$$
$$\frac{122 - 14\sqrt{73}}{12} - \frac{98 - 14\sqrt{73}}{12} - \frac{24}{12} = 0.$$

$$122 - 14\sqrt{73} - 98 + 14\sqrt{73} - 24 = 0.$$
$$0 = 0.$$

31. Solve for x: $7x^2 + 2x - 4 = 0.$

$$a = 7;$$
$$b = 2;$$
$$c = -4.$$

$$x = \frac{-b \pm \sqrt{b^2 - 4ac}}{2a}$$
$$= \frac{-2 \pm \sqrt{(2)^2 - 4(7)(-4)}}{2(7)}$$
$$= \frac{-2 \pm \sqrt{4 + 112}}{14}$$
$$= \frac{-2 \pm \sqrt{116}}{14}$$
$$= \frac{-2 \pm 2\sqrt{29}}{14}.$$

$$x = \frac{-1 + \sqrt{29}}{7}. \qquad x = \frac{-1 - \sqrt{29}}{7}.$$

Check:

$$7x^2 + 2x - 4 = 0.$$
$$7\left(\frac{-1 + \sqrt{29}}{7}\right)^2 + 2\left(\frac{-1 + \sqrt{29}}{7}\right) - 4 = 0.$$
$$7\left(\frac{1 - 2\sqrt{29} + 29}{49}\right) + \frac{-2 + 2\sqrt{29}}{7} - 4 = 0.$$
$$\frac{1 - 2\sqrt{29} + 29}{7} + \frac{-2 + 2\sqrt{29}}{7} - \frac{28}{7} = 0.$$
$$1 - 2\sqrt{29} + 29 - 2 + 2\sqrt{29} - 28 = 0.$$
$$0 = 0.$$

$$7x^2 + 2x - 4 = 0.$$
$$7\left(\frac{-1 - \sqrt{29}}{7}\right)^2 + 2\left(\frac{-1 - \sqrt{29}}{7}\right) - 4 = 0.$$
$$7\left(\frac{1 + 2\sqrt{29} + 29}{49}\right) + \frac{-2 - 2\sqrt{29}}{7} - \frac{28}{7} = 0.$$
$$\frac{1 + 2\sqrt{29} + 29}{7} + \frac{-2 - 2\sqrt{29}}{7} - \frac{28}{7} = 0.$$
$$1 + 2\sqrt{29} + 29 - 2 - 2\sqrt{29} - 28 = 0.$$
$$0 = 0.$$

32. Solve for x: $x^2 - 5x - 15 = 0.$

$$a = 1;$$
$$b = -5;$$
$$c = -15.$$

$$x = \frac{-b \pm \sqrt{b^2 - 4ac}}{2a} = 0$$
$$= \frac{-(-5) \pm \sqrt{(-5)^2 - 4(1)(-15)}}{2}$$
$$= \frac{5 \pm \sqrt{25 + 60}}{2}$$
$$= \frac{5 \pm \sqrt{85}}{2}.$$

$$x = \frac{5 + \sqrt{85}}{2}. \qquad x = \frac{5 - \sqrt{85}}{2}.$$

Check:

$$x^2 - 5x - 15 = 0.$$
$$\left(\frac{5 + \sqrt{85}}{2}\right)^2 - 5\left(\frac{5 + \sqrt{85}}{2}\right) - 15 = 0.$$
$$\frac{25 + 10\sqrt{85} + 85}{4} - \frac{25 + 5\sqrt{85}}{2} - 15 = 0.$$
$$\frac{25 + 10\sqrt{85} + 85}{4} - \frac{50 + 10\sqrt{85}}{4} - \frac{60}{4} = 0.$$
$$25 + 10\sqrt{85} + 85 - 50 - 10\sqrt{85} - 60 = 0.$$
$$0 = 0.$$

$$x^2 - 5x - 15 = 0.$$
$$\left(\frac{5 - \sqrt{85}}{2}\right)^2 - 5\left(\frac{5 - \sqrt{85}}{2}\right) - 15 = 0.$$
$$\frac{25 - 10\sqrt{85} + 85}{4} - \frac{25 - 5\sqrt{85}}{2} - 15 = 0.$$
$$\frac{25 - 10\sqrt{85} + 85}{4} - \frac{50 - 10\sqrt{85}}{4} - \frac{60}{4} = 0.$$
$$25 - 10\sqrt{85} + 85 - 50 + 10\sqrt{85} - 60 = 0.$$
$$0 = 0.$$

33. Solve for x: $x - 2x^2 = -6.$

Write the equation in standard form.

$$-2x^2 + x + 6 = 0.$$
$$a = -2;$$
$$b = 1;$$
$$c = 6.$$

$$x = \frac{-b \pm \sqrt{b^2 - 4ac}}{2a}$$
$$= \frac{-1 \pm \sqrt{(1)^2 - 4(-2)(6)}}{2(-2)}$$
$$= \frac{-1 \pm \sqrt{1 + 48}}{-4}$$
$$= \frac{-1 \pm \sqrt{49}}{-4}$$
$$= \frac{-1 \pm 7}{-4}.$$

$$x = \frac{-8}{-4} \qquad x = \frac{6}{-4}$$
$$= 2. \qquad = -\tfrac{3}{2}.$$

Check:

$$x - 2x^2 = -6.$$
$$2 - 2(2)^2 = -6.$$
$$2 - 2(4) = -6.$$
$$2 - 8 = -6.$$
$$-6 = -6.$$

$$x - 2x^2 = -6.$$
$$-\tfrac{3}{2} - 2(-\tfrac{3}{2})^2 = -6.$$
$$-\tfrac{3}{2} - 2(\tfrac{9}{4}) = -6.$$
$$-\tfrac{3}{2} - (\tfrac{9}{2}) = -6.$$
$$-\tfrac{12}{2} = -6.$$
$$-6 = -6.$$

34. Solve for x: $\frac{3x^2}{2} + \frac{2x}{3} - \frac{1}{4} = 0.$

Clear of fractions and then solve the new equation by use of the formula.

$$\frac{18x^2}{12} + \frac{8x}{12} - \frac{3}{12} = 0.$$
$$18x^2 + 8x - 3 = 0.$$

$$a = 18;$$
$$b = 8;$$
$$c = -3.$$

$$x = \frac{-b \pm \sqrt{b^2 - 4ac}}{2a}$$
$$= \frac{-8 \pm \sqrt{(8)^2 - 4(18)(-3)}}{2(18)}$$
$$= \frac{-8 \pm \sqrt{64 + 216}}{36}$$
$$= \frac{-8 \pm \sqrt{280}}{36}$$
$$= \frac{-8 \pm 2\sqrt{70}}{36}$$
$$= \frac{-4 \pm \sqrt{70}}{18}.$$

$$x = \frac{-4 + \sqrt{70}}{18} \qquad x = \frac{-4 - \sqrt{70}}{18}.$$

Check:

$$\frac{3x^2}{2} + \frac{2x}{3} - \frac{1}{4} = 0.$$
$$\frac{3}{2}\left(\frac{-4 + \sqrt{70}}{18}\right)^2 + \frac{2}{3}\left(\frac{-4 + \sqrt{70}}{18}\right) - \frac{1}{4} = 0.$$
$$\frac{3}{2}\left(\frac{16 - 8\sqrt{70} + 70}{324}\right) + \frac{-8 + 2\sqrt{70}}{54} - \frac{1}{4} = 0.$$
$$\frac{86 - 8\sqrt{70}}{216} + \frac{-8 + 2\sqrt{70}}{54} - \frac{1}{4} = 0.$$
$$\frac{43 - 4\sqrt{70}}{108} + \frac{-16 + 4\sqrt{70}}{108} - \frac{27}{108} = 0.$$
$$43 - 4\sqrt{70} - 16 + 4\sqrt{70} - 27 = 0.$$
$$0 = 0.$$

$$\frac{3x^2}{2} + \frac{2x}{3} - \frac{1}{4} = 0.$$
$$\frac{3}{2}\left(\frac{-4 - \sqrt{70}}{18}\right)^2 + \frac{2}{3}\left(\frac{-4 - \sqrt{70}}{18}\right) - \frac{1}{4} = 0.$$
$$\frac{3}{2}\left(\frac{16 + 8\sqrt{70} + 70}{324}\right) + \frac{-8 - 2\sqrt{70}}{54} - \frac{1}{4} = 0.$$
$$\frac{86 + 8\sqrt{70}}{216} + \frac{-8 - 2\sqrt{70}}{54} - \frac{1}{4} = 0.$$
$$\frac{43 + 4\sqrt{70}}{108} + \frac{-16 - 4\sqrt{70}}{108} - \frac{27}{108} = 0.$$
$$43 + 4\sqrt{70} - 16 - 4\sqrt{70} - 27 = 0.$$
$$0 = 0.$$

35. Solve for x: $\frac{x^2}{5} - \frac{x}{3} - \frac{7}{30} = 0.$

Clear of fractions.

$$\frac{6x^2}{30} - \frac{10x}{30} - \frac{7}{30} = 0.$$
$$6x^2 - 10x - 7 = 0.$$

$$a = 6;$$
$$b = -10;$$
$$c = -7.$$

$$x = \frac{-b \pm \sqrt{b^2 - 4ac}}{2a}$$
$$= \frac{-(-10) \pm \sqrt{(-10)^2 - 4(6)(-7)}}{2(6)}$$
$$= \frac{10 \pm \sqrt{100 + 168}}{12}$$
$$= \frac{10 \pm \sqrt{268}}{12} = \frac{10 \pm 2\sqrt{67}}{12}.$$

$$x = \frac{5 + \sqrt{67}}{6}. \qquad x = \frac{5 - \sqrt{67}}{6}.$$

Check:

$$\frac{x^2}{5} - \frac{x}{3} - \frac{7}{30} = 0.$$
$$\frac{1}{5}\left(\frac{5 + \sqrt{67}}{6}\right)^2 - \frac{1}{3}\left(\frac{5 + \sqrt{67}}{6}\right) - \frac{7}{30} = 0.$$
$$\frac{1}{5}\left(\frac{25 + 10\sqrt{67} + 67}{36}\right) - \frac{5 + \sqrt{67}}{18} - \frac{7}{30} = 0.$$
$$\frac{92 + 10\sqrt{67}}{180} - \frac{5 + \sqrt{67}}{18} - \frac{7}{30} = 0.$$
$$\frac{92 + 10\sqrt{67}}{180} - \frac{50 + 10\sqrt{67}}{180} - \frac{42}{180} = 0.$$
$$92 + 10\sqrt{67} - 50 - 10\sqrt{67} - 42 = 0.$$
$$0 = 0.$$

$$\frac{x^2}{5} - \frac{x}{3} - \frac{7}{30} = 0.$$
$$\frac{1}{5}\left(\frac{5 - \sqrt{67}}{6}\right)^2 - \frac{1}{3}\left(\frac{5 - \sqrt{67}}{6}\right) - \frac{7}{30} = 0.$$
$$\frac{1}{5}\left(\frac{25 - 10\sqrt{67} + 67}{36}\right) - \frac{5 - \sqrt{67}}{18} - \frac{7}{30} = 0.$$
$$\frac{92 - 10\sqrt{67}}{180} - \frac{50 - 10\sqrt{67}}{180} - \frac{42}{180} = 0.$$
$$92 - 10\sqrt{67} - 50 + 10\sqrt{67} - 42 = 0.$$
$$0 = 0.$$

THE ROOTS OF A QUADRATIC EQUATION

Forming a Quadratic Equation from Two Given Roots. Forming a quadratic equation when the two roots are given involves the reverse of finding the roots by factoring. The roots of the following equation are found by factoring.

$$x^2 - 3x - 40 = 0.$$
$$(x - 8)(x + 5) = 0.$$
$$x - 8 = 0. \qquad x + 5 = 0.$$
$$x = 8. \qquad x = -5.$$

Hence, the roots are 8 and -5.

To form the quadratic equation whose roots are 8 and -5, set x equal to each of these roots.

$$x = 8. \qquad x = -5.$$

Subtract the constants from each member of each equation to form the factors.

$$x - 8 = 0. \qquad x + 5 = 0.$$

Equate the product of these factors to zero.

$$(x - 8)(x + 5) = 0.$$

Multiply these factors to obtain the quadratic equation.

$$x^2 - 3x - 40 = 0.$$

1. Find the quadratic equation whose roots are 2 and 3.

$$x = 2. \qquad x = 3.$$
$$x - 2 = 0. \qquad x - 3 = 0.$$
$$(x - 2)(x - 3) = 0.$$
$$x^2 - 5x + 6 = 0.$$

2. Find the quadratic equation whose roots are 2 and -6.

$$x = 2. \qquad x = -6.$$
$$x - 2 = 0. \qquad x + 6 = 0.$$
$$(x - 2)(x + 6) = 0.$$
$$x^2 + 4x - 12 = 0.$$

3. Find the quadratic equation whose roots are -4 and -7.

$$x = -4. \qquad x = -7.$$
$$x + 4 = 0. \qquad x + 7 = 0.$$
$$(x + 4)(x + 7) = 0.$$
$$x^2 + 11x + 28 = 0.$$

4. Find the quadratic equation whose roots are $\frac{1}{2}$ and $\frac{3}{2}$.

$$x = \tfrac{1}{2}. \qquad x = \tfrac{3}{2}.$$

Clear of fractions in each of the above cases.

$$2x = 1. \qquad 2x = 3.$$
$$2x - 1 = 0. \qquad 2x - 3 = 0.$$
$$(2x - 1)(2x - 3) = 0.$$
$$4x^2 - 8x + 3 = 0.$$

5. Find the quadratic equation whose roots are $1 - \sqrt{3}$ and $1 + \sqrt{3}$.

$$x = 1 - \sqrt{3}. \qquad x = 1 + \sqrt{3}.$$
$$x - 1 + \sqrt{3} = 0. \qquad x - 1 - \sqrt{3} = 0.$$
$$(x - 1 + \sqrt{3})(x - 1 - \sqrt{3}) = 0.$$
$$x^2 - 2x - 2 = 0.$$

The Sum and Product of the Roots. If we add or multiply the two roots of a quadratic equation as found from the quadratic formula, we get a pair of useful relationships among the coefficients of the equation. Let r_1 and r_2 represent the roots, which would appear as follows:

$$r_1 = \frac{-b + \sqrt{b^2 - 4ac}}{2a}.$$
$$r_2 = \frac{-b - \sqrt{b^2 - 4ac}}{2a}.$$

If we add these two expressions we get

$$r_1 + r_2 = \frac{-b + \sqrt{b^2 - 4ac}}{2a} + \frac{-b - \sqrt{b^2 - 4ac}}{2a}$$
$$= \frac{-b + \sqrt{b^2 - 4ac} - b - \sqrt{b^2 - 4ac}}{2a}$$
$$= -\frac{2b}{2a}$$
$$= -\frac{b}{a}.$$

Therefore, the sum of the roots $(r_1 + r_2) = -\frac{b}{a}$.

If the roots are multiplied, their product will be

$$r_1 r_2 = \left(\frac{-b + \sqrt{b^2 - 4ac}}{2a}\right)\left(\frac{-b - \sqrt{b^2 - 4ac}}{2a}\right)$$
$$= \frac{b^2 + b\sqrt{b^2 - 4ac} - b\sqrt{b^2 - 4ac} - b^2 + 4ac}{4a^2}$$
$$= \frac{4ac}{4a^2}$$
$$= \frac{c}{a}.$$

Therefore, the product of the roots $(r_1 r_2) = \frac{c}{a}$.

6. In the equation $x^2 + 11x + 28 = 0$, one root is 3 more than the other root. What are the roots of the equation?

$$r_2 = r_1 + 3. \qquad (1)$$
$$r_1 + r_2 = -\tfrac{11}{1}. \qquad (2)$$

Substitute the value of r_2 from equation (1) in equation (2).

$$r_1 + r_1 + 3 = -11.$$
$$2r_1 = -14.$$
$$r_1 = -7. \qquad (3)$$

Substitute the value of r_1 from equation (3) in equation (1).

$$r_2 = r_1 + 3 \qquad (1)$$
$$= -7 + 3$$
$$= -4.$$

Hence, the roots are -7 and -4.

7. If one root is the negative of the other root in the equation $2x^2 - 18 = 0$, what are the roots of the equation?

$$r_1 r_2 = -\tfrac{18}{2}. \qquad (1)$$
$$r_2 = -r_1. \qquad (2)$$

Substitute the value of r_2 from equation (2) in equation (1).

$$(r_1)(-r_1) = -9.$$
$$-(r_1)^2 = -9.$$
$$(r_1)^2 = 9.$$
$$r_1 = \pm 3. \qquad (3)$$

Substitute the value of r_1 from equation (3) in equation (2).

$$r_2 = -r_1 \qquad (2)$$
$$= \mp 3.$$

Hence, the roots are 3 and -3.

8. In the equation $4x^2 + hx + 1 = 0$, the roots are equal. What is the value of h?

$$r_2 = r_1. \quad (1)$$
$$r_1 + r_2 = -\frac{h}{4}. \quad (2)$$
$$(r_1)(r_2) = \tfrac{1}{4}. \quad (3)$$

Substitute the value of r_2 from equation (1) in equation (2).

$$r_1 + r_1 = -\frac{h}{4}.$$
$$2r_1 = -\frac{h}{4}.$$
$$r_1 = -\frac{h}{8}. \quad (4)$$

Substitute the value of r_2 from equation (1) in equation (3).

$$(r_1)(r_1) = \tfrac{1}{4}.$$
$$(r_1)^2 = \tfrac{1}{4}.$$
$$r_1 = \pm\tfrac{1}{2}. \quad (5)$$

Substitute the value of r_1 from equation (4) in equation (5).

$$-\frac{h}{8} = \pm\frac{1}{2}.$$
$$h = \pm 4.$$

9. In the equation $hx^2 + 14x + 12 = 0$, one root is six times the other root. What is the value of h?

$$r_2 = 6r_1. \quad (1)$$
$$r_1r_2 = \frac{12}{h}. \quad (2)$$
$$r_1 + r_2 = -\frac{14}{h}. \quad (3)$$
$$(r_1)(6r_1) = \frac{12}{h}.$$
$$6(r_1)^2 = \frac{12}{h}.$$
$$(r_1)^2 = \frac{2}{h}. \quad (4)$$
$$r_1 + 6r_1 = -\frac{14}{h}.$$
$$7r_1 = -\frac{14}{h}.$$
$$r_1 = -\frac{2}{h}. \quad (5)$$

Substitute the value of r_1 from equation (5) in equation (4).

$$\left(-\frac{2}{h}\right)^2 = \frac{2}{h}.$$
$$\frac{4}{h^2} = \frac{2}{h}.$$
$$2h^2 = 4h.$$
$$h^2 = 2h.$$

Divide both members of the equation by h.

$$h = 2.$$

10. In the equation $3x^2 - 7x + h = 0$, one root is 3. What is the value of h?

$$\text{Let } r_2 = 3.$$
$$r_1 + r_2 = -\frac{b}{a}$$
$$= -(-\tfrac{7}{3}) - \tfrac{7}{3}.$$
$$r_1 + 3 = \tfrac{7}{3}.$$
$$r_1 = -\tfrac{9}{3} + \tfrac{7}{3} = -\tfrac{2}{3}.$$
$$(r_1)(r_2) = \frac{h}{3}.$$
$$(-\tfrac{2}{3})(3) = \frac{h}{3}.$$
$$\frac{h}{3} = -2.$$
$$h = -6.$$

EQUATIONS INVOLVING RADICALS

If an equation contains radicals of the second order (radicals indicating square roots), its roots may be found by squaring both members. If there are two radicals in the equation, they must be put on opposite sides before squaring. Should a radical remain, place it alone in one member of the equation and square both sides once more.

The procedure is similar for an equation with three or more radicals. Some radical(s) should appear in each member of the equation, and both members must be squared until all radicals have been eliminated.

When the roots have been found, they must be substituted in the *original* equation to see whether they will satisfy that equation. Any that do not must be discarded. Such values are called **extraneous roots** and may appear as a result of squaring both members of an equation.

$$\sqrt{x+1} - \sqrt{2x-2} = 0.$$
$$\sqrt{x+1} = \sqrt{2x-2}.$$
$$x + 1 = 2x - 2.$$
$$-x = -3.$$
$$x = 3.$$

Check:

$$\sqrt{x+1} - \sqrt{2x-2} = 0.$$
$$\sqrt{3+1} - \sqrt{2(3)-2} = 0.$$
$$\sqrt{4} - \sqrt{4} = 0.$$
$$0 = 0.$$

1. Find the roots of the equation $\sqrt{x+1} = 3$.

$$x + 1 = 9.$$
$$x = 8.$$

Check:

$$\sqrt{x+1} = 3.$$
$$\sqrt{8+1} = 3.$$
$$\sqrt{9} = 3.$$
$$3 = 3.$$

Hence, 8 is a root of the equation.

2. Find the roots of the equation $\sqrt{x+1} = \sqrt{7-x}$.

$$x + 1 = 7 - x.$$
$$2x = 6.$$
$$x = 3.$$

Check:

$$\sqrt{x+1} = \sqrt{7-x}.$$
$$\sqrt{3+1} = \sqrt{7-3}.$$
$$\sqrt{4} = \sqrt{4}.$$

Hence, 3 is a root of the equation.

3. Find the roots of the equation $\sqrt{x+10} - 3 = 0$.

Before squaring, add 3 to both members of the equation.

$$\sqrt{x+10} = 3.$$
$$x + 10 = 9.$$
$$x = -1$$

Check:

$$\sqrt{x+10} - 3 = 0.$$
$$\sqrt{-1+10} - 3 = 0.$$
$$\sqrt{9} - 3 = 0.$$
$$3 - 3 = 0.$$
$$0 = 0.$$

Hence, -1 is a root of the equation.

4. Find the roots of the equation $\sqrt{x^2+9} - 5 = 0$.

$$\sqrt{x^2+9} = 5.$$
$$x^2 + 9 = 25.$$
$$x^2 = 16.$$
$$x = \pm 4.$$

Check:

$$\sqrt{x^2+9} - 5 = 0. \qquad \sqrt{x^2+9} - 5 = 0.$$
$$\sqrt{(4)^2+9} - 5 = 0. \qquad \sqrt{(-4)^2+9} - 5 = 0.$$
$$\sqrt{16+9} - 5 = 0. \qquad \sqrt{16+9} - 5 = 0.$$
$$\sqrt{25} - 5 = 0. \qquad \sqrt{25} - 5 = 0.$$
$$5 - 5 = 0. \qquad 5 - 5 = 0.$$
$$0 = 0. \qquad 0 = 0.$$

Hence, 4 and -4 are roots of the equation.

5. Find the roots of the equation $\sqrt{x+1} - x + 1 = 0$.

Add $x - 1$ to both members of the equation.

$$\sqrt{x+1} = x - 1.$$
$$x + 1 = x^2 - 2x + 1.$$
$$-x^2 + 3x = 0.$$
$$x^2 - 3x = 0.$$
$$x(x - 3) = 0.$$
$$x = 0. \qquad x - 3 = 0.$$
$$x = 3.$$

Check:

$$\sqrt{x+1} - x + 1 = 0. \qquad \sqrt{x+1} - x + 1 = 0.$$
$$\sqrt{0+1} - 0 + 1 = 0. \qquad \sqrt{3+1} - 3 + 1 = 0.$$
$$\sqrt{1} + 1 = 0. \qquad \sqrt{4} - 3 + 1 = 0.$$
$$1 + 1 = 0. \qquad 2 - 3 + 1 = 0.$$
$$2 = 0. \qquad 0 = 0.$$

Since $x = 0$ does not check, it is an extraneous root and 3 is the only root of the original equation.

6. Find the roots of the equation $\sqrt{x+12} - x = 0$.

$$\sqrt{x+12} = x.$$
$$x + 12 = x^2.$$
$$-x^2 + x + 12 = 0.$$
$$x^2 - x - 12 = 0.$$
$$(x - 4)(x + 3) = 0.$$
$$x - 4 = 0. \qquad x + 3 = 0.$$
$$x = 4. \qquad x = -3.$$

Check:

$$\sqrt{x+12} - x = 0. \qquad \sqrt{x+12} - x = 0.$$
$$\sqrt{4+12} - 4 = 0. \qquad \sqrt{-3+12} + 3 = 0.$$
$$\sqrt{16} - 4 = 0. \qquad \sqrt{9} + 3 = 0.$$
$$0 = 0. \qquad 6 = 0.$$

Since -3 will not satisfy the original equation, it is an extraneous root and 4 is the only root of the equation.

7. Find the roots of the equation $\sqrt{x+7} = \sqrt{x-3}$.

$$x + 7 = x - 3.$$
$$x - x = -7 - 3.$$
$$0 = -10.$$

Since the x is eliminated, there is no value of x for which this equation is true.

8. Find the roots of the equation $\sqrt{x+4} - 1 = \sqrt{2x-6}$.

Square both members of the equation. The left member is a binomial, so use the rule for squaring a binomial.

$$x + 4 - 2\sqrt{x+4} + 1 = 2x - 6.$$

Using the axioms on pages 4-5, rearrange the terms so only the radical appears in the right member of the equation. Then simplify.

$$-x + 11 = 2\sqrt{x+4}.$$

Square both members of the equation again.

$$x^2 - 22x + 121 = 4(x + 4).$$
$$x^2 - 22x + 121 = 4x + 16.$$
$$x^2 - 26x + 105 = 0.$$
$$(x - 21)(x - 5) = 0.$$
$$x - 21 = 0. \qquad x - 5 = 0.$$
$$x = 21. \qquad x = 5.$$

Check:

$$\sqrt{x+4} - 1 = \sqrt{2x-6}. \qquad \sqrt{x+4} - 1 = \sqrt{2x-6}.$$
$$\sqrt{21+4} - 1 = \sqrt{42-6}. \qquad \sqrt{5+4} - 1 = \sqrt{10-6}.$$
$$\sqrt{25} - 1 = \sqrt{36}. \qquad \sqrt{9} - 1 = \sqrt{4}.$$
$$5 - 1 = 6. \qquad 3 - 1 = 2.$$
$$4 = 6. \qquad 2 = 2.$$

Since 21 will not satisfy the original equation, it is an extraneous root and the only root of the equation is 5.

9. Find the roots of the equation

$$\sqrt{2x+13} = \sqrt{x+3} + \sqrt{x+6}.$$

The right member of the equation is a binomial. Square it by use of the rule for squaring a binomial.

$$2x + 13 = x + 3 + 2\sqrt{(x+3)(x+6)} + x + 6.$$

Place the radical in the left member of the equation and all other terms in the right member and simplify.

$$-2\sqrt{x^2 + 9x + 18} = -4.$$
$$\sqrt{x^2 + 9x + 18} = 2.$$

Again square both members of the equation.

$$x^2 + 9x + 18 = 4.$$
$$x^2 + 9x + 14 = 0.$$
$$(x+7)(x+2) = 0.$$
$$x + 7 = 0. \qquad x + 2 = 0.$$
$$x = -7. \qquad x = -2.$$

Check:

$$\sqrt{2x+13} = \sqrt{x+3} + \sqrt{x+6}.$$
$$\sqrt{-14+13} = \sqrt{-7+3} + \sqrt{-7+6}.$$
$$\sqrt{-1} = \sqrt{-4} + \sqrt{-1}.$$
$$i = 2i + i.$$
$$i = 3i.$$

$$\sqrt{2x+13} = \sqrt{x+3} + \sqrt{x+6}.$$
$$\sqrt{-4+13} = \sqrt{-2+3} + \sqrt{-2+6}.$$
$$\sqrt{9} = \sqrt{1} + \sqrt{4}.$$
$$3 = 1 + 2.$$
$$3 = 3.$$

Since -7 will not satisfy the original equation, it is an extraneous root and -2 is the only root of the equation.

10. Find the roots of the equation $\sqrt{4x+1} + 1 = \sqrt{6x}$.

$$4x + 1 + 2\sqrt{4x+1} + 1 = 6x.$$
$$2\sqrt{4x+1} = 2x - 2.$$
$$\sqrt{4x+1} = x - 1.$$

Again square both members of the equation.

$$4x + 1 = x^2 - 2x + 1.$$
$$-x^2 + 6x = 0.$$
$$x^2 - 6x = 0.$$
$$x(x-6) = 0.$$
$$x = 0. \qquad x = 6.$$

Check:

$$\sqrt{4x+1} + 1 = \sqrt{6x}. \qquad \sqrt{4x+1} + 1 = \sqrt{6x}.$$
$$\sqrt{0+1} + 1 = \sqrt{0}. \qquad \sqrt{24+1} + 1 = \sqrt{36}.$$
$$\sqrt{1} + 1 = 0. \qquad \sqrt{25} + 1 = \sqrt{36}.$$
$$2 = 0. \qquad 6 = 6.$$

Since 0 will not satisfy the original equation, it is an extraneous root and 6 is the only root of the equation.

SYSTEMS OF QUADRATICS WITH TWO UNKNOWNS

Problems of this type are similar to those found in Chapter 6. The only difference is that some of the variables are raised to the second power. These problems may be solved by the same three elimination methods that were used to solve systems of linear equations, namely: addition, subtraction, or substitution.

1. Solve the following system of equations.

$$x^2 + y^2 = 10. \qquad (1)$$
$$x^2 - y^2 = 8. \qquad (2)$$

Add these equations to eliminate the y^2-terms.

$$x^2 + y^2 = 10 \qquad (1)$$
$$x^2 - y^2 = 8 \qquad (2)$$
$$2x^2 = 18$$
$$x^2 = 9.$$
$$x = \pm 3. \qquad (3)$$

Substitute the values of x from equation (3) in equation (1).

$$x^2 + y^2 = 10.$$
$$(\pm 3)^2 + y^2 = 10.$$
$$9 + y^2 = 10.$$
$$y^2 = 1.$$
$$y = \pm 1.$$

Check:

$$x^2 + y^2 = 10. \quad (1) \qquad x^2 - y^2 = 8. \quad (2)$$
$$(\pm 3)^2 + (\pm 1)^2 = 10. \qquad (\pm 3)^2 - (\pm 1)^2 = 8.$$
$$9 + 1 = 10. \qquad 9 - 1 = 8.$$
$$10 = 10. \qquad 8 = 8.$$

Hence, $x = \pm 3$ and $y = \pm 1$. This means that for either value of x, either value of y may be used.

2. Solve the following system of equations.

$$x^2 + 2y^2 = 6. \qquad (1)$$
$$x^2 + y^2 = 5. \qquad (2)$$

Subtract equation (2) from equation (1).

$$x^2 + 2y^2 = 6 \qquad (1)$$
$$x^2 + y^2 = 5 \qquad (2)$$
$$y^2 = 1$$
$$y = \pm 1. \qquad (3)$$

Substitute the values of y from equation (3) in equation (2).

$$x^2 + y^2 = 5. \qquad (2)$$
$$x^2 + (\pm 1)^2 = 5.$$
$$x^2 + 1 = 5.$$
$$x^2 = 4.$$
$$x = \pm 2.$$

Check:

$$x^2 + 2y^2 = 6. \quad (1)$$
$$(\pm 2)^2 + 2(\pm 1)^2 = 6.$$
$$4 + 2(1) = 6.$$
$$4 + 2 = 6.$$
$$6 = 6.$$

$$x^2 + y^2 = 5. \quad (2)$$
$$(\pm 2)^2 + (\pm 1)^2 = 5.$$
$$4 + 1 = 5.$$
$$5 = 5.$$

Hence, $x = \pm 2$ and $y = \pm 1$.

3. Solve the following system of equations.

$$2x^2 + 3y^2 = 59. \quad (1)$$
$$3x^2 - 2y^2 = 30. \quad (2)$$

Multiply equation (1) by 2 and equation (2) by 3.

$$4x^2 + 6y^2 = 118 \quad (1a)$$
$$9x^2 - 6y^2 = 90 \quad (2a)$$
$$13x^2 = 208$$

$$x^2 = 16.$$
$$x = \pm 4. \quad (3)$$

Substitute the values of x from equation (3) in equation (1).

$$2x^2 + 3y^2 = 59. \quad (1)$$
$$2(\pm 4)^2 + 3y^2 = 59.$$
$$2(16) + 3y^2 = 59.$$
$$32 + 3y^2 = 59.$$
$$3y^2 = 27.$$
$$y^2 = 9.$$
$$y = \pm 3.$$

Check:

$$2x^2 + 3y^2 = 59. \quad (1)$$
$$2(\pm 4)^2 + 3(\pm 3)^2 = 59.$$
$$2(16) + 3(9) = 59.$$
$$32 + 27 = 59.$$
$$59 = 59.$$

$$3x^2 - 2y^2 = 30. \quad (2)$$
$$3(\pm 4)^2 - 2(\pm 3)^2 = 30.$$
$$3(16) - 2(9) = 30.$$
$$48 - 18 = 30.$$
$$30 = 30.$$

Hence, $x = \pm 4$ and $y = \pm 3$.

4. Solve the following system of equations.

$$x^2 + y^2 = 20. \quad (1)$$
$$x + y^2 = 8. \quad (2)$$

Solve equation (2) for y^2.

$$y^2 = 8 - x. \quad (2a)$$

Substitute the value of y^2 from equation (2a) in equation (1).

$$x^2 + y^2 = 20. \quad (1)$$
$$x^2 + (8 - x) = 20.$$
$$x^2 + 8 - x = 20.$$
$$x^2 - x - 12 = 0.$$
$$(x - 4)(x + 3) = 0.$$

$$x - 4 = 0. \qquad x + 3 = 0.$$
$$x = 4. \quad (3) \qquad x = -3. \quad (4)$$

Substitute the value of x from equation (3) in equation (1).

$$x^2 + y^2 = 20. \quad (1)$$
$$(4)^2 + y^2 = 20.$$
$$16 + y^2 = 20.$$
$$y^2 = 4.$$
$$y = \pm 2.$$

Substitute the value of x from equation (4) in equation (1).

$$x^2 + y^2 = 20. \quad (1)$$
$$(-3)^2 + y^2 = 20.$$
$$9 + y^2 = 20.$$
$$y^2 = 11$$
$$y = \pm\sqrt{11}.$$

Check:

$$x^2 + y^2 = 20. \quad (1)$$
$$(4)^2 + (\pm 2)^2 = 20.$$
$$16 + 4 = 20.$$
$$20 = 20.$$

$$x + y^2 = 8. \quad (2)$$
$$4 + (\pm 2)^2 = 8.$$
$$4 + 4 = 8.$$
$$8 = 8.$$

$$x^2 + y^2 = 20. \quad (1)$$
$$(-3)^2 + (\pm\sqrt{11})^2 = 20.$$
$$9 + 11 = 20.$$
$$20 = 20.$$

$$x + y^2 = 8. \quad (2)$$
$$-3 + (\pm\sqrt{11})^2 = 8.$$
$$-3 + 11 = 8.$$
$$8 = 8.$$

Hence, the four solutions to this pair of equations are: $x = 4$, $y = 2$; $x = 4$, $y = -2$; $x = -3$, $y = \sqrt{11}$; and $x = -3$, $y = -\sqrt{11}$.

5. Solve the following system of equations.

$$x = y. \quad (1)$$
$$x^2 - 2y^2 = -9. \quad (2)$$

Substitute the value of y from equation (1) in equation (2).

$$x^2 - 2y^2 = -9. \quad (2)$$
$$x^2 - 2(x)^2 = -9.$$
$$x^2 - 2x^2 = -9.$$
$$-x^2 = -9.$$
$$x^2 = 9.$$
$$x = \pm 3. \quad (3)$$

Substitute the values of x from equation (3) in equation (1).

$$x = y. \quad (1)$$
$$\pm 3 = y.$$
$$y = \pm 3.$$

Check:

$$x = y. \quad (1)$$
$$3 = 3.$$

$$x = y. \quad (1)$$
$$-3 = -3.$$

$$x^2 - 2y^2 = -9. \quad (2)$$
$$(\pm 3)^2 - 2(\pm 3)^2 = -9.$$
$$9 - 18 = -9.$$
$$-9 = -9.$$

The signs must be the same for both variables. Hence, the solutions are: $x = 3$, $y = 3$; and $x = -3$, $y = -3$.

6. Solve the following system of equations.

$$3x^2 + y = 15. \quad (1)$$
$$x^2 - y^2 = -5. \quad (2)$$
$$x^2 = y^2 - 5. \quad (2a)$$

Substitute the value of x^2 from equation (2a) in equation (1).

$$3x^2 + y = 15. \quad (1)$$
$$3(y^2 - 5) + y = 15.$$
$$3y^2 - 15 + y = 15.$$
$$3y^2 + y - 30 = 0.$$
$$(3y + 10)(y - 3) = 0.$$

$$3y + 10 = 0.$$
$$3y = -10.$$
$$y = -\tfrac{10}{3}. \quad (3)$$

$$y - 3 = 0.$$
$$y = 3. \quad (4)$$

Substitute the value of y from equation (3) in equation (2a).

$$x^2 = y^2 - 5. \quad (2a)$$
$$= (-\tfrac{10}{3})^2 - 5$$
$$= \tfrac{100}{9} - 5$$
$$= \frac{100 - 45}{9}$$
$$= \tfrac{55}{9}.$$
$$x = \pm\frac{\sqrt{55}}{3}.$$

Substitute the value of y from equation (4) in equation (2a).

$$x^2 = y^2 - 5. \quad (2a)$$
$$= (3)^2 - 5$$
$$= 9 - 5$$
$$= 4.$$
$$x = \pm 2.$$

Check:

$$3x^2 + y = 15. \quad (1)$$
$$3\left(\pm\frac{\sqrt{55}}{3}\right)^2 - \frac{10}{3} = 15.$$
$$3(\tfrac{55}{9}) - \tfrac{10}{3} = 15.$$
$$\tfrac{55}{3} - \tfrac{10}{3} = 15.$$
$$\tfrac{45}{3} = 15.$$
$$15 = 15.$$

$$x^2 - y^2 = -5. \quad (2)$$
$$\left(\pm\frac{\sqrt{55}}{3}\right)^2 - \left(-\frac{10}{3}\right)^2 = -5.$$
$$\tfrac{55}{9} - \tfrac{100}{9} = -5.$$
$$-\tfrac{45}{9} = -5.$$
$$-5 = -5.$$

$$3x^2 + y = 15. \quad (1)$$
$$3(\pm 2)^2 + 3 = 15.$$
$$3(4) + 3 = 15.$$
$$12 + 3 = 15.$$
$$15 = 15.$$

$$x^2 - y^2 = -5. \quad (2)$$
$$(\pm 2)^2 - (3)^2 = -5$$
$$4 - 9 = -5.$$
$$-5 = -5.$$

Hence, the solutions are: $x = \pm\frac{\sqrt{55}}{3}$, $y = -\frac{10}{3}$; and $x = \pm 2$, $y = 3$.

7. Solve the following system of equations.

$$x^2 + y^2 = 6. \quad (1)$$
$$y^2 = x. \quad (2)$$

Substitute the value of y^2 from equation (2) in equation (1).

$$x^2 + y^2 = 6. \quad (1)$$
$$x^2 + x = 6.$$
$$x^2 + x - 6 = 0.$$
$$(x + 3)(x - 2) = 0.$$

$$x + 3 = 0.$$
$$x = -3. \quad (3)$$

$$x - 2 = 0.$$
$$x = 2. \quad (4)$$

Substitute the value of x from equation (3) in equation (2).

$$y^2 = x \quad (2)$$
$$= -3.$$
$$y = \pm\sqrt{-3}. \quad (5)$$

Substitute the value of x from equation (4) in equation (2).

$$y^2 = x \quad (2)$$
$$= 2.$$
$$y = \pm\sqrt{2}.$$

The values of y in equation (5) are square roots of negative numbers. Since we are interested in real values, these will be discarded. (See Chapter 9 for a discussion of square roots of negative numbers.)

Check:

$$x^2 + y^2 = 6. \quad (1)$$
$$(2)^2 + (\pm\sqrt{2})^2 = 6.$$
$$4 + 2 = 6.$$
$$6 = 6.$$

$$y^2 = x. \quad (2)$$
$$(\pm\sqrt{2})^2 = 2.$$
$$2 = 2.$$

Hence, the real solutions are: $x = 2$, $y = \pm\sqrt{2}$.

8. Solve the following system of equations.

$$xy = 12. \quad (1)$$
$$x^2 + y^2 = 25. \quad (2)$$
$$x = \frac{12}{y}. \quad (1a)$$

Substitute the value of x from equation (1a) in equation (2).

$$x^2 + y^2 = 25. \quad (2)$$
$$\left(\frac{12}{y}\right)^2 + y^2 = 25.$$
$$\frac{144}{y^2} + y^2 = 25.$$
$$\frac{144}{y^2} + \frac{y^4}{y^2} = \frac{25y^2}{y^2}.$$
$$144 + y^4 = 25y^2.$$
$$y^4 - 25y^2 + 144 = 0.$$

This is a fourth degree equation in the form of a quadratic equation, and it may be solved as a quadratic.

$$(y^2 - 9)(y^2 - 16) = 0.$$

$$y^2 - 9 = 0.$$
$$y^2 = 9.$$
$$y = \pm 3. \quad (3)$$

$$y^2 - 16 = 0.$$
$$y^2 = 16.$$
$$y = \pm 4. \quad (4)$$

Substitute the values of y from equation (3) in equation (1a).

$$x = \frac{12}{y} \quad (1a)$$
$$= \frac{12}{\pm 3}$$
$$= \pm 4.$$

Substitute the values of y from equation (4) in equation (1a).

$$x = \frac{12}{y}. \quad (1a)$$
$$= \frac{12}{\pm 4}.$$
$$= \pm 3.$$

It can be observed from equation (1) that, in order to satisfy the equation, both values must be either positive or negative. Hence, the only pairs of values to consider are those which have the same sign.

Check:

$$xy = 12. \quad (1)$$
$$(3)(4) = 12.$$
$$12 = 12.$$

$$x^2 + y^2 = 25. \quad (2)$$
$$(3)^2 + (4)^2 = 25.$$
$$9 + 16 = 25.$$
$$25 = 25.$$

$$xy = 12. \quad (1) \qquad x^2 + y^2 = 25. \quad (2)$$
$$(-3)(-4) = 12. \qquad (-3)^2 + (-4)^2 = 25.$$
$$12 = 12. \qquad 9 + 16 = 25.$$
$$25 = 25.$$

$$xy = 12. \quad (1) \qquad x^2 + y^2 = 25. \quad (2)$$
$$(4)(3) = 12. \qquad (4)^2 + (3)^2 = 25.$$
$$12 = 12. \qquad 16 + 9 = 25.$$
$$25 = 25.$$

$$xy = 12. \quad (1) \qquad x^2 + y^2 = 25. \quad (2)$$
$$(-4)(-3) = 12. \qquad (-4)^2 + (-3)^2 = 25.$$
$$12 = 12. \qquad 16 + 9 = 25.$$
$$25 = 25.$$

Hence, the solutions are: $x = 3$, $y = 4$; $x = -3$, $y = -4$; $x = 4$, $y = 3$; and $x = -4$, $y = -3$.

9. Solve the following system of equations.

$$x^2 + 2y^2 = 3. \quad (1)$$
$$\frac{3x}{2} - \frac{y}{2} = 2. \quad (2)$$

Clear equation (2) of fractions and we have

$$3x - y = 4. \quad (2a)$$
$$-y = -3x + 4.$$
$$y = 3x - 4. \quad (2b)$$

Substitute the value of y from equation (2b) in equation (1).

$$x^2 + 2y^2 = 3. \quad (1)$$
$$x^2 + 2(3x - 4)^2 = 3.$$
$$x^2 + 2(9x^2 - 24x + 16) = 3.$$
$$x^2 + 18x^2 - 48x + 32 = 3.$$
$$19x^2 - 48x + 29 = 0.$$
$$(19x - 29)(x - 1) = 0.$$

$$19x - 29 = 0. \qquad x - 1 = 0.$$
$$19x = 29. \qquad x = 1. \quad (4)$$
$$x = \tfrac{29}{19}. \quad (3)$$

Substitute the value of x from equation (3) in equation (2b).

$$y = 3x - 4 \quad (2b)$$
$$= 3(\tfrac{29}{19}) - 4$$
$$= \tfrac{87}{19} - 4$$
$$= \frac{87 - 76}{19}$$
$$= \tfrac{11}{19}.$$

Substitute the value of x from equation (4) in equation (2b).

$$y = 3x - 4 \quad (2b)$$
$$= 3(1) - 4$$
$$= 3 - 4$$
$$= -1.$$

Check:

$$x^2 + 2y^2 = 3. \quad (1) \qquad \frac{3x}{2} - \frac{y}{2} = 2. \quad (2)$$
$$(\tfrac{29}{19})^2 + 2(\tfrac{11}{19})^2 = 3. \qquad \tfrac{3}{2}(\tfrac{29}{19}) - \tfrac{1}{2}(\tfrac{11}{19}) = 2.$$
$$\tfrac{841}{361} + 2(\tfrac{121}{361}) = 3. \qquad \tfrac{87}{38} - \tfrac{11}{38} = 2.$$
$$\tfrac{841}{361} + \tfrac{242}{361} = 3. \qquad \tfrac{76}{38} = 2.$$
$$\tfrac{1083}{361} = 3. \qquad 2 = 2.$$
$$3 = 3.$$

$$x^2 + 2y^2 = 3. \quad (1) \qquad \frac{3x}{2} - \frac{y}{2} = 2. \quad (2)$$
$$(1)^2 + 2(-1)^2 = 3. \qquad \frac{3(1)}{2} - \frac{-1}{2} = 2.$$
$$1 + 2 = 3. \qquad \tfrac{3}{2} + \tfrac{1}{2} = 2.$$
$$3 = 3. \qquad \tfrac{4}{2} = 2.$$
$$2 = 2.$$

Hence, the solutions are: $x = 1, y = -1$; and $x = \tfrac{29}{19}$, $y = \tfrac{11}{19}$.

10. Solve the following system of equations.

$$\frac{xy}{3} = -2. \quad (1)$$
$$x^2 + y^2 = 13. \quad (2)$$

Clear equation (1) of fractions, and solve for x.

$$xy = -6. \quad (1a)$$
$$x = -\frac{6}{y}. \quad (1b)$$

Substitute the value of x from equation (1b) in equation (2).

$$x^2 + y^2 = 13. \quad (2)$$
$$\left(-\frac{6}{y}\right)^2 + y^2 = 13.$$
$$\frac{36}{y^2} + y^2 = 13.$$

Clear of fractions and we have

$$36 + y^4 = 13y^2.$$

Again, this is a fourth degree equation in the form of a quadratic equation, and it may be solved as a quadratic equation.

$$y^4 - 13y^2 + 36 = 0.$$
$$(y^2 - 9)(y^2 - 4) = 0.$$
$$y^2 - 9 = 0. \qquad y^2 - 4 = 0.$$
$$y^2 = 9. \qquad y^2 = 4.$$
$$y = \pm 3. \quad (3) \qquad y = \pm 2. \quad (4)$$

Substitute the values of y from equation (3) in equation (1b).

$$x = -\frac{6}{y} \quad (1b)$$
$$= -\frac{6}{\pm 3}$$
$$= \mp 2.$$

Substitute the values of y from equation (4) in equation (1b)

$$x = -\frac{6}{y} \quad (1b)$$
$$= -\frac{6}{\pm 2}$$
$$= \mp 3.$$

As in Problem 8, only certain combinations of these values will satisfy (1). In order to get the negative value in the right member of the equation, y will have to be negative when x is positive and y will have to be positive when x is negative.

Check:

$$\frac{xy}{3} = -2. \quad (1)$$
$$\frac{(2)(-3)}{3} = -2.$$
$$\frac{-6}{3} = -2.$$
$$-2 = -2.$$

$$x^2 + y^2 = 13. \quad (2)$$
$$(2)^2 + (-3)^2 = 13.$$
$$4 + 9 = 13.$$
$$13 = 13.$$

$$\frac{xy}{3} = -2. \quad (1)$$
$$\frac{(-2)(3)}{3} = -2.$$
$$\frac{-6}{3} = -2.$$
$$-2 = -2.$$

$$x^2 + y^2 = 13. \quad (2)$$
$$(-2)^2 + (3)^2 = 13.$$
$$4 + 9 = 13.$$
$$13 = 13.$$

$$\frac{xy}{3} = -2. \quad (1)$$
$$\frac{(3)(-2)}{3} = -2.$$
$$\frac{-6}{3} = -2.$$
$$-2 = -2.$$

$$x^2 + y^2 = 13. \quad (2)$$
$$(3)^2 + (-2)^2 = 13.$$
$$9 + 4 = 13.$$
$$13 = 13.$$

$$\frac{xy}{3} = -2. \quad (1)$$
$$\frac{(-3)(2)}{3} = -2.$$
$$\frac{-6}{3} = -2.$$
$$-2 = -2.$$

$$x^2 + y^2 = 13. \quad (2)$$
$$(-3)^2 + (2)^2 = 13.$$
$$9 + 4 = 13.$$
$$13 = 13.$$

Hence, the solutions are: $x = 2$, $y = -3$; $x = -2$, $y = 3$; $x = 3$, $y = -2$; and $x = -3$, $y = 2$.

VERBAL PROBLEMS INVOLVING QUADRATICS

The information given in some verbal problems results in the setting up of quadratic equations. These equations are then solved by the preceding methods.

1. The sum of two numbers is 2 and the sum of their squares is 34. What are the numbers?

Let x = one number;
y = the other number.

$$x + y = 2. \quad (1)$$
$$x^2 + y^2 = 34. \quad (2)$$
$$x = 2 - y. \quad (1a)$$

Substitute the value of x from equation (1a) in equation (2).

$$x^2 + y^2 = 34. \quad (2)$$
$$(2 - y)^2 + y^2 = 34.$$
$$4 - 4y + y^2 + y^2 = 34.$$
$$2y^2 - 4y - 30 = 0.$$
$$y^2 - 2y - 15 = 0.$$
$$(y - 5)(y + 3) = 0.$$

$$y - 5 = 0. \qquad y + 3 = 0.$$
$$y = 5. \quad (3) \qquad y = -3. \quad (4)$$

Substitute the value of y from equation (3) in equation (1a).

$$x = 2 - y \quad (1a)$$
$$= 2 - 5$$
$$= -3.$$

Substitute the value of y from equation (4) in equation (1a).

$$x = 2 - y \quad (1a)$$
$$= 2 - (-3)$$
$$= 2 + 3$$
$$= 5.$$

Check:

$$x + y = 2. \quad (1)$$
$$-3 + 5 = 2.$$
$$2 = 2.$$

$$x^2 + y^2 = 34. \quad (2)$$
$$(-3)^2 + (5)^2 = 34.$$
$$9 + 25 = 34.$$
$$34 = 34.$$

Hence, the two numbers are 5 and -3. The fact that there are two solutions does not give two distinct sets of answers. It means that if the first number is 5, the other one will be -3; if the first one is -3, the second one will be 5.

The same problem can be worked with only one variable in the following manner.

Let x = one number;
$2 - x$ = the other number.

$$x^2 + (2 - x)^2 = 34.$$
$$x^2 + 4 - 4x + x^2 = 34.$$
$$2x^2 - 4x - 30 = 0.$$
$$x^2 - 2x - 15 = 0.$$
$$(x + 3)(x - 5) = 0.$$

$$x + 3 = 0. \qquad x - 5 = 0.$$
$$x = -3. \qquad x = 5.$$

Hence, the same values are obtained as when the problem was solved with two variables.

2. The sum of two numbers is 8, and the difference of their squares is 32. What are the numbers?

Let x = one number;
y = the other number.

$$x + y = 8. \quad (1)$$
$$x^2 - y^2 = 32. \quad (2)$$
$$x = 8 - y. \quad (1a)$$

Substitute the value of x from equation (1a) in equation (2).

$$x^2 - y^2 = 32. \quad (2)$$
$$(8 - y)^2 - y^2 = 32.$$
$$64 - 16y + y^2 - y^2 = 32.$$
$$-16y = -32.$$
$$y = 2. \quad (3)$$

Substitute the value of y from equation (3) in equation (1a).

$$x = 8 - y$$
$$= 8 - 2$$
$$= 6.$$

Check:

$$x + y = 8. \quad (1) \qquad x^2 - y^2 = 32. \quad (2)$$
$$6 + 2 = 8. \qquad (6)^2 - (2)^2 = 32.$$
$$8 = 8. \qquad 36 - 4 = 32.$$
$$32 = 32.$$

Hence, the numbers are 6 and 2.

3. The length of a rectangle is 4 inches more than the width. The area is 77 square inches. What are the dimensions of the rectangle?

$$\text{Let } x = \text{the width;} \quad (1)$$
$$x + 4 = \text{the length;} \quad (2)$$
$$x(x + 4) = \text{the area.}$$
$$x(x + 4) = 77.$$
$$x^2 + 4x = 77.$$
$$x^2 + 4x - 77 = 0.$$
$$(x - 7)(x + 11) = 0.$$
$$x - 7 = 0. \qquad x + 11 = 0.$$
$$x = 7. \qquad x = -11.$$
$$x + 4 = 11.$$

We discard the -11 since the width cannot be a negative number.

Check:

$$(x + 4) - x = 4. \qquad x(x + 4) = 77.$$
$$7 + 4 - 7 = 4. \qquad 7(7 + 4) = 77.$$
$$4 = 4. \qquad 7(11) = 77.$$
$$77 = 77.$$

Hence, the width is 7 inches and the length is 11 inches.

4. The product of two consecutive positive numbers is 210. What are the numbers?

$$\text{Let } x = \text{the smaller number;}$$
$$x + 1 = \text{the larger number;}$$
$$x(x + 1) = \text{the product.}$$
$$x(x + 1) = 210.$$
$$x^2 + x = 210.$$
$$x^2 + x - 210 = 0.$$
$$(x - 14)(x + 15) = 0.$$
$$x - 14 = 0. \qquad x + 15 = 0.$$
$$x = 14. \qquad x = -15.$$
$$x + 1 = 15.$$

Since it was stated that the numbers were to be positive, the value of -15 must be discarded.

Check:

$$x(x + 1) = 210.$$
$$14(14 + 1) = 210.$$
$$(14)(15) = 210.$$
$$210 = 210.$$

Hence, the numbers are 14 and 15.

5. The product of two consecutive even numbers is 288. What are the numbers?

$$\text{Let } 2x = \text{the smaller number;}$$
$$2x + 2 = \text{the larger number;}$$
$$2x(2x + 2) = \text{the product.}$$
$$2x(2x + 2) = 288.$$
$$4x^2 + 4x = 288.$$
$$x^2 + x = 72.$$
$$x^2 + x - 72 = 0.$$
$$(x - 8)(x + 9) = 0.$$
$$x - 8 = 0. \qquad x + 9 = 0.$$
$$x = 8. \qquad x = -9.$$
$$2x = 16. \qquad 2x = -18.$$
$$2x + 2 = 18. \qquad 2x + 2 = -16.$$

Check:

$$2x(2x + 2) = 288. \qquad 2x(2x + 2) = 288.$$
$$16(16 + 2) = 288. \qquad -18([-18] + 2) = 288.$$
$$(16)(18) = 288. \qquad (-18)(-16) = 288.$$
$$288 = 288. \qquad 288 = 288.$$

Hence, the numbers are 16 and 18 or -18 and -16.

6. A rectangular fish pool 4 feet wide and 9 feet long has a walk of uniform width around it. The area of the walk is 68 square feet. How wide is the walk? (See Fig. 6.)

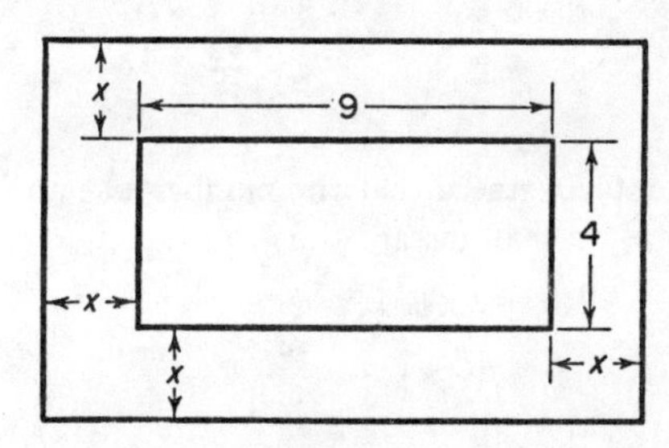

FIG. 6

$$\text{Let } x = \text{width of walk;}$$
$$2x + 9 = \text{length of rectangle formed by pool and walk;}$$
$$2x + 4 = \text{width of rectangle formed by pool and walk;}$$
$$(2x + 9)(2x + 4) = \text{area of rectangle formed by pool and walk.}$$
$$(2x + 9)(2x + 4) = (4)(9) + 68.$$
$$4x^2 + 26x + 36 = 104.$$
$$4x^2 + 26x - 68 = 0.$$
$$2x^2 + 13x - 34 = 0.$$
$$(x - 2)(2x + 17) = 0.$$
$$x - 2 = 0. \qquad 2x + 17 = 0.$$
$$x = 2. \qquad 2x = -17.$$
$$x = -\tfrac{17}{2}.$$

The width must be a positive value and therefore the $-\frac{17}{2}$ will be discarded.

Check:

The rectangle formed by the pool and the walk is 4 feet longer and 4 feet wider than the pool. Hence, that rectangle is 8 feet by 13 feet. The difference between the area of this large rectangle and the area of the pool will be equal to the area of the walk.

$$(8)(13) - (4)(9) = 68.$$
$$104 - 36 = 68.$$
$$68 = 68.$$

Hence, the width of the walk is 2 feet.

7. The sum of an integer and twice its reciprocal is $\frac{19}{3}$. What is the number?

Let $x =$ the number;

$\frac{1}{x} =$ its reciprocal.

$$x + \frac{2}{x} = \frac{19}{3}.$$

Clear of fractions in the above equation.

$$\frac{3x^2}{3x} + \frac{6}{3x} = \frac{19x}{3x}.$$
$$3x^2 + 6 = 19x.$$
$$3x^2 - 19x + 6 = 0.$$
$$(x - 6)(3x - 1) = 0.$$
$$x - 6 = 0. \qquad 3x - 1 = 0.$$
$$x = 6. \qquad 3x = 1.$$
$$x = \tfrac{1}{3}.$$

The problem stated that the number was an integer. Therefore, we shall discard the $\frac{1}{3}$.

Check:

$$x + \frac{2}{x} = \frac{19}{3}.$$
$$6 + \tfrac{2}{6} = \tfrac{19}{3}.$$
$$\tfrac{18}{3} + \tfrac{1}{3} = \tfrac{19}{3}.$$
$$\tfrac{19}{3} = \tfrac{19}{3}.$$

Hence, the number is 6.

8. Separate 63 into two parts whose product is 936.

Let $x =$ one part;

$63 - x =$ the other part;

$x(63 - x) =$ the product.

$$x(63 - x) = 936.$$
$$63x - x^2 = 936.$$
$$-x^2 + 63x - 936 = 0.$$
$$x^2 - 63x + 936 = 0.$$
$$(x - 24)(x - 39) = 0.$$
$$x - 24 = 0. \qquad x - 39 = 0.$$
$$x = 24. \qquad x = 39.$$
$$63 - x = 39. \qquad 63 - x = 24.$$

Check:

$$24 + 39 = 63. \qquad (24)(39) = 936.$$
$$63 = 63. \qquad 936 = 936.$$

Hence, the numbers are 24 and 39.

9. If a fraction is multiplied by one more than its reciprocal, the product is $\frac{5}{3}$. What is the fraction?

Let $x =$ the fraction;

$\frac{1}{x} =$ its reciprocal.

$$x\left(\frac{1}{x} + 1\right) = \frac{5}{3}.$$
$$x\left(\frac{1 + x}{x}\right) = \frac{5}{3}.$$
$$\frac{x + x^2}{x} = \frac{5}{3}.$$

Clear of fractions and we have

$$3x + 3x^2 = 5x.$$
$$3x^2 - 2x = 0.$$
$$x(3x - 2) = 0.$$
$$x = 0. \qquad 3x - 2 = 0.$$
$$3x = 2.$$
$$x = \tfrac{2}{3}.$$

The 0 will be discarded as there could be no product other than 0 if one of the factors is equal to 0.

Check:

$$\tfrac{2}{3}(\tfrac{3}{2} + 1) = \tfrac{5}{3}.$$
$$\tfrac{2}{3}(\tfrac{3}{2} + \tfrac{2}{2}) = \tfrac{5}{3}.$$
$$\tfrac{2}{3}(\tfrac{5}{2}) = \tfrac{5}{3}.$$
$$\tfrac{5}{3} = \tfrac{5}{3}.$$

Hence, the fraction is $\frac{2}{3}$.

10. The length of a rectangular piece of tin is twice its width. In each corner a 2-inch square is cut out so the sides can be turned up to form a box. If the box has a volume of 60 cubic inches, what were the dimensions of the piece of tin from which the box was made? (See Fig. 7.)

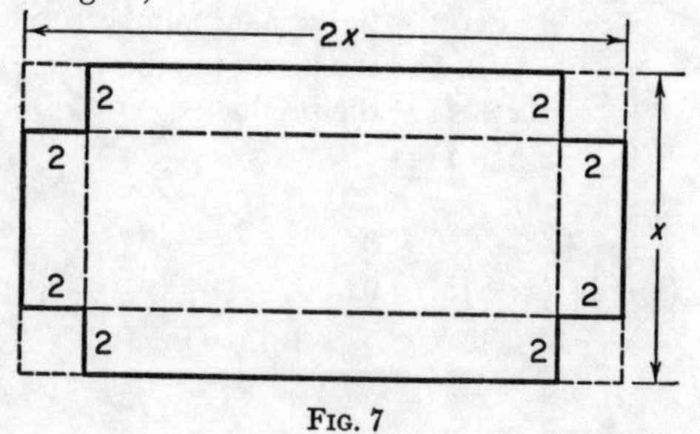

FIG. 7

Let $x =$ width of original piece;

$2x =$ length of original piece;

$x - 4 =$ width of box;

$2x - 4 =$ length of box;

$2 =$ height of box.

$$(2x - 4)(x - 4)(2) = \text{volume of box.}$$
$$2(2x - 4)(x - 4) = 60.$$
$$2x^2 - 12x + 16 = 30.$$
$$2x^2 - 12x - 14 = 0.$$
$$x^2 - 6x - 7 = 0.$$
$$(x - 7)(x + 1) = 0.$$

$$x - 7 = 0. \qquad x + 1 = 0.$$
$$x = 7. \qquad x = -1.$$
$$2x = 14.$$

The -1 will be discarded, since the length must be a positive value.

Check:

The length and width of the box will be 4 inches less than the length and width of the piece of tin.

$$(14 - 4)(7 - 4)(2) = 60.$$
$$(10)(3)(2) = 60.$$
$$60 = 60.$$

Hence, the dimensions of the piece of tin were 7 inches by 14 inches.

11. The sum of the areas of two squares is 136 square feet. If the side of one square is 4 feet more than that of the other square, what is the side of each square?

Let x = side of smaller square;
$x + 4$ = side of larger square.

$$x^2 + (x + 4)^2 = 136.$$
$$x^2 + x^2 + 8x + 16 = 136.$$
$$2x^2 + 8x - 120 = 0.$$
$$x^2 + 4x - 60 = 0.$$
$$(x - 6)(x + 10) = 0.$$
$$x - 6 = 0. \qquad x + 10 = 0.$$
$$x = 6. \qquad x = -10.$$
$$x + 4 = 10.$$

We discard the negative value because the length of a side must be positive.

Check:

$$(6)^2 + (10)^2 = 136.$$
$$36 + 100 = 136.$$
$$136 = 136.$$

Hence, the sides of the squares are 6 feet and 10 feet respectively.

12. The diagonal of a rectangle is 17 feet. If the length is 7 feet more than the width, what are the dimensions of the rectangle? (See Fig. 8.)

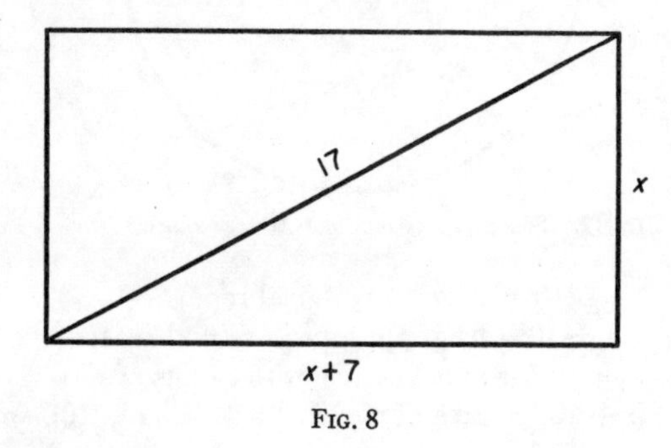

Fig. 8

Let x = width;
$x + 7$ = length.

By the Pythagorean Theorem,

$$x^2 + (x + 7)^2 = (17)^2.$$
$$x^2 + x^2 + 14x + 49 = 289.$$
$$2x^2 + 14x - 240 = 0.$$
$$x^2 + 7x - 120 = 0.$$
$$(x - 8)(x + 15) = 0.$$
$$x - 8 = 0. \qquad x + 15 = 0.$$
$$x = 8. \qquad x = -15.$$
$$x + 7 = 15.$$

Discard the negative value because the dimensions of a rectangle must be positive.

Check:

$$15 - 8 = 7. \qquad 8^2 + 15^2 = 17^2.$$
$$7 = 7. \qquad 64 + 225 = 289.$$
$$289 = 289.$$

Hence, the dimensions of the rectangle are 8 feet by 15 feet.

13. The base of a triangle is twice its altitude, and its area is 81 square inches. Find the base and altitude of the triangle.

Let x = altitude;
$2x$ = base.

$$\tfrac{1}{2}(2x)(x) = 81.$$
$$x^2 = 81.$$
$$x = \pm 9.$$

Discard the negative value, since the sides of a triangle must be positive.

$$2x = 18.$$

Check:

$$2(9) = 18. \qquad \frac{9(18)}{2} = 81.$$
$$18 = 18. \qquad (9)(9) = 81.$$
$$81 = 81.$$

Hence, the altitude of the triangle is 9 inches and its base is 18 inches.

14. A farmer had a certain amount of fencing with which to make a pen. In order to make a larger pen he decided to use one side of his barn for a side of the pen and use the fencing for the other three sides. By his first plan, the pen would contain 150 square yards. He discovered that if he doubled the distance between the barn and the opposite side of the pen, the pen would include 200 square yards. How many yards of fencing did he have? (See Figs. 9 and 10 on p. 104.)

Let x = distance from barn to opposite side in original plan;
y = length of side opposite barn in original plan;
$2x + y$ = total amount of fencing;
$2x$ = distance from barn to opposite side in second plan;

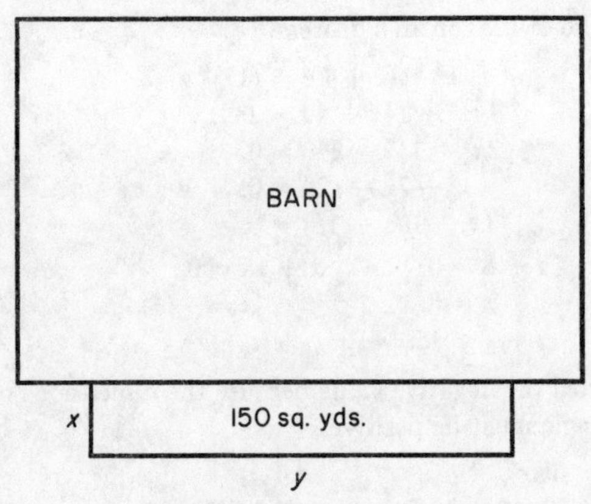

FIG. 9. Original plan.

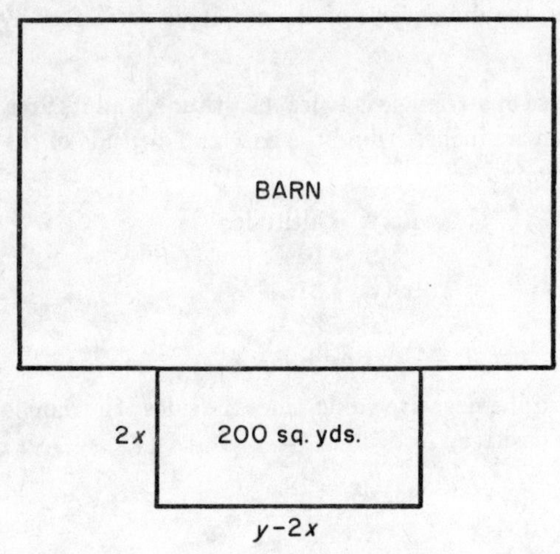

FIG. 10. Second plan.

$(2x + y) - 2(2x) =$ length of side opposite barn in second plan or

$y - 2x =$ length of side opposite barn in second plan.

$$xy = 150. \quad (1)$$

$$2x(y - 2x) = 200. \quad (2)$$

$$xy - 2x^2 = 100. \quad (2a)$$

Subtract equation (2a) from equation (1).

$$\begin{aligned} xy &= 150 \\ xy - 2x^2 &= 100 \\ \hline 2x^2 &= 50 \\ x^2 &= 25. \\ x &= \pm 5. \quad (3) \end{aligned}$$

Discard the negative value, since the sides of a pen must be positive.

Substitute the positive value of x from equation (3) in equation (1).

$$xy = 150. \quad (1)$$

$$5y = 150.$$

$$y = 30.$$

$$2x + y = 2(5) + 30.$$

$$= 40.$$

Check:

$$xy = 150. \qquad 2x(y - 2x) = 200.$$

$$(5)(30) = 150. \qquad 10(30 - 10) = 200.$$

$$150 = 150. \qquad 10(20) = 200.$$

$$200 = 200.$$

Hence, the amount of fencing was 40 yards.

15. A farmer staked his cow at the corner of a building 30 feet long and 25 feet wide so she could eat the grass near that corner of the building. He found that if the rope by which the cow was tied was lengthened by 10 feet, she could graze over four times as much area. How long was the original rope if in neither case it was as much as 25 feet long? (See Figs. 11 and 12.)

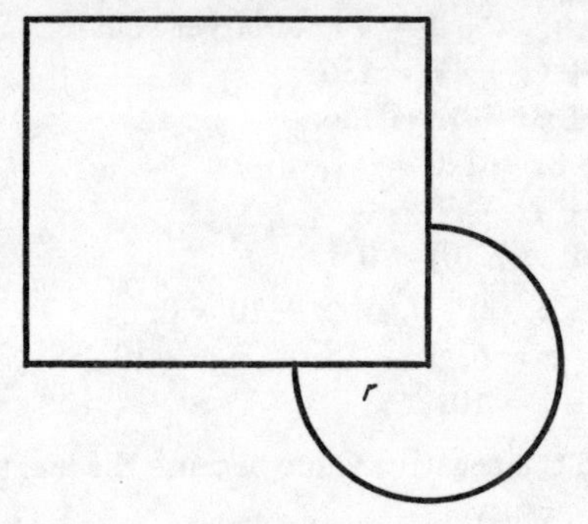

FIG. 11. First case (rope r feet long).

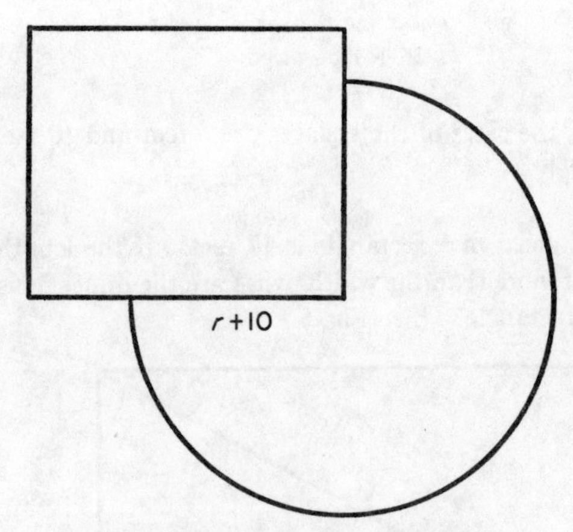

FIG. 12. Second case (rope $r + 10$ feet long).

Let $r =$ length of original rope;

$r + 10 =$ length of rope in second case;

$\pi r^2 =$ area of circle with radius r;

$\pi(r + 10)^2 =$ area of circle with radius $r + 10$;

$\frac{3}{4}\pi r^2 =$ grazing area in first case;

$\frac{3}{4}\pi(r + 10)^2 =$ grazing area in second case.

$$\tfrac{3}{4}\pi(r + 10)^2 = 4(\tfrac{3}{4}\pi r^2).$$

Divide both members of the equation by $\frac{3}{4}\pi$.

$$(r+10)^2 = 4r^2.$$
$$r^2 + 20r + 100 = 4r^2.$$
$$-3r^2 + 20r + 100 = 0.$$
$$3r^2 - 20r - 100 = 0.$$
$$(r-10)(3r+10) = 0.$$

$r - 10 = 0.$ $\quad 3r + 10 = 0.$

$r = 10.$ $\quad 3r = -10.$

$r = -\frac{10}{3}.$

The length of the rope is a positive value, so the negative root will be discarded.

Check:

$$\frac{3\pi}{4}(r+10)^2 = (4)\frac{3\pi r^2}{4}.$$
$$(r+10)^2 = 4r^2.$$
$$(10+10)^2 = 4(10)^2.$$
$$(20)^2 = 4(100).$$
$$400 = 400.$$

Hence, the original rope was 10 feet long.

SUPPLEMENTARY PROBLEMS

Solve the following equations.

1. $x^2 + 5x - 24 = 0.$
2. $x^2 - 13x + 36 = 0.$
3. $x^2 + 16x + 55 = 0.$
4. $2x^2 - 9x - 18 = 0.$
5. $12x^2 + x - 6 = 0.$
6. $18x^2 + 21x + 5 = 0.$
7. $x^2 - 7x + 3 = 0.$
8. $2x^2 + 4x - 9 = 0.$
9. $3x^2 - 8x + 2 = 0.$
10. $x^2 - 9x + 20 = 0.$
11. $2x^2 - 5x + 3 = 0.$
12. $3x^2 + 4x + 5 = 0.$
13. $\frac{x^2}{2} + \frac{3x}{4} = \frac{7}{2}.$
14. $\frac{x^2}{3} - \frac{2x}{5} - \frac{3}{15} = 0.$
15. $\frac{x^2}{4} - \frac{2x}{3} - \frac{7}{12} = 0.$
16. $\frac{2x^2}{3} + \frac{2x}{5} - \frac{3}{10} = 0.$
17. $\sqrt{2x+3} + \sqrt{4+3x} = \sqrt{x+5}.$
18. $\sqrt{4x+1} - \sqrt{2x-3} - \sqrt{2x} = 0.$
19. $\sqrt{5x-1} - \sqrt{11-x} - \sqrt{4-2x} = 0.$
20. $\sqrt{5x+1} - 3 = \sqrt{x-2}.$

Form the quadratic equations whose roots are given in the following problems.

21. $x = 2.$ $\quad x = -3.$
22. $x = \frac{2}{5}.$ $\quad x = -\frac{1}{2}.$
23. $x = -\frac{3}{2}.$ $\quad x = -\frac{2}{3}.$
24. $x = i\sqrt{3}.$ $\quad x = -i\sqrt{3}.$
25. $x = \frac{3+i\sqrt{2}}{2}.$ $\quad x = \frac{3-i\sqrt{2}}{2}.$

Find the value or values of h for which the roots of the following equations are equal.

26. $x^2 - hx + 16 = 0.$
27. $x^2 - 3hx + 9 = 0.$
28. $hx^2 - 24x + 9 = 0.$
29. $x^2 - 3x + h = 0.$
30. $3x^2 - 12x + 3h = 0.$

Chapter 11

Functions and Graphs

RELATIONS AND FUNCTIONS

Ordered Pairs. In an earlier discussion we introduced the concept of ordered pairs. From a specific universal set, U, it is possible to list or indicate all of the ordered pairs that may be selected from that set. The symbolism $U \times U$ (read: **U cross U**) is the set of all ordered pairs which can be formed from the elements of U. Thus, if $U = \{1, 2, 3\}$, then

$$U \times U = \{(1,1), (1,2), (1,3), (2,1), (2,2), (2,3), (3,1), (3,2), (3,3)\}.$$

The number of ordered pairs that can be formed in $U \times U$ is always equal to the square of the number of elements in U. Thus, if U has two members, $U \times U$ will contain four ordered pairs. Likewise, if U contains five members, $U \times U$ will consist of twenty-five ordered pairs.

Recall the work done with two variables in Chapter 6. In that chapter, systems of linear equations were discussed; and it was pointed out that each ordered pair of a solution set consists of a value of the first variable, say x, and a corresponding value of the second variable, say y, which satisfy both equations of the system. In the subsequent section we shall extend the discussion of such ordered pairs.

Relations. If we state that $x + y = 3$, we show a relationship between values of x and values of y. Any equality, inequality, or other mathematical statement concerning x and y shows a certain relationship between those two variables. The statement $\{(x, y) \mid x - y = 3\}$ relates x and y in such a way that any given value of y must differ by 3 from the corresponding value of x. The statement $x - y = 3$, then, places a condition upon the values of x and y which may constitute ordered pairs in the solution set. That is, in any ordered pair of the solution set, the value of x must be related to the value of y in the way stated in $\{(x, y) \mid x - y = 3\}$.

If, for a given statement concerning x and y, the universal set (U) is the set of real numbers, then the solution set for the statement is the set of all ordered pairs of real numbers which satisfy the conditions specified by the statement. And this solution set defines the relation. Specifically, then, a relation is a subset of the set of ordered pairs of the real numbers, $U \times U$.

Now since the set of real numbers is infinite, and since between any two real numbers there are an infinite number of real numbers, it is possible that the set of ordered pairs defining a given relation will also be infinite. This is not always the case, however; and even when it is the case if the real numbers constitute the universe, it need not be the case if there is some other universe. In some cases, then, the ordered pairs may be listed by reason of a restriction on U. As an example, if $x + y = 6$ and both x and y are natural numbers, the solution set is finite and is the following set of ordered pairs: $\{(1, 5), (2, 4), (3, 3), (4, 2), (5, 1)\}$. If the set of real numbers were the universal set for this problem, the solution set would be infinite.

If, from the above set of ordered pairs, we formulate a set of numbers consisting of all the first members of the ordered pairs, then we obtain the set $\{1, 2, 3, 4, 5\}$. Any set whose elements are the first members of all ordered pairs of a relation is called the **domain** of that relation. In like manner, from the above set of ordered pairs we can form a set whose elements are all the second members of those ordered pairs. In this case, the set would be $\{5, 4, 3, 2, 1\}$. Any set whose elements are all the second members of the ordered pairs of a relation is called the **range** of that relation.

An example of a given relation, its domain, and its range is the following.

Relation:	$\{(1,3),(2,6),(3,9)\}$
Domain:	$\{1,2,3\}$
Range:	$\{3,6,9\}$

In the following problems let $U = \{1,2,3,4,5,6,7,8,9\}$, and assume that the members of (x,y) are elements of U. List the relation, the domain, and the range for each problem. It must be kept in mind that all numbers used in the relation, the domain, and the range must be elements of U.

1. $\{(x,y) \mid y = 2x\}$.
   Relation: $\{(1,2),(2,4),(3,6),(4,8)\}$
   Domain: $\{1,2,3,4\}$
   Range: $\{2,4,6,8\}$

2. $\{(x,y) \mid y = \frac{1}{2}x\}$.
   Relation: $\{(2,1),(4,2),(6,3),(8,4)\}$
   Domain: $\{2,4,6,8\}$
   Range: $\{1,2,3,4\}$
3. $\{(x,y) \mid y = x + 1\}$.
   Relation: $\{(1,2)(2,3),(3,4),(4,5),(5,6),(6,7),(7,8),(8,9)\}$
   Domain: $\{1,2,3,4,5,6,7,8\}$
   Range: $\{2,3,4,5,6,7,8,9\}$
4. $\{(x,y) \mid y = 2x - 1\}$.
   Relation: $\{(1,1),(2,3),(3,5),(4,7),(5,9)\}$
   Domain: $\{1,2,3,4,5\}$
   Range: $\{1,3,5,7,9\}$
5. $\{(x,y) \mid y = x - 1\}$.
   Relation: $\{(2,1),(3,2),(4,3),(5,4),(6,5),(7,6),(8,7),(9,8)\}$
   Domain: $\{2,3,4,5,6,7,8,9\}$
   Range: $\{1,2,3,4,5,6,7,8\}$

Solve problems 6 through 10 in a similar manner with the condition that $U = \{1, 2, 3, 4, \ldots\}$. (Recall that the three dots mean "and so on indefinitely"). In other words, U is the set of natural numbers.

6. $\{(x,y) \mid y = 4 - x\}$.
   Relation: $\{(1,3),(2,2),(3,1)\}$
   Domain: $\{1,2,3\}$
   Range: $\{3,2,1\}$
7. $\{(x,y) \mid y = 20 - x^2\}$.
   Relation: $\{(1,19),(2,16),(3,11),(4,4)\}$
   Domain: $\{1,2,3,4\}$
   Range: $\{19,16,11,4\}$
8. $\{(x,y) \mid y = 4\}$.
   (This, of course, means that for any and all values of x, $y = 4$).
   Relation: $\{(1,4),(2,4),(3,4), \ldots\}$
   Domain: $\{1,2,3, \ldots\}$
   Range: $\{4\}$
9. $\{(x,y) \mid y = x^2\}$.
   Relation: $\{(1,1),(2,4),(3,9),(4,16), \ldots\}$
   Domain: $\{1,2,3,4, \ldots\}$
   Range: $\{1,4,9,16, \ldots\}$
10. $\{(x,y) \mid y - x = 0\}$.
   Relation: $\{(1,1),(2,2),(3,3), \ldots\}$
   Domain: $\{1,2,3, \ldots\}$
   Range: $\{1,2,3, \ldots\}$

In problems 11 through 15 proceed as in the above problems, but use $U = \{-5, -4, -3, -2, -1, 0, 1, 2, 3, 4, 5\}$.

11. $\{(x,y) \mid y = x + 1\}$.
   Relation: $\{(-5,-4),(-4,-3),(-3,-2),(-2,-1),(-1,0),(0,1),(1,2),(2,3),(3,4),(4,5)\}$
   Domain: $\{-5,-4,-3,-2,-1,0,1,2,3,4\}$
   Range: $\{-4,-3,-2,-1,0,1,2,3,4,5\}$
12. $\{(x,y) \mid y = x^2 + x\}$.
   Relation: $\{(-2,2),(-1,0),(0,0),(1,2)\}$
   Domain: $\{-2,-1,0,1\}$ Range: $\{2,0\}$
13. $\{(x,y) \mid y = 2x - 3\}$.
   Relation: $\{(-1,-5),(0,-3),(1,-1),(2,1),(3,3),(4,5)\}$
   Domain: $\{-1,0,1,2,3,4\}$
   Range: $\{-5,-3,-1,1,3,5\}$
14. $\{(x,y) \mid y = x^2 + 3\}$.
   Relation: $\{(-1,4),(0,3),(1,4)\}$
   Domain: $\{-1,0,1\}$
   Range: $\{4,3\}$
15. $\{(x,y) \mid y = x^2 - x + 1\}$.
   Relation: $\{(-1,3),(0,1),(1,1),(2,3)\}$
   Domain: $\{-1,0,1,2\}$
   Range: $\{1,3\}$

Functions. A special kind of relation, which we call a function, has an important role in the study of mathematics. A **function** is a relation in which there is *one and only one* value in the range for each value in the domain. In other words, a function is a relation in which no two of the ordered pairs have the same first element.

The lower case letter f is used to denote a function; and, in terms of the variable x, is symbolized $f(x)$ (read "function of x" or merely "f of x"); and may be thought of as the value of the relation at x. This value is the ***unique*** value of y at a given value of x. For this reason, it is customary, throughout the field of mathematics, to state a functional relationship between x and y as $y = f(x)$ (read "y is equal to a function of x").

Since x is used as a variable, it may assume different values—either a finite or infinite number of values, depending upon what elements are in the universal set.

To find the value of y which corresponds to a given value of x, we merely substitute for x in the function and find the corresponding value of $f(x)$.

The following illustrates this procedure.

$y = f(x) = x^2 - x + 3.$

Find the value of y which corresponds to each of the following values of x: $-2, -1, 0, 1, 2$.

$$y = f(x) = x^2 - x + 3.$$

If $x = -2$,
$$y = f(x) = f(-2) = (-2)^2 - (-2) + 3 = 4 + 2 + 3 = 9.$$

If $x = -1$,
$$y = f(x) = f(-1) = (-1)^2 - (-1) + 3 = 1 + 1 + 3 = 5.$$

If $x = 0$,
$$y = f(x) = f(0) = (0)^2 - (0) + 3 = 3.$$

If $x = 1$,
$$y = f(x) = f(1) = (1)^2 - (1) + 3 = 1 - 1 + 3 = 3.$$

If $x = 2$,
$$y = f(x) = f(2) = (2)^2 - (2) + 3 = 4 - 2 + 3 = 5.$$

In the following problems find the value of $f(x)$ for each of the given values of x. The given set of values of x is, of course, the domain of the function.

16. $f(x) = x^2 - 5$.

Domain: $\{2, 3, 0 -1\}$

$$\begin{aligned} f(2) &= (2)^2 - 5 \\ &= 4 - 5 \\ &= -1. \\ f(3) &= (3)^2 - 5 \\ &= 9 - 5 \\ &= 4. \\ f(0) &= (0)^2 - 5 \\ &= 0 - 5 \\ &= -5. \\ f(-1) &= (-1)^2 - 5 \\ &= 1 - 5 \\ &= -4. \end{aligned}$$

17. $f(x) = 2x^2 + 3$.

Domain: $\{-2, 2, 4\}$

$$\begin{aligned} f(-2) &= 2(-2)^2 + 3 \\ &= 2(4) + 3 \\ &= 8 + 3 \\ &= 11. \\ f(2) &= 2(2)^2 + 3 \\ &= 2(4) + 3 \\ &= 8 + 3 \\ &= 11. \\ f(4) &= 2(4)^2 + 3 \\ &= 2(16) + 3 \\ &= 32 + 3 \\ &= 35. \end{aligned}$$

18. $f(x) = x^3 - 2x^2 + 1$.

Domain: $\{-2, -1, 0, 1, 2\}$

$$\begin{aligned} f(-2) &= (-2)^3 - 2(-2)^2 + 1 \\ &= -8 - 2(4) + 1 \\ &= -8 - 8 + 1 \\ &= -15. \\ f(-1) &= (-1)^3 - 2(-1)^2 + 1 \\ &= -1 - 2 + 1 \\ &= -2. \\ f(0) &= (0)^3 - 2(0)^2 + 1 \\ &= 0 - 0 + 1 \\ &= 1. \\ f(1) &= (1)^3 - 2(1)^2 + 1 \\ &= 1 - 2(1) + 1 \\ &= 1 - 2 + 1 \\ &= 0. \\ f(2) &= (2)^3 - 2(2)^2 + 1 \\ &= 8 - 2(4) + 1 \\ &= 8 - 8 + 1 \\ &= 1. \end{aligned}$$

19. $f(x) = 4 - x + 2x^2$.

Domain: $\{2, 4, 6\}$

$$\begin{aligned} f(2) &= 4 - (2) + 2(2)^2 \\ &= 4 - 2 + 2(4) \\ &= 4 - 2 + 8 \\ &= 10. \\ f(4) &= 4 - (4) + 2(4)^2 \\ &= 4 - 4 + 2(16) \\ &= 32. \\ f(6) &= 4 - (6) + 2(6)^2 \\ &= 4 - 6 + 2(36) \\ &= 4 - 6 + 72 \\ &= 70. \end{aligned}$$

20. $f(x) = 1 + x + 2x^2 - x^3$.

Domain: $\{-2, -1, 0, 1, 2\}$

$$\begin{aligned} f(-2) &= 1 + (-2) + 2(-2)^2 - (-2)^3 \\ &= 1 - 2 + 2(4) - (-8) \\ &= 1 - 2 + 8 + 8 \\ &= 15. \\ f(-1) &= 1 + (-1) + 2(-1)^2 - (-1)^3 \\ &= 1 - 1 + 2(1) - (-1) \\ &= 1 - 1 + 2 + 1 \\ &= 3. \\ f(0) &= 1 + (0) + 2(0)^2 - (0)^3 \\ &= 1 + 0 + 2(0) - 0 \\ &= 1. \\ f(1) &= 1 + (1) + 2(1)^2 - (1)^3 \\ &= 1 + 1 + 2(1) - (1) \\ &= 1 + 1 + 2 - 1 \\ &= 3. \\ f(2) &= 1 + (2) + 2(2)^2 - (2)^3 \\ &= 1 + 2 + 2(4) - (8) \\ &= 1 + 2 + 8 - 8 \\ &= 3. \end{aligned}$$

21. $f(x) = x^4 - 3x^3 + 2x^2 - x - 3$.

Domain: $\{-2, 0, \frac{1}{2}, 2, 4\}$

$$\begin{aligned} f(-2) &= (-2)^4 - 3(-2)^3 + 2(-2)^2 - (-2) - 3 \\ &= 16 - 3(-8) + 2(4) + 2 - 3 \\ &= 16 + 24 + 8 + 2 - 3 \\ &= 47. \\ f(0) &= (0)^4 - 3(0)^3 + 2(0)^2 - (0) - 3 \\ &= 0 - 0 + 0 - 0 - 3 \\ &= -3. \\ f(\tfrac{1}{2}) &= (\tfrac{1}{2})^4 - 3(\tfrac{1}{2})^3 + 2(\tfrac{1}{2})^2 - (\tfrac{1}{2}) - 3 \\ &= \tfrac{1}{16} - 3(\tfrac{1}{8}) + 2(\tfrac{1}{4}) - \tfrac{1}{2} - 3 \\ &= \tfrac{1}{16} - \tfrac{3}{8} + \tfrac{1}{2} - \tfrac{1}{2} - 3 \\ &= -\tfrac{53}{16} \\ &= -3\tfrac{5}{16}. \\ f(2) &= (2)^4 - 3(2)^3 + 2(2)^2 - (2) - 3 \\ &= 16 - 3(8) + 2(4) - 2 - 3 \\ &= 16 - 24 + 8 - 2 - 3 \\ &= -5. \end{aligned}$$

$$\begin{aligned} f(4) &= (4)^4 - 3(4)^3 + 2(4)^2 - (4) - 3 \\ &= 256 - 3(64) + 2(16) - 4 - 3 \\ &= 256 - 192 + 32 - 4 - 3 \\ &= 89. \end{aligned}$$

22. $f(x) = 3x^2 - 2x + 5.$

Domain: $\{-a, 0, \frac{1}{2}a, a, 2a\}$

$$\begin{aligned} f(-a) &= 3(-a)^2 - 2(-a) + 5 \\ &= 3a^2 + 2a + 5 \\ f(0) &= 3(0)^2 - 2(0) + 5 \\ &= 0 - 0 + 5 \\ &= 5. \\ f(\tfrac{1}{2}a) &= 3(\tfrac{1}{2}a)^2 - 2(\tfrac{1}{2}a) + 5 \\ &= 3(\tfrac{1}{4}a^2) - a + 5 \\ &= \frac{3a^2}{4} - a + 5. \\ f(a) &= 3(a)^2 - 2(a) + 5 \\ &= 3a^2 - 2a + 5. \\ f(2a) &= 3(2a)^2 - 2(2a) + 5 \\ &= 3(4a^2) - 4a + 5 \\ &= 12a^2 - 4a + 5. \end{aligned}$$

23. $f(x) = 2x^2 + 3x - 3.$

Domain: $\{-a, 0\ h, 2h, 4\}$

$$\begin{aligned} f(-a) &= 2(-a)^2 + 3(-a) - 3 \\ &= 2a^2 - 3a - 3. \\ f(0) &= 2(0)^2 + 3(0) - 3 \\ &= 0 + 0 - 3 \\ &= -3. \\ f(h) &= 2(h)^2 + 3(h) - 3 \\ &= 2h^2 + 3h - 3. \\ f(2h) &= 2(2h)^2 + 3(2h) - 3 \\ &= 2(4h^2) + 6h - 3 \\ &= 8h^2 + 6h - 3. \\ f(4) &= 2(4)^2 + 3(4) - 3 \\ &= 2(16) + 12 - 3 \\ &= 32 + 12 - 3 \\ &= 41. \end{aligned}$$

24. $f(x) = x^2 - 6x + 7.$

Domain: $\{-1,0,1,h,h+1\}$

$$\begin{aligned} f(-1) &= (-1)^2 - 6(-1) + 7 \\ &= 1 + 6 + 7 \\ &= 14. \\ f(0) &= (0)^2 - 6(0) + 7 \\ &= 0 - 0 + 7 \\ &= 7. \\ f(1) &= (1)^2 - 6(1) + 7 \\ &= 1 - 6 + 7 \\ &= 2. \\ f(h) &= (h)^2 - 6(h) + 7 \\ &= h^2 - 6h + 7. \\ f(h+1) &= (h+1)^2 - 6(h+1) + 7 \\ &= h^2 + 2h + 1 - 6h - 6 + 7 \\ &= h^2 - 4h + 2. \end{aligned}$$

25. $f(x) = 2x^2 - 3x + 1.$

Domain: $\{a, h, a+h, a-h\}$

$$\begin{aligned} f(a) &= 2(a)^2 - 3(a) + 1 \\ &= 2a^2 - 3a + 1. \\ f(h) &= 2(h)^2 - 3(h) + 1 \\ &= 2h^2 - 3h + 1. \\ f(a+h) &= 2(a+h)^2 - 3(a+h) + 1 \\ &= 2(a^2 + 2ah + h^2) - 3a - 3h + 1 \\ &= 2a^2 + 4ah + 2h^2 - 3a - 3h + 1. \\ f(a-h) &= 2(a-h)^2 - 3(a-h) + 1 \\ &= 2(a^2 - 2ah + h^2) - 3a + 3h + 1 \\ &= 2a^2 - 4ah + 2h^2 - 3a + 3h + 1. \end{aligned}$$

GRAPHS OF LINEAR EQUATIONS

The Rectangular Coordinate System. The most commonly used system for graphing algebraic equations is known as the **rectangular coordinate** system. In this plan of graphing, a plane is represented by a grid, and points in the plane are located on that grid by dots. These dots are connected by a line which is the graph of the equation.

On the grid, two lines perpendicular to each other are constructed. The point of intersection of the two lines is known as the **origin.** The horizontal line is called the $X'X$-axis and the vertical line is known as the $Y'Y$-axis. These are commonly referred to as the **X-axis** and the **Y-axis,** and they divide the grid into four parts known as **quadrants.** The quadrants are numbered I to IV, starting with Quadrant I in the upper right-hand corner and moving counter-clockwise. This is shown in Fig. 13. Points on the graph are located with respect to their position from the origin. Positive numbers on the X-axis are to the right of the origin, and negative numbers are to the left of the origin. Positive numbers on the Y-axis are above the origin, and negative numbers are below the origin. Arrows are placed at both ends of each axis to show that the plane extends indefinitely in all directions.

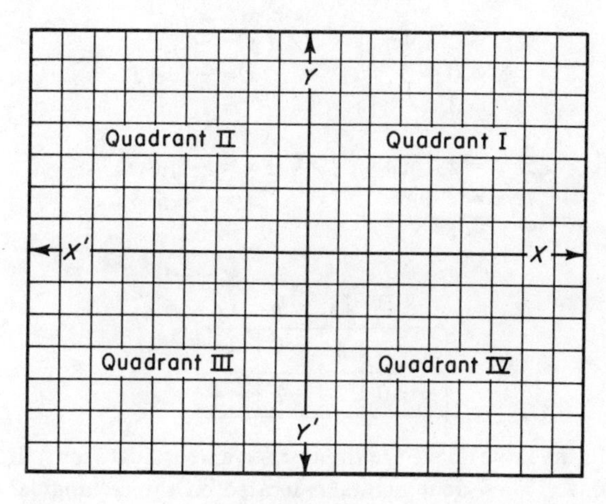

FIG. 13

All points on the graph are represented by ordered pairs in the form $P(x,y)$. The first number, x, known as the **abscissa** of P, is the perpendicular distance from the point to the Y-axis. The second number, y, called the **ordinate** of P, is the perpendicular distance from the point P to the X-axis. The two are called the **coordinates** of the point. Values of x are to the right or left of the origin, and values of y are located above or below the origin. Figure 14 shows the position of the points $P_1(3,5)$, $P_2(-4,2)$, $P_3(-6,-5)$, and $P_4(4,-4)$ with respect to the coordinate axes.

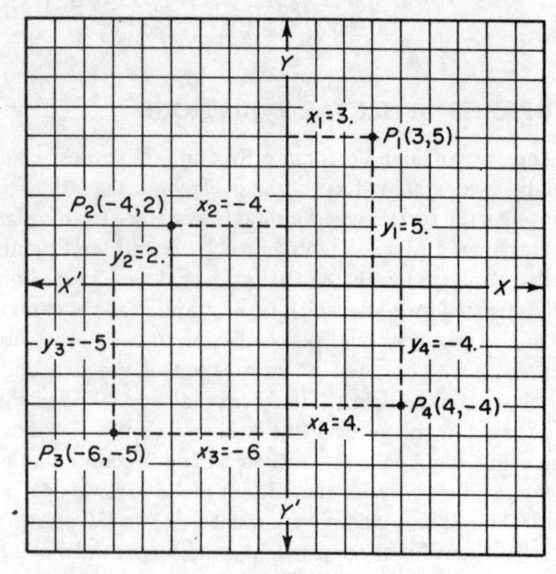

FIG. 14

Graphs of Linear Equations. In any algebraic equation containing two variables, a variety of values can be assigned to one of the variables and the corresponding values of the other variable can be determined.

In the equation $x = y + 3$, corresponding pairs of values are found as follows:

If $y = 0$, then
$x = 0 + 3$
$= 3.$

If $y = -2$, then
$x = -2 + 3$
$= 1.$

If $y = 2$, then
$x = 2 + 3$
$= 5.$

If $y = -4$, then
$x = -4 + 3$
$= -1.$

	P_1	P_2	P_3	P_4
x	3	5	1	−1
y	0	2	−2	−4

In the table above are the abscissa and ordinate for four points. These four points, if located on the rectangular coordinate system, will appear as shown in Fig. 15.

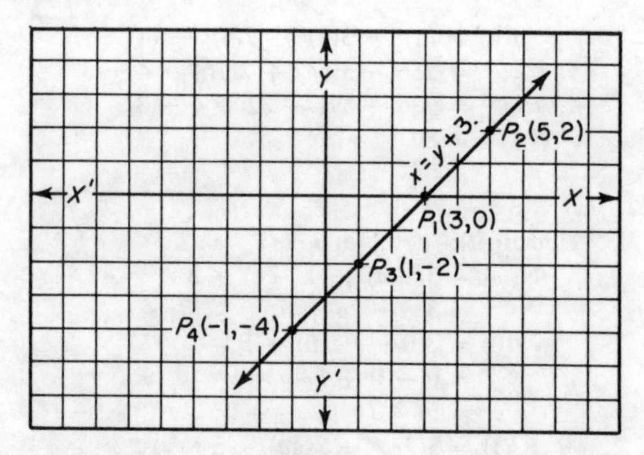

FIG. 15

The graph is the line joining all these points. In this case the line is a straight line and can be extended indefinitely in either direction. All first degree equations in two unknowns represent a straight line graph. This straight line can be extended indefinitely in either direction because an infinite number of points satisfy $x = y + 3$ as x is either increased or decreased.

Since two points determine a straight line, the location of two points will correctly determine the graph. However, it is usually desirable to locate three points as this will probably expose any error that has been made in locating the first two points.

Intercepts. The **x-intercept** is the point at which the graph intersects the X-axis. At this point the value of y is zero. Likewise, the **y-intercept** is the point at which the graph intersects the Y-axis, and the corresponding value of x is zero. If one assigns the value 0 to x and then to y, the corresponding values of y and x are the y- and x-intercepts. The location of these points is of special interest in a graph and usually should be determined.

Graph the following first degree equations.

1. $y = x + 5.$

If $x = 0$,
$y = 0 + 5$
$= 5.$

If $x = -2$,
$y = -2 + 5$
$= 3.$

If $x = -5$,
$y = -5 + 5$
$= 0.$

x	0	−2	−5
y	5	3	0

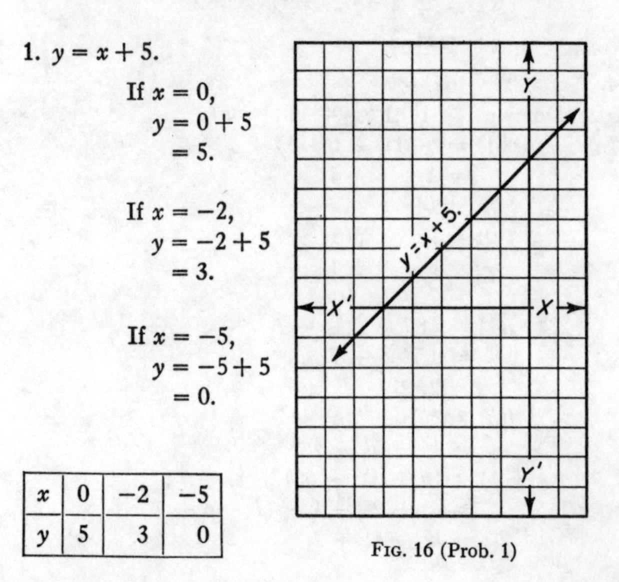

FIG. 16 (Prob. 1)

2. $x + y = 5$.

For convenience in determining the values of x, use the subtraction axiom and subtract y from each member of the equation.

$$x + y = 5.$$
$$x = 5 - y.$$

Find the intercepts and one "extra point".

If $y = 0$,
$x = 5 - 0$
$= 5$.
If $y = -1$,
$x = 5 - (-1)$
$= 6$.
If $x = 0$,
$0 = 5 - y$.
$y = 5$.

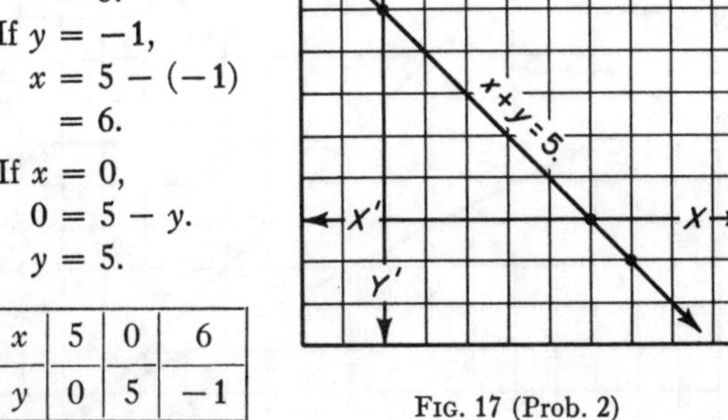

x	5	0	6
y	0	5	-1

FIG. 17 (Prob. 2)

3. $x = 2y - 3$.

If $y = 0$,
$x = 2(0) - 3$
$= -3$.

If $y = 1$,
$x = 2(1) - 3$
$= 2 - 3$
$= -1$.

Since the x-intercept is fractional and difficult to locate accurately on the grid, we will find a different third point.

If $y = -1$,
$x = 2(-1) - 3$
$= -2 - 3$
$= -5$.

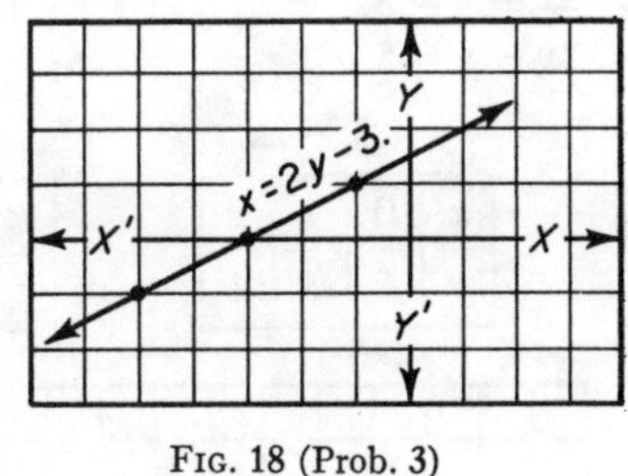

x	-3	-5	-1
y	0	-1	1

FIG. 18 (Prob. 3)

4. $2x = y + 2$.

Solve the equation for y.

$$2x = y + 2.$$
$$y = 2x - 2.$$

Find the y-intercept and one "extra" point.

If $x = 0$,
$y = 2(0) - 2$
$= -2$.
If $x = 2$,
$y = 2(2) - 2$
$= 4 - 2$
$= 2$.

Find the x-intercept.

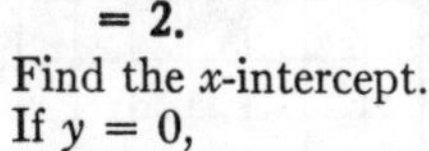

If $y = 0$,
$0 = 2x - 2$

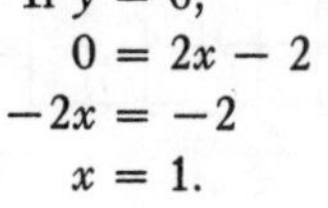

$-2x = -2$
$x = 1$.

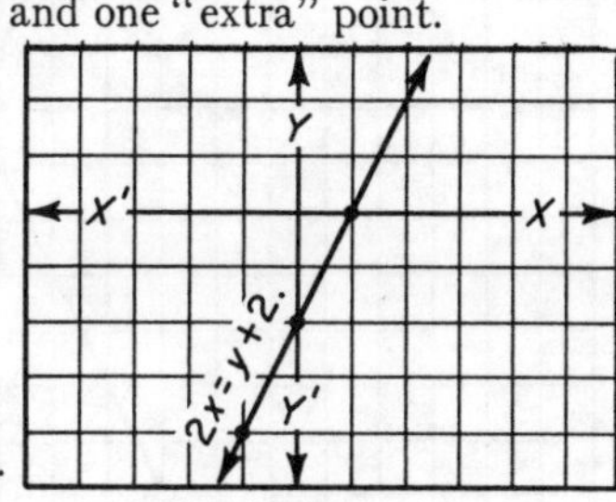

FIG. 19 (Prob. 4)

x	0	2	1
y	-2	2	0

5. $2x = 3y + 1$.

$$2x = 3y + 1.$$

Divide both members of the equation by 2.

$$x = \frac{3y + 1}{2}.$$

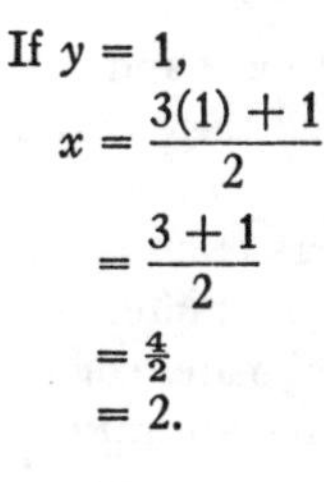

If $y = 1$,
$x = \frac{3(1) + 1}{2}$
$= \frac{3 + 1}{2}$
$= \frac{4}{2}$
$= 2$.

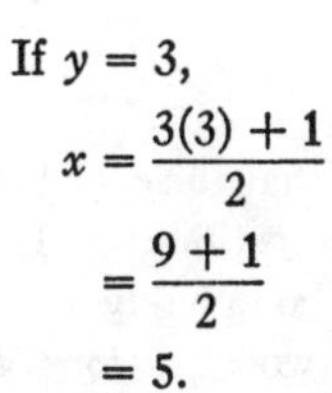

If $y = 3$,
$x = \frac{3(3) + 1}{2}$
$= \frac{9 + 1}{2}$
$= 5$.

If $y = -1$,
$x = \frac{3(-1) + 1}{2}$
$= \frac{-3 + 1}{2}$
$= -1$.

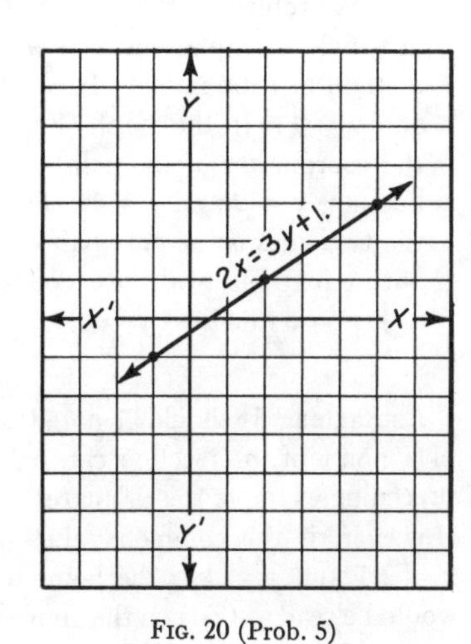

FIG. 20 (Prob. 5)

x	2	5	-1
y	1	3	-1

Systems of Linear Equations. The graphs of two or more linear equations may be drawn on the same coordinate system. When this is done, the point or points at which the graphs intersect is the solution set of the system of equations. Thus, a graphical method may be used for solving systems of linear equations such as those found in Chapter 6.

When two or more equations are graphed on one coordinate system, each equation is taken separately and graphed as were the preceding problems in this chapter.

Graph the following pairs of equations on the same set of axes.

6. $x + y = 3$.
$x - y = 1$.

$$x + y = 3.$$

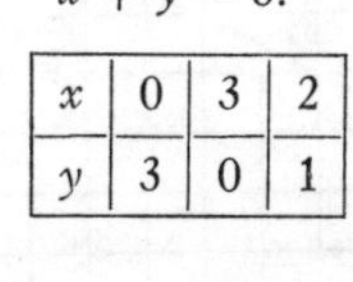

x	0	3	2
y	3	0	1

$$x - y = 1.$$

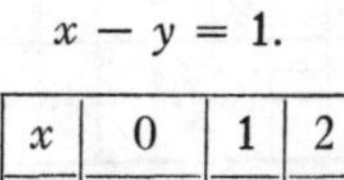

x	0	1	2
y	-1	0	1

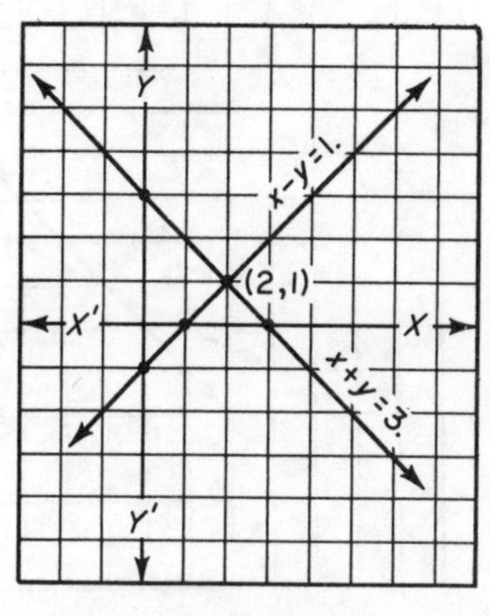

FIG. 21 (Prob. 6)

It can be seen that the two straight lines intersect. The coordinates of the point of intersection are (2, 1). This means that only one point (the point of intersection) is common to both lines or to both equations; and that the solution set of the system of equations is $\{(2, 1)\}$.

This problem is the same as Problem 1 in Chapter 6 in which the equations were solved for x and y by the method of elimination. It was found there that $x = 2$ and $y = 1$. In the graph (Fig. 21), it can be seen that the coordinates of the point of intersection of the two lines are $x = 2$ and $y = 1$. The point of intersection of the two lines is the only pair of coordinates common to both equations, and these coordinates will satisfy both $x + y = 3$ and $x - y = 1$.

Estimates. It should be noted here that the coordinates of a point of intersection can only be estimated. In the preceding example the estimates were completely accuate. However, if the common values of two equations were $x = 2.1$ and $y = 1.1$, the point of intersection probably would be read as (2,1) on the above set of axes. Sometimes the conditions of a problem can help us to establish the accuracy of our estimates. Graphic estimates can be checked by use of the algebraic methods of solving systems of equations that were used in Chapter 6.

7. $x - 2y = 3.$
   $x = y + 1.$

$x = 2y + 3.$

x	3	5	1
y	0	1	-1

$x = y + 1.$

x	1	2	0
y	0	1	-1

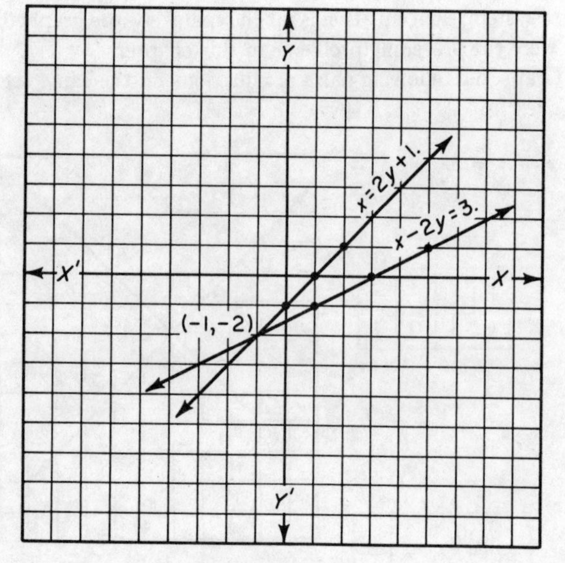

FIG. 22 (Prob. 7)

8. $x = 3 - y.$
   $x + 2y = 0.$

$x = 3 - y.$

x	3	2	1
y	0	1	2

$x = -2y.$

x	0	-2	2
y	0	1	-1

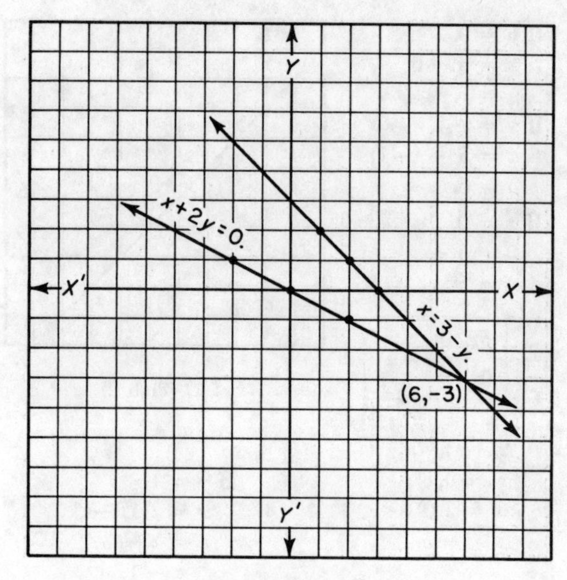

FIG. 23 (Prob. 8)

9. $2x + y = 5.$
   $2x - y = 5.$

$y = 5 - 2x.$

x	0	1	2
y	5	3	1

$y = 2x - 5.$

x	0	2	3
y	-5	-1	1

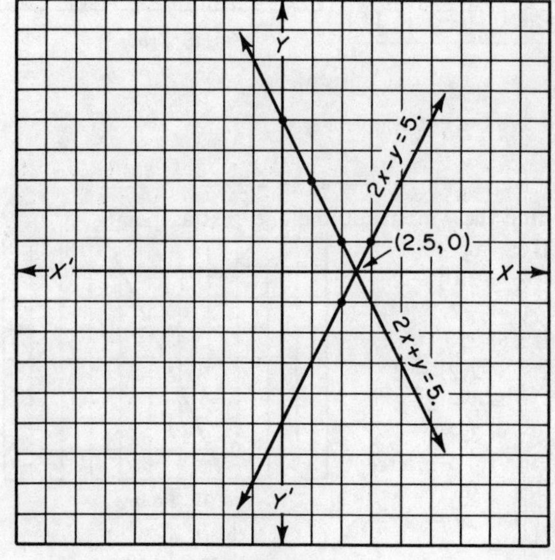

FIG. 24 (Prob. 9)

10. $x - y = 0.$
$x = 2y.$

$x = y.$

x	0	1	-1
y	0	1	-1

$x = 2y.$

x	0	2	-2
y	0	1	-1

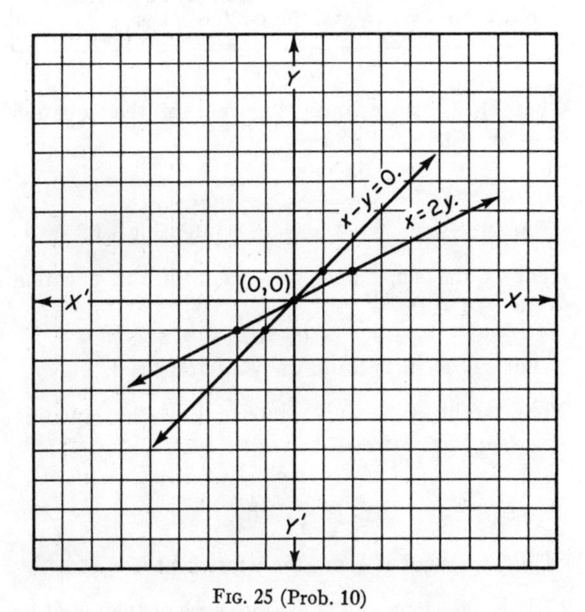

FIG. 25 (Prob. 10)

SPECIAL FORMS OF THE LINEAR EQUATION

The Slope of a Line. By the slope of a line we mean its inclination with respect to the X-axis; or, in other words, how steep it is with respect to the X-axis. To find the slope of a line we need the coordinates of two points on the line.

The slope of a line, which we shall call m, is defined by the formula

$$m = \frac{y_1 - y_2}{x_1 - x_2}.$$

The slope of a line is a positive number if the line rises as we move from left to right on the coordinate system (as in Problem 1 above). The slope is a negative number if the line falls as we move from left to right (as in Problem 2).

Find the slope of the line which passes through the points whose coordinates are (3,1) and (5,4).

$$\begin{aligned} m &= \frac{y_1 - y_2}{x_1 - x_2} \\ &= \frac{4 - 1}{5 - 3} \\ &= \frac{3}{2}. \end{aligned}$$

It does not matter which ordered pair is assigned the subscripts 1 and which the subscripts 2. However, it is mandatory that both members of any ordered pair have the same subscript.

1. Find the slope of the line passing through the points whose coordinates are $(-3,-1)$ and $(3,-5)$.

$$\begin{aligned} m &= \frac{y_1 - y_2}{x_1 - x_2} \\ &= \frac{-5 - (-1)}{3 - (-3)} \\ &= \frac{-5 + 1}{3 + 3} \\ &= \frac{-4}{6} \\ &= -\tfrac{2}{3}. \end{aligned}$$

2. What is the slope of the line passing through the points whose coordinates are $(-5,0)$ and $(0,-3)$?

$$\begin{aligned} m &= \frac{y_1 - y_2}{x_1 - x_2} \\ &= \frac{-3 - 0}{0 - (-5)} \\ &= \frac{-3}{5} \\ &= \tfrac{3}{5}. \end{aligned}$$

3. What is the slope of the line whose equation is $y = x + 5$?

In order to solve the problem, at this time, it will be necessary to find two ordered pairs that satisfy the equation. Those two ordered pairs, of course, will represent two points on the graph of the equation.

If $x = 3,$
$$\begin{aligned} y &= 3 + 5 \\ &= 8. \end{aligned}$$

If $x = -2,$
$$\begin{aligned} y &= -2 + 5 \\ &= 3. \end{aligned}$$

Thus, the two ordered pairs are (3,8) and $(-2,3)$ and the slope is

$$\begin{aligned} m &= \frac{y_1 - y_2}{x_1 - x_2} \\ &= \frac{8 - 3}{3 - (-2)} \\ &= \frac{5}{3 + 2} \\ &= \tfrac{5}{5} \\ &= 1. \end{aligned}$$

4. What is the slope of the line whose equation is

$$3y = 2x - 1?$$

$$\begin{aligned} 3y &= 2x - 1. \\ y &= \frac{2x - 1}{3}. \end{aligned}$$

If $x = -1$,

$$y = \frac{2(-1) - 1}{3} = \frac{-2 - 1}{3} = \frac{-3}{3} = -1.$$

If $x = 5$,

$$y = \frac{2(5) - 1}{3} = \frac{10 - 1}{3} = \frac{9}{3} = 3.$$

Thus, the two ordered pairs which we shall use are $(-1,-1,)$ and $(5,3)$ and the slope is

$$m = \frac{y_1 - y_2}{x_1 - x_2} = \frac{3 - (-1)}{5 - (-1)} = \frac{3 + 1}{5 + 1} = \tfrac{4}{6} = \tfrac{2}{3}.$$

5. What is the slope of the line whose equation is

$$2y = \frac{3x}{4} + 2?$$

$$2y = \frac{3x}{4} + 2.$$

$$y = \frac{3x}{8} + 1.$$

If $x = 1$,

$$y = \frac{3(1)}{8} + 1 = \tfrac{3}{8} + \tfrac{8}{8} = \tfrac{11}{8}.$$

If $x = -2$,

$$y = \frac{3(-2)}{8} + 1 = -\tfrac{6}{8} + \tfrac{8}{8} = \tfrac{1}{4}.$$

Hence, the two ordered pairs which we shall use are $(1, \tfrac{11}{8})$ and $(-2, \tfrac{1}{4})$. The slope is

$$m = \frac{y_1 - y_2}{x_1 - x_2} = \frac{\tfrac{11}{8} - \tfrac{1}{4}}{1 - (-2)} = \frac{\tfrac{11}{8} - \tfrac{2}{8}}{1 + 2} = \frac{\tfrac{9}{8}}{3} = \tfrac{3}{8} \cdot 1 = \tfrac{3}{8}.$$

The Slope-Intercept Form of the Equation of a Line. When the equation of a line is written in slope-intercept form, y is expressed in terms of the slope m and the point at which the line intersects the y-axis. The form of this way of writing the equation of a line is

$$y = mx + b$$

where m is the slope of the line and b is the y-intercept.

What is the slope and y-intercept of the equation $x - 2y = 3$?

Solve for y and we have

$$x - 2y = 3.$$
$$-2y = -x + 3.$$
$$y = \tfrac{1}{2}x - \tfrac{3}{2}.$$

Thus, the slope is $\tfrac{1}{2}$ and the y-intercept is $-\tfrac{3}{2}$.

6. What is the slope and y-intercept of the equation $3y + 3x = 0$?

$$3y + 3x = 0.$$
$$3y = -3x. \quad y = -x.$$

Thus, the slope is -1 and the y-intercept is 0.

7. What is the slope and y-intercept of the equation $y - x - 3 = 0$?

$$y - x - 3 = 0. \quad y = x + 3.$$

Thus, the slope is 1 and the y-intercept is 3.

8. What is the slope and y-intercept of the equation $4 = x - 3y$?

$$4 = x - 3y.$$
$$3y = x - 4.$$
$$y = \tfrac{1}{3}x - \tfrac{4}{3}.$$

Thus, the slope is $\tfrac{1}{3}$ and the y-intercept is $-\tfrac{4}{3}$.

9. What is the slope and the y-intercept of the equation $\frac{x}{3} + \frac{y}{4} = 1$?

$$\frac{x}{3} + \frac{y}{4} = 1.$$
$$\frac{y}{4} = -\frac{x}{3} + 1.$$
$$4\left(\frac{y}{4}\right) = 4\left(\frac{-x}{3}\right) + 4(1).$$
$$y = -\tfrac{4}{3}x + 4.$$

Hence, the slope is $-\tfrac{4}{3}$ and the y-intercept is 4.

10. What is the slope and y-intercept of the equation $\frac{x+1}{3} - \frac{y-1}{2} = \frac{1}{6}$?

$$\frac{x+1}{3} - \frac{y-1}{2} = \frac{1}{6}.$$

Clear of fractions and solve for y.

$$\frac{2(x+1)}{6} - \frac{3(y-1)}{6} = \frac{1}{6}.$$
$$6\left[\frac{2(x+1)}{6}\right] - 6\left[\frac{3(y-1)}{6}\right] = 6\left[\frac{1}{6}\right].$$
$$2x + 2 - 3y + 3 = 1.$$
$$2x - 3y + 5 = 1.$$
$$-3y = -2x - 4.$$
$$3y = 2x + 4.$$
$$y = \tfrac{2}{3}x + \tfrac{4}{3}.$$

Hence, the slope is $\tfrac{2}{3}$ and the y-intercept is $\tfrac{4}{3}$.

The Point-Slope Form of the Equation of a Line. If two points on a line are designated P and P_1; and if their coordinates are (x, y) and (x_1, y_1) respectively, the formula for the slope of that line would read

$$m = \frac{y - y_1}{x - x_1}.$$

If both members of this equation were multiplied by $(x - x_1)$, we would obtain

$$m(x - x_1) = \frac{(y - y_1)}{(x - x_1)} \cdot (x - x_1).$$

$$m(x - x_1) = y - y_1.$$

By the symmetric relationship (page 4), this becomes

$$y - y_1 = m(x - x_1).$$

where x_1 and y_1 are the given coordinates of a point on the line and where m is the slope of that line. This is known as the point-slope form of the equation of a line.

Find the equation of the line whose slope is 2 and which passes through the point whose coordinates are (3, 1).

$$y - y_1 = m(x - x_1).$$
$$y - 1 = 2(x - 3).$$
$$y - 1 = 2x - 6.$$
$$-2x + y + 5 = 0.$$
$$2x - y - 5 = 0.$$

11. Write the equation of the line passing through the point whose coordinates are $(-2,3)$ and having the slope -1.

$$y - y_1 = m(x - x_1).$$
$$y - 3 = -1[x - (-2)].$$
$$y - 3 = -1(x + 2).$$
$$y - 3 = -x - 2.$$
$$x + y - 1 = 0.$$

12. Write the equation of the line passing through the point whose coordinates are $(4,-3)$ and having a slope of 3.

$$y - y_1 = m(x - x_1).$$
$$y - (-3) = 3(x - 4).$$
$$y + 3 = 3x - 12.$$
$$-3x + y + 15 = 0.$$
$$3x - y - 15 = 0.$$

13. Write the equation of the line passing through the point whose coordinates are $(-1,-2)$ and having a slope of $-\frac{2}{3}$.

$$y - y_1 = m(x - x_1).$$
$$y - (-2) = -\tfrac{2}{3}[x - (-1)].$$
$$y + 2 = -\tfrac{2}{3}(x + 1).$$
$$3y + 6 = -2(x + 1).$$
$$3y + 6 = -2x - 2.$$
$$2x + 3y + 8 = 0.$$

14. Write the equation of the line passing through the point whose coordinates are $(-2,\frac{2}{3})$ and having a slope of $\frac{3}{2}$.

$$y - y_1 = m(x - x_1).$$
$$y - \tfrac{2}{3} = \tfrac{3}{2}[x - (-2)].$$
$$\frac{3y - 2}{3} = \frac{3(x + 2)}{2}.$$
$$\frac{2(3y - 2)}{6} = \frac{9(x + 2)}{6}.$$
$$6\left[\frac{2(3y - 2)}{6}\right] = 6\left[\frac{9(x + 2)}{6}\right].$$
$$6y - 4 = 9x + 18.$$
$$-9x + 6y - 22 = 0.$$
$$9x - 6y + 22 = 0.$$

15. Write the equation of the line passing through the point whose coordinates are $(\frac{2}{3}, \frac{3}{4})$ and having a slope of $\frac{1}{3}$.

$$y - y_1 = m(x - x_1).$$
$$y - \tfrac{3}{4} = \tfrac{1}{3}(x - \tfrac{2}{3}).$$
$$\frac{4y - 3}{4} = \frac{1}{3}\left(\frac{3x - 2}{3}\right).$$
$$\frac{4y - 3}{4} = \frac{3x - 2}{9}.$$
$$\frac{9(4y - 3)}{36} = \frac{4(3x - 2)}{36}.$$
$$36\left[\frac{9(4y - 3)}{36}\right] = 36\left[\frac{4(3x - 2)}{36}\right].$$
$$36y - 27 = 12x - 8.$$
$$-12x + 36y - 19 = 0.$$
$$12x - 36y + 19 = 0.$$

The Two-Point Form of the Equation of a Line. Let three points on a line be designated by the letters P, P_1, and P_2. Their coordinates are (x,y), (x_1y_1), and (x_2y_2) respectively. Using points P and P_1 we could indicate the slope of the line by the equation

$$m_1 = \frac{y - y_1}{x - x_1}.$$

Then by using the points P_1 and P_2, the slope could be written

$$m_2 = \frac{y_1 - y_2}{x_1 - x_2}.$$

Since both slopes are for the same line $m_1 = m_2$ and

$$\frac{y - y_1}{x - x_1} = \frac{y_1 - y_2}{x_1 - x_2}.$$

This equation is known as the two-point equation of a line; and it enables us to write the equation of any line if two ordered pairs that satisfy the equation of that line are given.

Write the linear equation satisfied by the ordered pairs (2,5) and (−3,1).

$$\frac{y-y_1}{x-x_1}=\frac{y_1-y_2}{x_1-x_2}.$$
$$\frac{y-5}{x-2}=\frac{5-1}{2-(-3)}.$$
$$\frac{y-5}{x-2}=\frac{5-1}{2+3}.$$
$$\frac{y-5}{x-2}=\frac{4}{5}.$$

Clear of fractions and we get

$$5y-25=4x-8.$$
$$-4x+5y-17=0.$$
$$4x-5y+17=0.$$

16. Write the linear equation satisfied by the ordered pairs (2,0) and (3,−1).

$$\frac{y-y_1}{x-x_1}=\frac{y_1-y_2}{x_1-x_2}.$$
$$\frac{y-(-1)}{x-3}=\frac{-1-0}{3-2}.$$
$$\frac{y+1}{x-3}=\frac{-1}{1}.$$
$$\frac{y+1}{x-3}=-1.$$
$$y+1=-x+3.$$
$$x+y-2=0.$$

17. Write the linear equation satisfied by the ordered pairs (2,2) and (−1,4).

$$\frac{y-y_1}{x-x_1}=\frac{y_1-y_2}{x_1-x_2}.$$
$$\frac{y-2}{x-2}=\frac{2-4}{2-(-1)}.$$
$$\frac{y-2}{x-2}=\frac{-2}{2+1}.$$
$$\frac{y-2}{x-2}=\frac{-2}{3}.$$
$$3y-6=-2x+4.$$
$$2x+3y-10=0.$$

18. Write the equation of the line which passes through the points whose coordinates are $(\frac{1}{2},\frac{2}{3})$ and (0,−2).

$$\frac{y-y_1}{x-x_1}=\frac{y_1-y_2}{x_1-x_2}.$$
$$\frac{y-\frac{2}{3}}{x-\frac{1}{2}}=\frac{\frac{2}{3}-(-2)}{\frac{1}{2}-0}.$$
$$\frac{\frac{3y-2}{3}}{\frac{2x-1}{2}}=\frac{\frac{2}{3}+2}{\frac{1}{2}}.$$
$$\frac{3y-2}{3}\cdot\frac{2}{2x-1}=\frac{8}{3}\cdot\frac{2}{1}.$$
$$\frac{6y-4}{6x-3}=\frac{16}{3}.$$
$$18y-12=96x-48.$$
$$-96x+18y+36=0.$$
$$16x-3y-6=0.$$

19. Write the equation of the line which passes through the points whose coordinates are (5,−2) and $(\frac{1}{2},0)$.

$$\frac{y-y_1}{x-x_1}=\frac{y_1-y_2}{x_1-x_2}.$$
$$\frac{y-(-2)}{x-5}=\frac{-2-0}{5-\frac{1}{2}}.$$
$$\frac{y+2}{x-5}=\frac{-2}{\frac{9}{2}}.$$
$$\frac{y+2}{x-5}=\frac{-4}{9}.$$
$$9y+18=-4x+20.$$
$$4x+9y-2=0.$$

20. Write the equation of the line which passes through the points whose coordinates are $(\frac{2}{3},-\frac{1}{3})$ and $(\frac{1}{4},\frac{3}{4})$.

$$\frac{y-y_1}{x-x_1}=\frac{y_1-y_2}{x_1-x_2}.$$
$$\frac{y-(-\frac{1}{3})}{x-\frac{2}{3}}=\frac{-\frac{1}{3}-\frac{3}{4}}{\frac{2}{3}-\frac{1}{4}}.$$
$$\frac{y+\frac{1}{3}}{x-\frac{2}{3}}=\frac{\frac{-4-9}{12}}{\frac{8-3}{12}}.$$
$$\frac{\frac{3y+1}{3}}{\frac{3x-2}{3}}=\frac{\frac{-13}{12}}{\frac{5}{12}}.$$
$$\frac{3y+1}{3}\cdot\frac{3}{3x-2}=-\frac{13}{12}\cdot\frac{12}{5}.$$
$$\frac{3y+1}{3x-2}=\frac{-13}{5}.$$
$$15y+5=-39x+26.$$
$$39x+15y-21=0.$$
$$13x+5y-7=0.$$

The Two-Intercept Form of the Equation of a Line. The x-intercept of an equation is the point at which the graph of the line intersects the x-axis. Likewise, the y-intercept is the point at which it crosses the y-axis. If a is the x-intercept and b the y-intercept, then two points on that line have the coordinates $(a,0)$ and $(0,b)$. By use of the two-point form of the equation of a line, and using those coordinates for the two points we obtain

$$\frac{y-y_1}{x-x_1}=\frac{y_1-y_2}{x_1-x_2}.$$
$$\frac{y-0}{x-a}=\frac{0-b}{a-0}.$$
$$\frac{y}{x-a}=\frac{-b}{a}.$$

$$ay = -bx + ab.$$
$$bx + ay = ab.$$

Divide both members of the above equation by ab and we have

$$\frac{bx}{ab} + \frac{ay}{ab} = \frac{ab}{ab}.$$
$$\frac{x}{a} + \frac{y}{b} = 1, \qquad (1)$$

where a is the x-intercept and b is the y-intercept.

Equation (1) is called the two-intercept form of the equation of a line.

21. Write the equation $2x + 3y = 6$ in the two-intercept form.

$$2x + 3y = 6.$$
$$\frac{2x}{6} + \frac{3y}{6} = \frac{6}{6}.$$
$$\frac{x}{3} + \frac{y}{2} = 1.$$

Hence, the x-intercept is at (3,0) and the y-intercept is at (0,2).

22. Write the equation $x - 5y = 2$ in the two-intercept form.

$$x - 5y = 2.$$
$$\frac{x}{2} - \frac{5y}{2} = \frac{2}{2}.$$
$$\frac{x}{2} + \frac{y}{-\frac{2}{5}} = 1.$$

Hence, the x-intercept is at (2,0) and the y-intercept is at $(0, -\frac{2}{5})$.

23. Write the equation $y = \frac{2}{3}x - 3$ in the two-intercept form.

$$y = \tfrac{2}{3}x - 3.$$
$$3y = 2x - 9.$$
$$-2x + 3y = -9.$$
$$2x - 3y = 9.$$
$$\frac{2x}{9} - \frac{3y}{9} = \frac{9}{9}.$$
$$\frac{2x}{9} - \frac{y}{3} = 1.$$
$$\frac{x}{\frac{9}{2}} + \frac{y}{-3} = 1.$$

Hence, the x-intercept is at $(\frac{9}{2}, 0)$ and the y-intercept is at $(0, -3)$.

24. Write the linear equation satisfied by the coordinates $(-1,3)$ and $(2,-1)$ in the two-intercept form.

$$\frac{y - y_1}{x - x_1} = \frac{y_1 - y_2}{x_1 - x_2}.$$
$$\frac{y - 3}{x - (-1)} = \frac{3 - (-1)}{-1 - 2}.$$
$$\frac{y - 3}{x + 1} = \frac{3 + 1}{-3}.$$
$$\frac{y - 3}{x + 1} = \frac{4}{-3}.$$
$$-3y + 9 = 4x + 4.$$
$$-4x - 3y = -5.$$
$$4x + 3y = 5.$$
$$\frac{4x}{5} + \frac{3y}{5} = \frac{5}{5}.$$
$$\frac{x}{\frac{5}{4}} + \frac{y}{\frac{5}{3}} = 1.$$

25. Write, in two-intercept form, the equation of the line which passes through the point $(0, -3)$ and whose slope is $-\frac{2}{3}$.

First write the slope-intercept form of the equation.

$$y - y_1 = m(x - x_1).$$
$$y - (-3) = -\tfrac{2}{3}(x - 0).$$
$$y + 3 = \frac{-2x}{3}.$$
$$3y + 9 = -2x.$$
$$2x + 3y = -9.$$
$$\frac{2x}{-9} + \frac{3y}{-9} = \frac{-9}{-9}.$$
$$\frac{x}{-\frac{9}{2}} + \frac{y}{-3} = 1.$$

GRAPHS OF HIGHER-DEGREE EQUATIONS

Higher-Degree Equations. The study of mathematics involves much work with equations which contain variables raised to powers higher than the first. All such equations are known as higher-degree equations. Some examples are: $x^2 + 2x + 9 = 0$, $x^3 - 3x^2y + 3xy^2 = -9$, and $x^4 = 96$. Many times the graphing of these equations is either necessary or helpful in completing the solutions involved.

The method used for graphing linear equations may be used for higher-degree equations as well; but until one has had considerable experience in such graphing, it is necessary to locate quite a number of points before drawing the graph. One must be especially careful to use the correct signs when graphing these equations.

The table of square roots and cube roots in the Appendix will be helpful in working with second- and third-degree equations.

Graph the following equations.

1. $x^2 + y^2 = 25$.

Assign values to one of the variables and solve for the other one in the same manner as was done with the linear equations. Be sure to use both plus and minus signs when they appear.

Let $x = 0$.
$0^2 + y^2 = 25$.
$y^2 = 25$.
$y = \pm 5$.

Let $y = 0$.
$x^2 + 0^2 = 25$.
$x^2 = 25$.
$x = \pm 5$.

Let $y = \pm 3$.
$x^2 + (\pm 3)^2 = 25$.
$x^2 + 9 = 25$.
$x^2 = 16$.
$x = \pm 4$.

Let $y = \pm 2$.
$x^2 + (\pm 2)^2 = 25$.
$x^2 + 4 = 25$.
$x^2 = 21$.
$x = \pm 4.6$.

Let $y = \pm 4$.
$x^2 + (\pm 4)^2 = 25$.
$x^2 + 16 = 25$.
$x^2 = 9$.
$x = \pm 3$.

Let $x = \pm 2$.
$(\pm 2)^2 + y^2 = 25$.
$4 + y^2 = 25$.
$y^2 = 21$.
$y = \pm 4.6$.

x	0	±5	±4	±3	±4.6	±2
y	±5	0	±3	±4	±2	±4.6

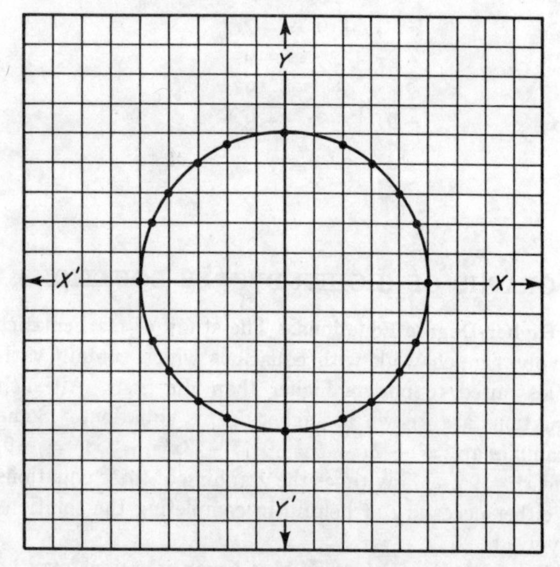

FIG. 26 (Prob. 1)

Let $y = \pm 6$.
$x^2 + (\pm 6)^2 = 25$.
$x^2 + 36 = 25$.
$x^2 = -11$.
$x = \pm\sqrt{-11}$
$= \pm i\sqrt{11}$.

This is an imaginary number, and it will not appear on the graph. It can be seen that any number numerically greater than 5, if substituted in the equation for one of the variables, will produce an imaginary value for the other variable. Consequently, no values numerically greater than 5 will be used.

When the points are plotted, the graph turns out to be a **circle.** It can be proven that any equation indicating the sum of x^2 and y^2 will produce a circle when a graph is made of it. If the coefficients of x^2 and y^2 are each 1, the square root of the constant will be the radius of a circle whose center is at the origin. If the coefficients of the x^2 and y^2 terms are the same but not equal to 1, divide all terms in both members of the equation by that coefficient, and the square root of the constant will be the radius of a circle whose center is at the origin.

2. $y = x^2$.

Assign values to x and determine the corresponding values of y.

Let $x = 0$.
$y = 0$.

Let $x = \pm 1$.
$y = 1$.

Let $x = \pm 2$.
$y = 4$.

Let $x = \pm 3$.
$y = 9$.

Let $x = \pm 4$.
$y = 16$.

x	0	±1	±2	±3	±4
y	0	1	4	9	16

There is no limit to the number of values that can be used. As x increases, y increases very rapidly.

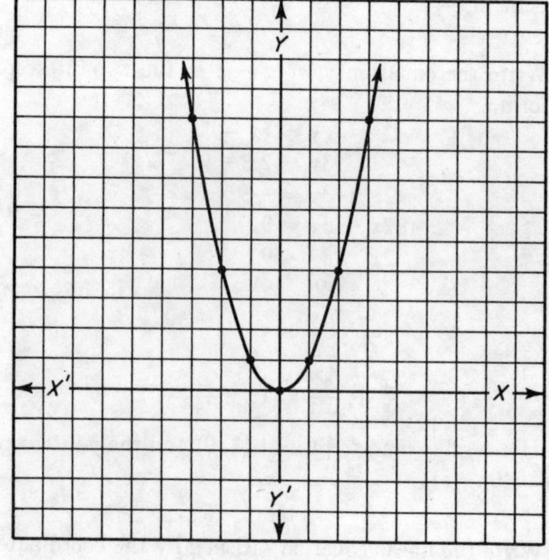

FIG. 27 (Prob. 2)

This is known as a **parabola** and will be the type of graph for all equations in which y is to the first power and x is to the second power.

3. $x = y^2$.

x	0	1	4	9	16
y	0	± 1	± 2	± 3	± 4

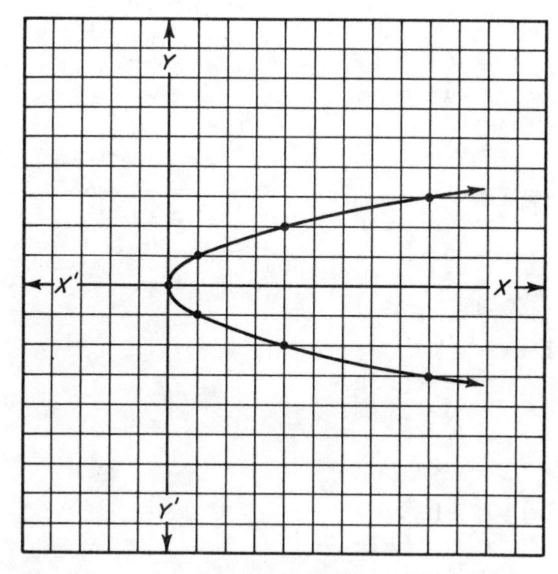

FIG. 28 (Prob. 3)

This equation is also a parabola, but it opens to the right rather than upward. In the graphs of $-y = x^2$ and $y = -x^2$, the parabolas will be similar to that in Problem 2 except that they will open downward. The graphs of $-x = y^2$ and $x = -y^2$ will differ from the parabola in Problem 3 only in that they will open to the left rather than to the right.

4. $y = x^2 + 2$.

Let $x = 0$.
$y = 2$.

Let $x = \pm 1$.
$y = 1 + 2$
$= 3$.

Let $x = \pm 2$.
$y = 4 + 2$
$= 6$.

Let $x = \pm 3$.
$y = 9 + 2$
$= 11$.

Let $x = \pm 4$.
$y = 16 + 2$
$= 18$.

x	0	± 1	± 2	± 3	± 4
y	2	3	6	11	18

Observe that Problem 4 is the same as Problem 2 except for the 2 that is added to x^2. This results in a parabola similar to the one in Problem 2 except that the vertex is two units above the origin. In the graph of $y = x^2 - 2$, the vertex is on the Y-axis and two units below the origin.

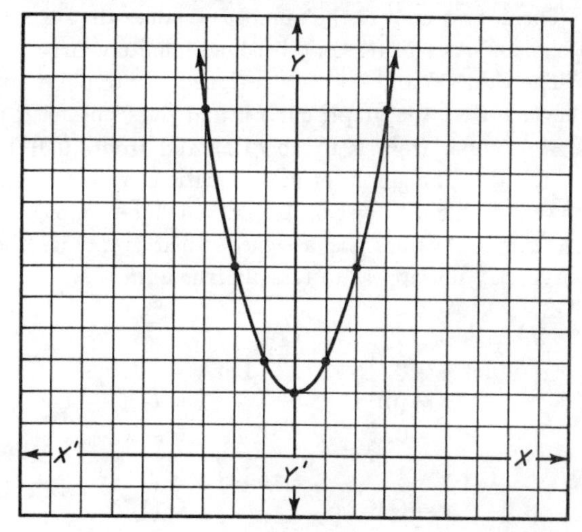

FIG. 29 (Prob. 4)

5. $y = x^3$.

Let $x = 0$.
$y = (0)^3$
$= 0$.

Let $x = 1$.
$y = (1)^3$
$= 1$.

Let $x = -\frac{1}{2}$.
$y = (-\frac{1}{2})^3$
$= -\frac{1}{8}$.

Let $x = \frac{1}{2}$.
$y = (\frac{1}{2})^3$
$= \frac{1}{8}$.

Let $x = 2$.
$y = (2)^3$
$= 8$.

Let $x = -2$.
$y = -8$.

Let $x = -1$.
$y = -1$.

x	0	$\frac{1}{2}$	1	$-\frac{1}{2}$	2	-2	-1
y	0	$\frac{1}{8}$	1	$-\frac{1}{8}$	8	-8	-1

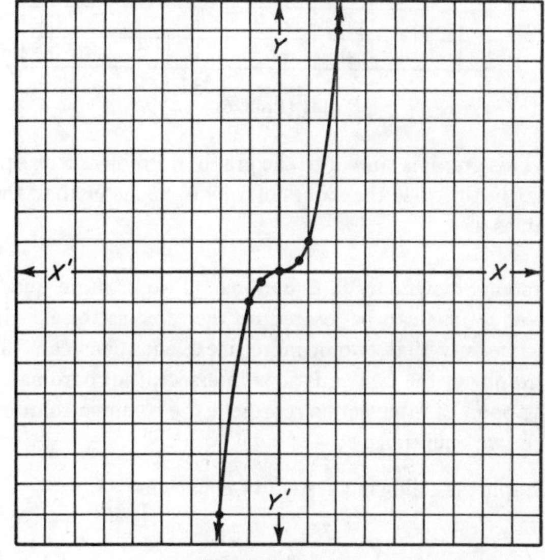

FIG. 30 (Prob. 5)

There is no limit to the number of values that can be assigned. As x increases, y becomes infinitely large.

The values $\frac{1}{2}$ and $-\frac{1}{2}$ were assigned to x to point out the fact that the graph curves, and does not form a straight line, from (0,0) to (1,1) and from (0,0) to (−1,−1). The graphs of most equations of the second power, or higher powers, are curved lines. Unless a great many values are assigned when graphing, the curve may not appear to take its true form.

6. $x = y^3$.

Let $y = 0$.
$x = (0)^3$
$= 0$.

Let $y = \frac{1}{2}$.
$x = (\frac{1}{2})^3$
$= \frac{1}{8}$.

Let $y = -\frac{1}{2}$.
$x = (-\frac{1}{2})^3$
$= -\frac{1}{8}$.

Let $y = 1$.
$x = (1)^3$
$= 1$.

Let $y = -1$.
$x = (-1)^3$
$= -1$.

Let $y = 2$.
$x = (2)^3$
$= 8$.

Let $y = -2$
$x = (-2)^3$
$= -8$.

x	0	$\frac{1}{8}$	$-\frac{1}{8}$	1	−1	8	−8
y	0	$\frac{1}{2}$	$-\frac{1}{2}$	1	−1	2	−2

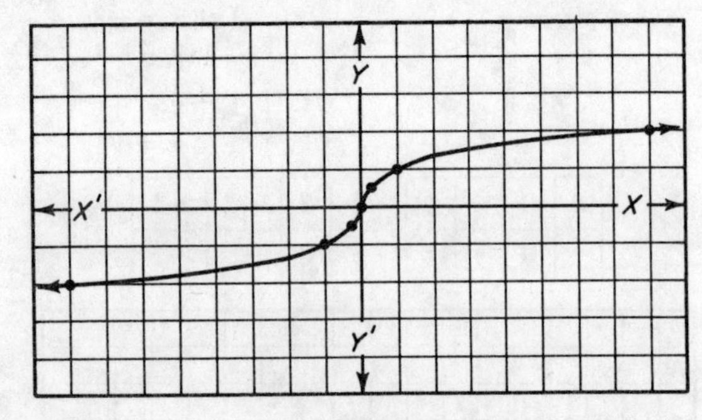

FIG. 31 (Prob. 6)

This graph is similar to the graph in Problem 5 except that in this case the values of x increase faster than the values of y.

Systems of Quadratic Equations. Two or more quadratic equations can be located on the same set of axes in the same way that two or more linear equations can. If the graphs of the two equations intersect, the coordinates of the point of intersection represent the common solution of the two equations.

7. Graph the following system of equations.

$$x^2 + y^2 = 36.$$
$$x - 2y = 6.$$

$x^2 = 36 - y^2$.

Let $y = 0$.
$x^2 = 36 - 0$
$= 36$.
$x = \pm 6$.

Let $y = \pm 6$.
$x^2 = 36 - (\pm 6)^2$
$= 36 - 36$
$= 0$.
$x = 0$.

Let $y = \pm 3$.
$x^2 = 36 - (\pm 3)^2$
$= 36 - 9$
$= 27$.
$x = \pm 5.2$.

Let $y = \pm 5$.
$x^2 = 36 - (\pm 5)^2$
$= 36 - 25$
$= 11$.
$x = \pm 3.3$.

Let $y = \pm 1$.
$x^2 = 36 - (\pm 1)^2$
$= 36 - 1$
$= 35$.
$x = \pm 5.9$.

$x = 2y + 6$.

Let $y = 0$.
$x = 2(0) + 6$
$= 6$.

Let $y = -3$.
$x = (2)(-3) + 6$
$= 6 - 6$
$= 0$.

Let $y = -1$.
$x = 2(-1) + 6$
$= -2 + 6$
$= 4$.

x	±6	0	±5.2	±3.3	±5.9
y	0	±6	±3	±5	±1

x	6	0	4
y	0	−3	−1

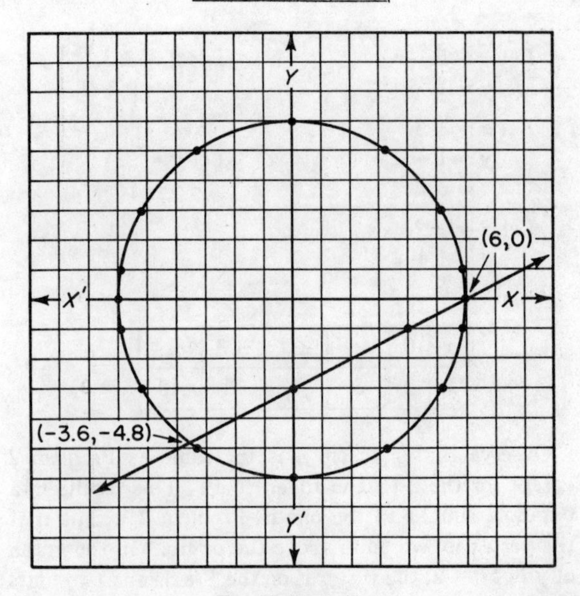

FIG. 32 (Prob. 7)

8. Graph the following system of equations.

$$x^2 + 4y^2 = 36$$
$$x^2 + y^2 = 25.$$

$x^2 = 36 - 4y^2$

Let $y = 0$.
$x^2 = 36 - 4(0)^2$
$= 36 - 0$
$= 36.$
$x = \pm 6.$

Let $y = \pm 3$.
$x^2 = 36 - 4(\pm 3)^2$
$= 36 - 4(9)$
$= 36 - 36.$
$x = 0.$

Let $y = \pm 2$.
$x^2 = 36 - 4(\pm 2)^2$
$= 36 - 4(4)$
$= 36 - 16$
$= 20.$
$x = \pm 4.5.$

Let $y = \pm 1$.
$x^2 = 36 - 4(\pm 1)^2$
$= 36 - 4$
$= 32.$
$x = \pm 5.7.$

$x^2 = 25 - y^2.$

Let $y = 0$.
$x^2 = 25 - 0$
$= 25.$
$x = \pm 5.$

Let $x = 0$.
$0 = 25 - y^2.$
$y^2 = 25.$
$y = \pm 5.$

Let $y = \pm 3$.
$x^2 = 25 - (\pm 3)^2$
$= 25 - 9$
$= 16.$
$x = \pm 4.$

Let $y = \pm 4$.
$x^2 = 25 - (\pm 4)^2$
$= 25 - 16$
$= 9.$
$x = \pm 3.$

x	± 6	0	± 4.5	± 5.7
y	0	± 3	± 2	± 1

x	± 5	0	± 4	± 3
y	0	± 5	± 3	± 4

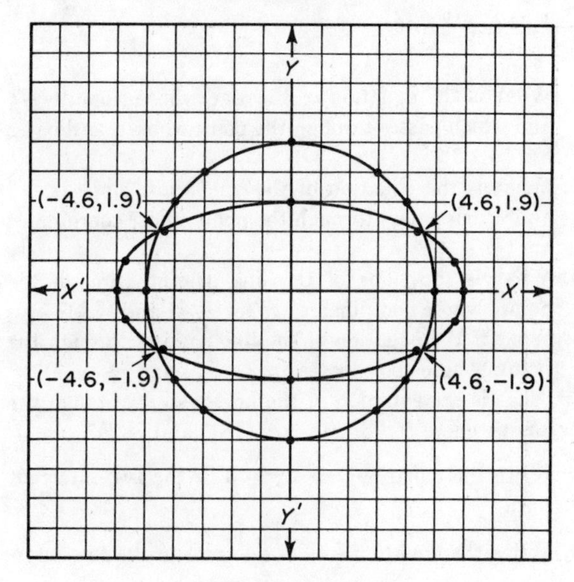

FIG. 33 (Prob. 8)

9. Graph the following system of equations.

$$y^2 = 3x.$$
$$x^2 = 3y.$$

$y^2 = 3x.$

Let $x = 0$.
$y^2 = 3(0)$
$= 0.$
$y = 0.$

Let $x = 1$.
$y^2 = 3(1)$
$= 3.$
$y = \pm 1.7.$

Let $x = 2$.
$y^2 = 3(2)$
$= 6.$
$y = \pm 2.4.$

Let $x = 3$.
$y^2 = 3(3)$
$= 9.$
$y = \pm 3.$

Let $x = 4$.
$y^2 = 3(4)$
$= 12.$
$y = \pm 3.5.$

Let $x = 5$.
$y^2 = 3(5)$
$= 15.$
$y = \pm 3.9.$

$x^2 = 3y.$

Let $y = 0$.
$x^2 = 3(0)$
$= 0.$
$x = 0.$

Let $y = 1$.
$x^2 = 3(1)$
$= 3.$
$x = \pm 1.7.$

Let $y = 2$.
$x^2 = 3(2)$
$= 6.$
$x = \pm 2.4.$

Let $y = 3$.
$x^2 = 3(3)$
$= 9.$
$x = \pm 3.$

Let $y = 4$.
$x^2 = 3(4)$
$= 12.$
$x = \pm 3.5.$

Let $y = 5$.
$x^2 = 3(5)$
$= 15.$
$x = \pm 3.9.$

x	0	1	2	3	4	5
y	0	± 1.7	± 2.4	± 3	± 3.5	± 3.9

x	0	± 1.7	± 2.4	± 3	± 3.5	± 3.9
y	0	1	2	3	4	5

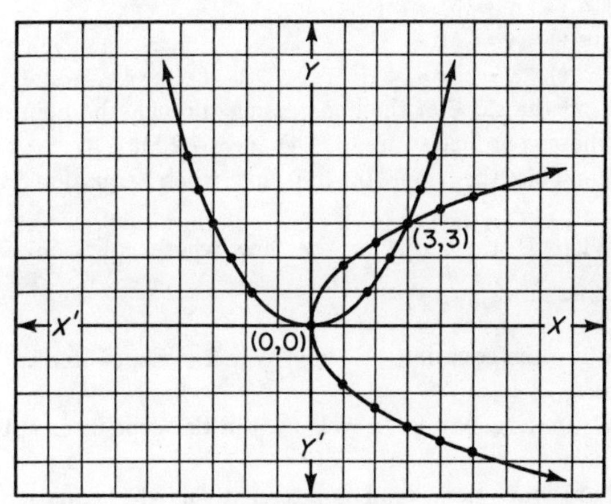

FIG. 34 (Prob. 9)

10. Graph the following system of equations.

$$x^2 - y^2 = 6.$$
$$x^2 + y^2 = 16.$$

$x^2 = y^2 + 6.$

Let $y = 0.$
$x^2 = (0)^2 + 6$
$= 6.$
$x = \pm 2.4.$

Let $y = \pm 1.$
$x^2 = (\pm 1)^2 + 6$
$= 1 + 6$
$= 7.$
$x = \pm 2.6.$

Let $y = \pm 2.$
$x^2 = (\pm 2)^2 + 6$
$= 4 + 6$
$= 10.$
$x = \pm 3.2.$

Let $y = \pm 3.$
$x^2 = (\pm 3)^2 + 6$
$= 9 + 6$
$= 15.$
$x = \pm 3.9.$

Let $y = \pm 4.$
$x^2 = (\pm 4)^2 + 6$
$= 16 + 6$
$= 22.$
$x = \pm 4.7.$

Let $y = \pm 5.$
$x^2 = (\pm 5)^2 + 6$
$= 25 + 6$
$= 31.$
$x = \pm 5.6.$

$x^2 = 16 - y^2.$

Let $y = 0.$
$x^2 = 16 - (0)^2$
$= 16.$
$x = \pm 4.$

Let $x = 0.$
$0 = 16 - y^2.$
$y^2 = 16.$
$y = \pm 4.$

Let $y = \pm 1.$
$x^2 = 16 - 1$
$= 15.$
$x = \pm 3.9.$

Let $y = \pm 2.$
$x^2 = 16 - (\pm 2)^2$
$= 16 - 4$
$= 12.$
$x = \pm 3.5.$

x	± 2.4	± 2.6	± 3.2	± 3.9	± 4.7	± 5.6
y	0	± 1	± 2	± 3	± 4	± 5

x	± 4	0	± 3.9	± 3.5
y	0	± 4	± 1	± 2

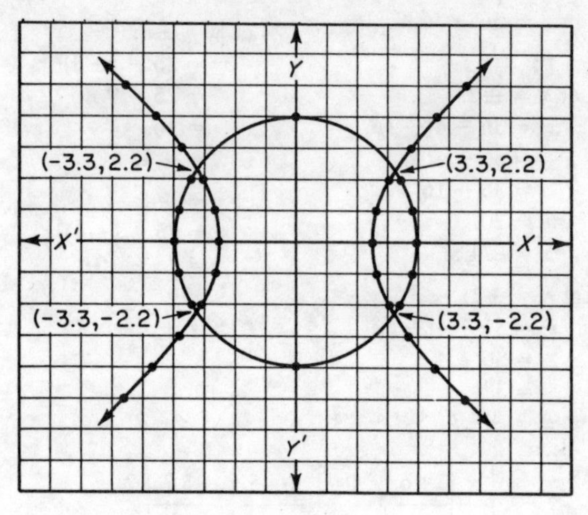

FIG. 35 (Prob. 10)

SUPPLEMENTARY PROBLEMS

In Problems 1 – 5, let $U = \{-4, -3, -2, -1, 0, 1, 2, 3, 4, 5\}$. Find the relation, the domain, and the range for each.

1. $\{(x,y) \mid y = 3x + 2\}$.

2. $\{(x,y) \mid y - x = 4\}$.

3. $\{(x,y) \mid y = 8 - x^2\}$.

4. $\{(x,y) \mid y = \frac{1}{2}x + 3\}$.

5. $\{(x,y) \mid y - 2x = 5\}$.

6. Find the slope of the line passing through the points whose coordinates are $(3,-4)$ and $(-2,5)$.

7. What is the slope of the line whose equation is $x = 3y - 5$?

8. What is the slope of the line whose equation is $\frac{3x}{4} - \frac{5y}{6} = 3$?

9. Write the equation $4 - x = 3y$ in the slope-intercept form.

10. Write the equation $\frac{3}{4}x = 1 - 5y$ in the slope-intercept form.

11. Find the slope and y-intercept of the equation $x = \frac{2}{3}y + 3$.

12. Find the slope and y-intercept of the equation $\frac{x-1}{3} + \frac{y+2}{4} = 1$.

13. What is the equation of the line whose slope is $-\frac{1}{3}$ and which passes through the point whose coordinates are $(-1,3)$?

14. What is the equation of the line whose slope is -5 and which passes through the point whose coordinates are $(\frac{2}{3}, -\frac{1}{2})$?

15. What is the slope of the line passing through the points whose coordinates are $(3, -\frac{1}{2})$ and $(\frac{2}{3},\frac{3}{2})$?

16. What is the equation of the line passing through the points whose coordinates are $(-5,3)$ and $(\frac{1}{2}, -2)$?

17. What is the equation of the line passing through the points whose coordinates are $(\frac{2}{3}, \frac{1}{3})$ and $(\frac{5}{3} \frac{4}{3})$?

18. Write the equation $7x - y = 3$ in the two-intercept form.

19. Write the equation $\frac{x}{3} - \frac{y+1}{2} = 4$ in the two-intercept form.

20. What are the x- and y-intercepts of the equation $\frac{y-1}{2} = \frac{x+1}{4} + \frac{1}{3}$?

21. Graph the equation $x + 3y = 7$.

22. Graph the system of equations.

$$x + 2y = 6.$$
$$2x - y = 8.$$

23. Graph the equation $x + 2 = y^2$.

24. Graph the system of equations.

$$x + 5 = y^2.$$
$$y = x^2 - 4.$$

25. Graph the system of equations.

$$y^2 - x^2 = 2.$$
$$3x = y^2.$$

Chapter 12

Inequalities

SOLVING INEQUALITIES

Of two real numbers a and b, either $a = b$ or $a \neq b$. If we wish to give further information about the inequality, we may use the symbols $>$ and $<$. If a is greater than b, we denote this by the symbolism $a > b$ (read "a is greater than b"), or $b < a$ (read "b is less than a").

If the symbols for two inequalities point in the same direction they are said to be unequal in the same **order** or the same **sense.**

As with equalities (or equations), we speak of that part to the left of $>$ or $<$ as the **left member of the inequality** while that part to the right of the symbol is called the **right member of the inequality.**

The following properties hold true for all inequalities.

1. If a, b, and c are any real numbers and $a > b$ and $b > c$, then $a > c$.
2. If a, b, and c are any real numbers and $a > b$, then $a + c > b + c$. In other words if any real number, either positive or negative, is added to both members of an inequality it remains an inequality in the same sense. Since subtraction is the same as adding the negative of a number, this means that adding or subtracting a number to both members of an inequality maintains an inequality in the same sense.
3. If a and b are any real numbers and c is a positive real number and $a > b$, then $ac > bc$ and $\frac{a}{c} > \frac{b}{c}$. Thus, if both members of an inequality are multiplied or divided by a positive number, the products or the quotients are unequal in the same sense.
4. If a and b are any real numbers and c is a negative real number and $a > b$, then $ac < bc$ and $\frac{a}{c} < \frac{b}{c}$. Hence, if both members of an inequality are multiplied or divided by a negative real number, the sense of the inequality is reversed.
5. If a, b, c, and d are any real numbers and $a > b$ and $c > d$, then $(a + c) > (b + d)$.
6. If a is any real number and $a < 0$, then $(-a) > 0$. Likewise, if $a > 0$, then $(-a) < 0$.

Solve for x: $x + 3 > 4$.

$x + 3 > 4.$

Add -3 to both members of the inequality (Property 2).

$x > 1.$

Hence, $x + 3 > 4$ will be true if x is any real number greater than 1.

Solve for x: $3x + 1 < 2$.

$3x + 1 < 2.$

Add -1 to both members of the inequality (Property 2).

$3x < 1.$

Divide both members of the inequality by 3 (Property 3).

$x < \frac{1}{3}.$

Hence, $3x + 1 < 2$ will be true if x is any real number less than $\frac{1}{3}$.

Solve for x: $-\frac{1}{3}x + 2 < 5$.

$-\frac{1}{3}x + 2 < 5.$

Add -2 to both members of the inequality (Property 2).

$-\frac{1}{3}x < 3.$

Multiply both members of the inequality by 3 (Property 3).

$-x < 9.$

Multiply both members of the inequality by -1 (Property 4).

$x > -9.$

Hence, $-\frac{1}{3}x + 2 < 5$ will be true if x is any real number greater than -9.

Solve for x: $-\frac{2}{3}x - 2 < 3$.

$-\frac{2}{3}x - 2 < 3.$

Multiply both members of the inequality by $-\frac{3}{2}$ (Property 4).

$x + 3 > -\frac{9}{2}.$

Subtract 3 from both members of the inequality (Property 2).

$x > -\frac{15}{2}.$

Hence, $-\frac{2}{3}x - 2 < 3$ will be true if x is any real number greater than $-\frac{15}{2}$.

Solve the following inequalities.

1. $x - 5 < -2$.

$$x - 5 < -2.$$

Add 5 to both members of the inequality.

$$x < 3.$$

2. $2x + 1 < -2$.

$$2x + 1 < -2.$$

Subtract 1 from both members of the inequality.

$$2x < -3.$$

Divide both members of the inequality by 2.

$$x < -\tfrac{3}{2}.$$

3. $\frac{1}{2}x + 2 > -5$.

$$\tfrac{1}{2}x + 2 > -5.$$

Subtract 2 from both members of the inequality.

$$\tfrac{1}{2}x > -7.$$

Multiply both members of the inequality by 2.

$$x > -14.$$

4. $3 - 2x > 4$.

$$3 - 2x > 4.$$

Subtract 3 from both members of the inequality.

$$-2x > 1.$$

Divide both members of the inequality by -2.

$$x < -\tfrac{1}{2}.$$

5. $2 - \frac{2}{3}x < -6$.

$$2 - \tfrac{2}{3}x < -6.$$

Subtract 2 from both members of the inequality.

$$-\tfrac{2}{3}x < -8.$$

Multiply both members of the inequality by $-\frac{3}{2}$.

$$x > 12.$$

6. $3x - 1 > x + 3$.

$$3x - 1 > x + 3.$$

Add 1 to both members of the inequality.

$$3x > x + 4.$$

Subtract x from both members of the inequality.

$$2x > 4.$$

Divide both members of the inequality by 2.

$$x > 2.$$

7. $2x + 1 > \frac{1}{3}x + 1$.

$$2x + 1 > \tfrac{1}{3}x + 1.$$

Subtract 1 from both members of the inequality.

$$2x > \tfrac{1}{3}x.$$

Subtract $\frac{1}{3}x$ from each member of the inequality.

$$\tfrac{5}{3}x > 0.$$

Multiply both members of the inequality by $\frac{3}{5}$.

$$x > 0.$$

8. $2.1x - 1 < 1.3x + 2$.

$$2.1x - 1 < 1.3x + 2.$$

Add 1 to both members of the inequality.

$$2.1x < 1.3x + 3.$$

Subtract $1.3x$ from both members of the inequality.

$$0.8x < 3.$$

Multiply both members of the inequality by 10.

$$8x < 30.$$

Divide both members of the inequality by 8.

$$x < \tfrac{15}{4}.$$

9. $\frac{2}{3} - \frac{1}{2}x < \frac{1}{4}x + 2$.

$$\tfrac{2}{3} - \tfrac{1}{2}x < \tfrac{1}{4}x + 2.$$

Subtract $\frac{2}{3}$ from both members of the inequality.

$$-\tfrac{1}{2}x < \tfrac{1}{4}x + \tfrac{4}{3}.$$

Subtract $\frac{1}{4}x$ from both members of the inequality.

$$-\tfrac{3}{4}x < \tfrac{4}{3}.$$

Multiply both members of the inequality by $-\frac{4}{3}$.

$$x > -\tfrac{16}{9}.$$

10. $3 - \frac{2}{3}x > 1.1x - 2$.

This will be more easily solved if $1.1x$ is written as $\frac{11}{10}x$.

$$3 - \tfrac{2}{3}x > \tfrac{11}{10}x - 2.$$

Subtract $\frac{11}{10}x$ from both members of the inequality.

$$3 - \tfrac{53}{30}x > -2.$$

Subtract 3 from both members of the inequality.

$$-\tfrac{53}{30}x > -5.$$

Multiply both members of the inequality by $-\frac{30}{53}$.

$$x < \tfrac{150}{53}.$$

Inequalities in Set-Builder Form. Like equalities, inequalities may be written in set-builder form. In such cases, the inequality is the condition imposed upon the variable. Thus, if U is the set of natural numbers, then $\{x \mid x - 3 < 4\}$ means that the value of the variable x may be any natural number such that $x - 3 < 4$. The set of such natural numbers is the solution set of the inequality. The following example illustrates the steps involved in arriving at the solution set.

> Find the solution set of $\{x \mid x - 3 < 4\}$ if $U = \{1,2,3, \ldots\}$.
>
> $$\begin{aligned}\{x \mid x - 3 < 4\} &= \{x \mid x < 7\} \\ &= \{1,2,3,4,5,6\}.\end{aligned}$$

11. Find the solution set of $\{x \mid 3x - 1 > 11\}$ if $U = \{1,3,5, \ldots,15\}$.

$$\begin{aligned}\{x \mid 3x - 1 > 11\} &= \{x \mid 3x > 12\} \\ &= \{x \mid x > 4\} \\ &= \{5,7,9,11,13,15\}.\end{aligned}$$

12. Find the solution set of $\{x \mid 2x - 3 > x + 1\}$ if $U = \{1,2,3, \ldots,12\}$.

$$\begin{aligned}\{x \mid 2x - 3 > x + 1\} &= \{x \mid x - 3 > 1\} \\ &= \{x \mid x > 4\} \\ &= \{5,6,7,8,9,10,11,12\}.\end{aligned}$$

13. Find the solution set of $\{x \mid x + 2 > 3x - 5\}$ if $U = \{1,2,3, \ldots\}$.

$$\begin{aligned}\{x \mid x + 2 > 3x - 5\} &= \{x \mid -2x + 2 > -5\} \\ &= \{x \mid -2x > -7\} \\ &= \{x \mid x < \tfrac{7}{2}\} \\ &= \{1,2,3\}.\end{aligned}$$

14. Find the solution set of $\{\frac{2}{3}x - 1 \geq -\frac{1}{3}x + 2\}$ if U is the set of one-digit natural numbers. (The symbol $\geq$ is read "is greater than or equal to". Likewise, the symbol $\leq$ is read "is less than or equal to".)

$$\{x \mid \tfrac{2}{3}x - 1 \geq -\tfrac{1}{3}x + 2\} = \{x \mid x - 1 \geq 2\}$$
$$= \{x \mid x \geq 3\}$$
$$= \{3,4,5,6,7,8,9\}.$$

15. Find the solution set of $\{x \mid 3 < 2x + 1 < 11\}$ if U is the set of one digit natural numbers.

$$\{x \mid 3 < 2x + 1 < 11\} = \{x \mid 2 < 2x < 10\}$$

In the above step, 1 was subtracted from each member of the inequality. For the next step in the solution we divide each of the three members of the inequality by 2 and obtain

$$= \{x \mid 1 < x < 5\}$$
$$= \{2,3,4\}.$$

Inequalities Involving Variable Denominators. Mathematical expressions employing the symbols $<$, $>$, $\leq$, and $\geq$ can sometimes be more easily understood if we use parentheses and brackets to indicate the interval within which or outside of which values of the variable are located. For example, if $3 \geq x \geq 0$, the **open interval** is used to express the permissible values of x. The interval $(0, 3)$ is called an **open** interval because neither 0 nor 3 can be the value of x, but all real numbers greater than 0 and less than 3 are permissible values of x. If we were using the expression $3 \geq x \geq 0$, both 0 and 3 are included in the set of values which x may assume. We denote this by use of brackets, and the interval within which values of x are located would be written $[0, 3]$. This notation symbolizes a **closed interval** of values that x may assume.

Sometimes only part of an expression involves the closed interval. Consider the expression $0 < x \leq 3$. Such an expression is referred to as **half-open** on the left, and is expressed by the symbolism $(0, 3]$. This, of course, means that the set of values of x does not include 0, but that x can assume all values in our real number system greater than zero and up to and including 3. In a similar manner, we would call the expression $0 \leq x < 3$ an interval half open on the right, and would denote it by the symbolism $[0, 3)$.

If we are working with a variable, say x, we do not know whether x is positive or negative unless this information is given. This poses a problem which we have not encountered before; namely, how to use Properties 3 and 4 when the value by which we wish to multiply or divide is a variable. To arrive at solutions of problems in which this situation occurs, it is necessary to consider both possibilities. First, we must solve the inequality assuming that $x > 0$; and, secondly, we must solve the inequality assuming that $x < 0$. If x is a denominator, it cannot have the value 0; and x cannot assume any value which would make a denominator equal to zero.

16. Solve for x $(x \neq 0)$: $\dfrac{2}{x} > 3$.

Case 1. $x > 0$. (x is positive.)

Multiply both members of the inequality by x. Thus

$$\frac{2}{x} > 3.$$
$$2 > 3x.$$
$$3x < 2.$$
$$x < \tfrac{2}{3}.$$

In Case 1 we assumed that $x > 0$, and found that when $x > 0$, x must be less than $\frac{2}{3}$. In order for these two statements to be consistent, the permissible values of x will be those included in the open interval $(0,\frac{2}{3})$.

Case 2. $x < 0$. (x is negative.)

Multiply both members of the original inequality by x. Since $x < 0$, reverse the inequality found for the solution in Case 1. Thus, we have $x > \frac{2}{3}$. This is not consistent with our basic assumption in Case 2 (that x is negative). Hence, x cannot be negative in this inequality.

In problems of this sort we combine the answers found in Case 1 and Case 2 to arrive at our final solution. In this problem, Case 2 results in incompatable solutions and, therefore, it is impossible for x to be negative.

Our solution to the inequality, then, is that found in Case 1. Consequently, x may assume all real-number values greater than zero and less than $\frac{2}{3}$.

17. Solve for x $(x \neq 0)$: $\dfrac{2}{x} < \dfrac{5}{6}$.

$$\frac{2}{x} < \frac{5}{6}.$$

Case 1. $x > 0$.

$$12 < 5x.$$
$$5x > 12.$$
$$x > \tfrac{12}{5}.$$

If x is positive and $x > \frac{12}{5}$ the interval in which x is found is $(\frac{12}{5}, +\infty)$.

Case 2. $x < 0$.

If x is negative the solution of Case 1 is reversed and

$$x < \tfrac{12}{5}.$$

Now since x is negative, it cannot have values in $[0,\frac{12}{5})$. Therefore, the interval which satisfies both $x < 0$ and $x < \frac{12}{5}$ is $(-\infty,0)$.

Combining Case 1 and Case 2, x is any value greater than $\frac{12}{5}$ and any value less than zero. A more concise way of stating this is that the solution consists of all x *not* in the interval $[0,\frac{12}{5}]$.

18. Solve for x $(x \neq 4)$: $\frac{x-2}{x-4} > 3$.

$$\frac{x-2}{x-4} > 3.$$

Case 1. If $x - 4 > 0$, then $x > 4$.

Multiply both members of the inequality by $x - 4$ and we have

$$x - 2 > 3x - 12.$$
$$-2x - 2 > -12.$$
$$-2x > -10.$$
$$x < 5.$$

Thus, if $x - 4$ is positive x must be greater than 4 and less than 5. These possible values for x are included in the interval (4,5).

Case 2. If $x - 4 < 0$, then $x < 4$.

This reverses the solution found in Case 1 and we have

$$x > 5.$$

The two statements, $x < 4$ and $x > 5$, are inconsistent. Thus, $x - 4$ cannot be negative.

Combining the results in Case 1 and Case 2, we find that the only values x can assume are those values found in the interval (4,5).

19. Solve for x $(x \neq 3)$: $\frac{x+2}{x-3} < 3$.

$$\frac{x+2}{x-3} < 3.$$

Case 1. If $x - 3 > 0$, then $x > 3$.

$$x + 2 < 3x - 9.$$
$$-2x < -11.$$
$$x > \tfrac{11}{2}.$$

Therefore, when $x - 3$ is positive the possible values of x will fall in the interval $(\frac{11}{2}, +\infty)$.

Case 2. If $x - 3 < 0$, then $x < 3$.

If the denominator is negative then the inequality will be the reverse of that found in Case 1 and

$$x < \tfrac{11}{2}.$$

Thus, if the denominator is negative, $x < 3$ and $x < \frac{11}{2}$. All values of x in the interval $(-\infty, 3)$ will satisfy both of these statements.

Combining the results obtained in Case 1 and Case 2, the solution is: all x *not* in the interval $[3, \frac{11}{2}]$.

20. Solve for x $(x \neq 1)$: $\frac{3-2x}{x-1} < \frac{5}{2}$.

$$\frac{3-2x}{x-1} < \frac{5}{2}.$$

Case 1. If $x - 1 > 0$, then $x > 1$.

$$6 - 4x < 5x - 5.$$
$$-9x < -11.$$
$$x > \tfrac{11}{9}.$$

Then, if $x - 1$ is positive and $x > \frac{11}{9}$, x will fall in the interval $(\frac{11}{9}, +\infty)$.

Case 2. If $x - 1 < 0$, then $x < 1$.

If $x - 1$ is negative, the inequality will be the reverse of that found in Case 1 and

$$x < \tfrac{11}{9}.$$

Thus, if $x - 1$ is negative, then $x < 1$ and $x < \frac{11}{9}$. Therefore, x will fall in the interval $(-\infty, 1)$.

Combining the results found in Case 1 and Case 2, we find that the inequality is true when x assumes any value from our real number system *not* in the interval $[1, \frac{11}{9}]$.

Inequalities Involving Absolute Value. If x is any value that satisfies the condition $|x| < 2$, then x could be any number greater than -2 and less than 2. This could also be stated: x may be any value in the open interval $(-2,2)$, which we would write as

$$-2 < x < 2.$$

Likewise, $|x + 1| < 7$ means that $x + 1$ is greater than -7 and less than 7 or $x + 1$ lies in the open interval $(-7,7)$. This, too, can be stated as an inequality in the form

$$-7 < x + 1 < 7.$$

By subtracting 1 from each of the three members of the inequality, we would get

$$-8 < x < 6$$

and x could be any value in the open interval $(-8,6)$.

The inequality $|x| \leq 9$ is satisfied by values of x within the closed interval $[-9,9]$.

The general principles associated with the previous inequalities apply to those involving absolute values. The following illustrates the procedure for solving these inequalities.

> Solve for x: $|x - 1| < 3$.
> This problem may be written as
> $$-3 < x - 1 < 3.$$
> Add 1 to each of the three members of the inequality.
> $$-2 < x < 4.$$
> The solution, then, consists of all values of x in the interval $(-2,4)$.

21. Solve for x: $|2x + 1| < 11$.

$$-11 < 2x + 1 < 11.$$

Subtract 1 from all three members of the inequality to get

$$-12 < 2x < 10.$$

Divide all three members of the inequality by 2.

$$-6 < x < 5.$$

The solution, then, consists of all values of x in the open interval $(-6,5)$.

22. Solve for x: $|3 - x| < 7$.

$$-7 < 3 - x < 7.$$

Subtract 3 from each of the three members of the inequality and we have

$$-10 < -x < 4.$$

Multiply all three members of the inequality by -1 in order to obtain a positive x and we get

$$10 > x > -4.$$

The solution, then, consists of all x in the open interval $(-4,10)$.

23. Solve for x: $\left|\frac{x+3}{x-1}\right| < 5$.

This, as in the previous problems, may be written

$$-5 < \frac{x+3}{x-1} < 5.$$

Here we do not know whether $x - 1$ is positive or negative so both cases must be considered.

Case 1. If $x - 1 > 0$, then $x > 1$. This implies that the denominator is positive; therefore, the sense or direction of the inequality is maintained when we multiply or divide all members of the inequality by the denominator. By multiplying each of the three members of the inequality by $x - 1$, we get

$$-5(x-1) < x + 3 < 5(x-1).$$
$$-5x + 5 < x + 3 < 5x - 5.$$

The left inequality (using only the first two members) then states

$$-5x + 5 < x + 3.$$

By subtracting $x + 5$ from both members of this inequality we have

$$-6x < -2.$$

Divide both members of the inequality by -6 to obtain

$$x > \tfrac{1}{3}.$$

Then we use the right inequality (the last two of the three members of the same inequality) and we have

$$x + 3 < 5x - 5.$$
$$-4x < -8.$$
$$x > 2.$$

From Case 1, then, we have three statements of inequality: from the fact that we assumed the denominator to be positive we have $x - 1 > 0$ or $x > 1$; and from the two solutions we have $x > \frac{1}{3}$ and $x > 2$. These three conditions all hold true if $x > 2$. Consequently, the solution for Case 1 is all x in the open interval $(2, +\infty)$.

Case 2. If $x - 1 < 0$, then $x < 1$. (The denominator is negative.) If the denominator is negative and we multiply all three members of the inequality by $x - 1$, the sense or direction of the inequality is reversed and we have

$$-5(x-1) > x + 3 > 5(x-1).$$
$$-5x + 5 > x + 3 > 5x - 5.$$

Using the left inequality we have

$$-5x + 5 > x + 3.$$

Subtract $x + 5$ from both members of the inequality and we have

$$-6x > -2.$$

Divide both members of the inequality by -6 and we get

$$x < \tfrac{1}{3}.$$

Using the right inequality we have

$$x + 3 > 5x - 5.$$

Subtract $5x + 3$ from both members of the inequality and we get

$$-4x > -8.$$

Divide both members of the inequality by -4 to obtain

$$x < 2.$$

From Case 2 we have three conditions $x < 1$, $x < 2$, and $x < \frac{1}{3}$. Consequently, Case 2 permits the use of all values of x less than $\frac{1}{3}$, and this would be expressed with the open interval $(-\infty, \frac{1}{3})$.

Combining the results of Case 1 and Case 2 for a final solution tells us that the original problem holds true for all x less than $\frac{1}{3}$ and greater than 2. This also may be stated as all x *not* in the closed interval $[\frac{1}{3},2]$.

24. Solve for x: $|3x - 1| \leq 5$.

$$|3x - 1| \leq 5.$$

We write this inequality as

$$-5 \leq 3x - 1 \leq 5.$$

Add 1 to each of the three members of the inequality and we have

$$-4 \leq 3x \leq 6.$$

Divide each member of the inequality by 3 and we have

$$-\tfrac{4}{3} \leq x \leq 2.$$

Thus, the solution consists of all x in the closed interval $[-\frac{4}{3}, 2]$.

25. Solve for x: $\left|\frac{4x+3}{1-x}\right| \leq 3$.

$$\left|\frac{4x+3}{1-x}\right| \leq 3.$$

Rewrite the inequality in the form

$$-3 \leq \frac{4x+3}{1-x} \leq 3.$$

Again we do not know whether the denominator is positive or negative so we must consider both cases.

Case 1. $1 - x > 0$.

$$1 - x > 0.$$
$$-x > -1.$$
$$x < 1.$$

Thus, if the denominator is positive, $x < 1$.

$$-3 \leq \frac{4x+3}{1-x} \leq 3.$$

Multiply each of the three members of the inequality by the denominator and we have

$$-3(1-x) \leq 4x + 3 \leq 3(1-x).$$
$$-3 + 3x \leq 4x + 3 \leq 3 - 3x.$$

Using the left inequality we have

$$-3 + 3x \leq 4x + 3.$$

Add $3 - 4x$ to both members of the inequality and we get

$$-x \leq 6.$$
$$x \geq -6.$$

Using the right inequality we have

$$4x + 3 \leq 3 - 3x.$$

Add $3x - 3$ to both members of the inequality to get

$$7x \leq 0.$$
$$x \leq 0.$$

Thus, in Case 1 we have the three inequalities: $x < 1$, $x \geq -6$, and $x \leq 0$. Hence the solution for Case 1 is all x in the interval $[-6,0]$.

Case 2. $1 - x < 0$, then

$$-x < -1.$$
$$x > 1.$$

Hence, if the denominator is negative x must be greater than 1. With the denominator negative the two inequalities found in Case 1 will be reversed and we have $x \leq -6$ and $x \geq 0$. These two values are incompatible and, consequently, the denominator cannot be negative.

Combining the results of Case 1 and Case 2 we find that the solution consists of all x in the interval $[-6,0]$.

GRAPHING INEQUALITIES

Simple inequalities (inequalities in one variable) may be graphed on the number line. In such cases, the variable involved may represent either no number (the solution set is ϕ), one number, or several numbers in our real number system.

Consider the following problem.

$$A = \{x \mid x - 1 > 2\},\ U = \{1,2,3,4,5\}.$$

The solution set associated with the problem gives us the values to be graphed on the number line. Solving the problem, we have

$$\{x \mid x - 1 > 2\} = \{x \mid x > 3\}$$
$$= \{4,5\}.$$

Thus, the only values from our universal set which satisfy the condition imposed on the variable x are 4 and 5. We indicate those values on the number line by heavy dots as shown in Fig. 36.

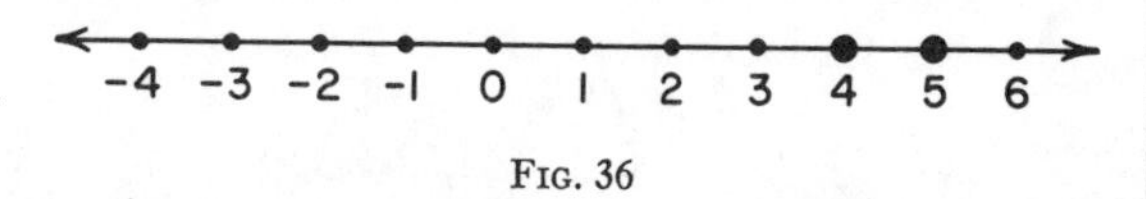

Fig. 36

When the graph consists of specific rational numbers between which are other real numbers, we denote those values by heavy dots as shown in Fig. 36. If all real numbers between two rational numbers satisfy the condition imposed on the variable, this is shown by a heavy line (see Fig. 41). If the two endpoints are included, they are shown by the use of the heavy dots (see Fig. 46). The graph then represents a closed interval. If one or both of those endpoints are not included in the solution set, we use an open circle rather than the dot (see Fig. 41). This means that all values up to but not including the endpoint are in the solution set; and the graph then represents a half-open or open interval. When the possible values of the solution set continue indefinitely in a positive direction, an arrow is drawn to the right end of the number line to so indicate (see Fig. 42). Likewise, if the solution set contains all values in the negative direction (to the left of a certain point), we draw an arrow to the left end of the number line to indicate this fact (see Fig. 47). In certain problems the graph may involve two such arrows (see Fig. 49).

1. Graph the set A: $A = \{x \mid x - 1 < 4\}$, U = the set of positive integers.

$$\{x \mid x - 1 < 4\} = \{x \mid x < 5\}$$
$$= \{1,2,3,4\}.$$

-4 -3 -2 -1 0 1 2 3 4 5 6

Fig. 37 (Prob. 1)

2. Graph the set X: $X = \{x \mid 3 + x > 10\}$, $U = \{2,4,6,\ldots,14\}$.

$$\{x \mid 3 + x > 10\} = \{x \mid x > 7\}$$
$$= \{8,10,12,14\}.$$

-4 -2 0 2 4 6 8 10 12 14 16

Fig. 38 (Prob. 2)

3. Graph the set C: $C = \{x \mid 2x - 3 < 5\}$, $U = \{1,2,3,\ldots,15\}$.

$$\{x \mid 2x - 3 < 5\} = \{x \mid 2x < 8\} = \{x \mid x < 4\} = \{1,2,3\}.$$

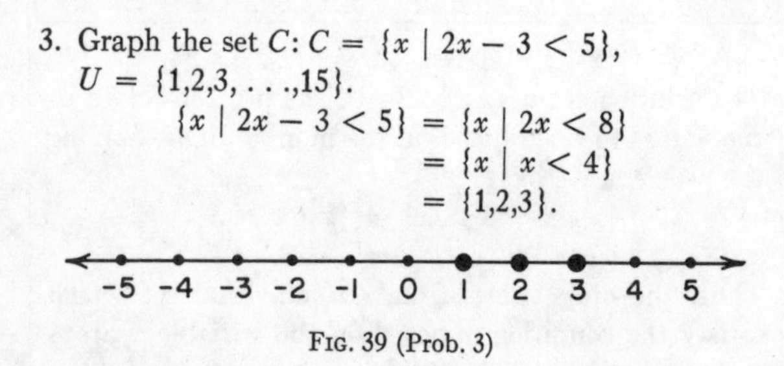

FIG. 39 (Prob. 3)

4. Graph the set B: $B = \{x \mid x + 3 \geq 7\}$, $U =$ the set of all one digit natural numbers.

$$\{x \mid x + 3 \geq 7\} = \{x \mid x \geq 4\} = \{4,5,6,7,8,9\}.$$

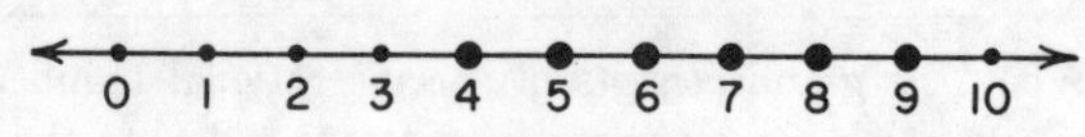

FIG. 40 (Prob. 4)

5. Graph the set X: $X = \{x \mid -3 < x < 2\}$, $U =$ the set of all real numbers.

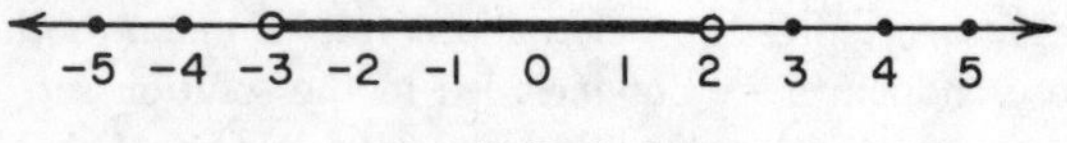

FIG. 41 (Prob. 5)

6. Graph the set A: $A = \{x \mid x \geq -5\}$, $U =$ the set of all real numbers.

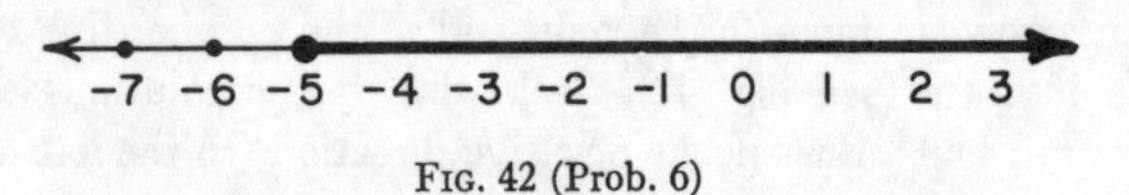

FIG. 42 (Prob. 6)

7. Graph the set C: $C = \{x \mid 2x - 1 \geq -3\}$, $U =$ the set of all real numbers.

$$\{x \mid 2x - 1 \geq -3\} = \{x \mid 2x \geq -2\} = \{x \mid x \geq -1\}.$$

FIG. 43 (Prob. 7)

The **intersection** of two sets is the set of all elements that are common to those two given sets. The symbol $\cap$ is used to indicate the intersection of two sets. Thus, if $A = \{x \mid x + 1 < 3\}$ and $B = \{x \mid x - 3 > -5\}$ then $\{x \mid x + 1 < 3\} \cap \{x \mid x - 3 > -5\}$ is a third set whose elements are found both in set A and in set B.

$$A = \{x \mid x + 1 < 3\} = \{x \mid x < 2\}.$$

Hence, A consists of all values in the interval $(-\infty, 2)$.

$$B = \{x \mid x - 3 > -5\} = \{x \mid x > -2\}.$$

Hence, B consists of all values in the interval $(-2, +\infty)$.

It can be seen from these solutions that the intersection of sets A and B will include all numbers greater than -2 and all numbers less than 2. The graph of this intersection is shown in Fig. 44.

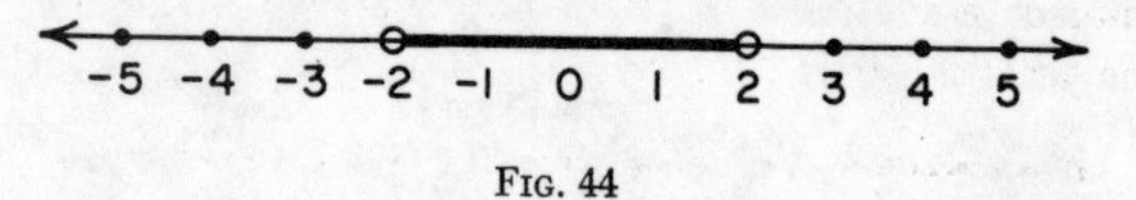

FIG. 44

8. Graph: $\{x \mid x > 1\} \cap \{x \mid x > 4\}$.

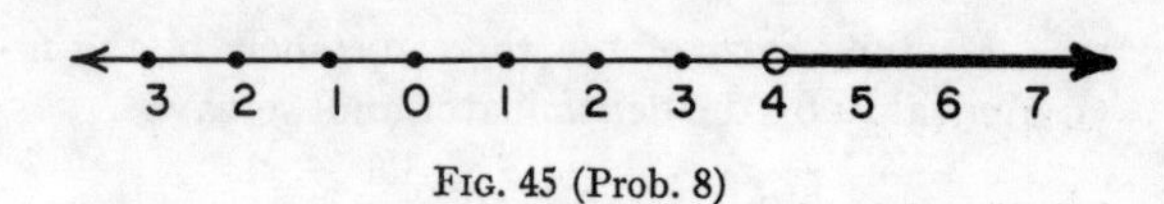

FIG. 45 (Prob. 8)

9. Graph: $\{x \mid x + 3 \geq -1\} \cap \{x \mid x - 1 \leq 2\}$.

$$\{x \mid x + 3 \geq -1\} = \{x \mid x \geq -4\}$$

and

$$\{x \mid x - 1 \leq 2\} = \{x \mid x \leq 3\}.$$

Thus, we have

$$\{x \mid x \geq 4\} \cap \{x \mid x \leq 3\}.$$

FIG. 46 (Prob. 9)

10. Graph: $\{x \mid 2x + 3 < 5\} \cap \{x \mid x + 1 < -2\}$.

$$\{x \mid 2x + 3 < 5\} = \{x \mid 2x < 2\} = \{x \mid x < 1\}$$

and

$$\{x \mid x + 1 < -2\} = \{x \mid x < -3\}$$

Thus, we have

$$\{x \mid x < 1\} \cap \{x \mid x < -3\}.$$

FIG. 47 (Prob. 10)

11. Graph: $\{x \mid 2x \geq 6\} \cap \{x \mid 2x > -4\}$

$$\{x \mid 2x \geq 6\} = \{x \mid x \geq 3\}$$

and

$$\{x \mid 2x > -4\} = \{x \mid x > -2\}.$$

Thus, we have

$$\{x \mid x \geq 3\} \cap \{x \mid x > -2\}.$$

FIG. 48 (Prob. 11)

The **union** of two sets is the set each of whose elements is a member of one or the other of the given sets. The symbol $\cup$ indicates the union of two sets. Thus if $A = \{x \mid x > 3\}$ and $B = \{x \mid x \leq 1\}$ then $\{x \mid x > 3\} \cup \{x \mid x \leq 1\}$ is the set that contains all of the elements in set A and all of the elements in set B. To graph the union of sets A and B, we place both graphs on the number line, as is shown in Fig. 49.

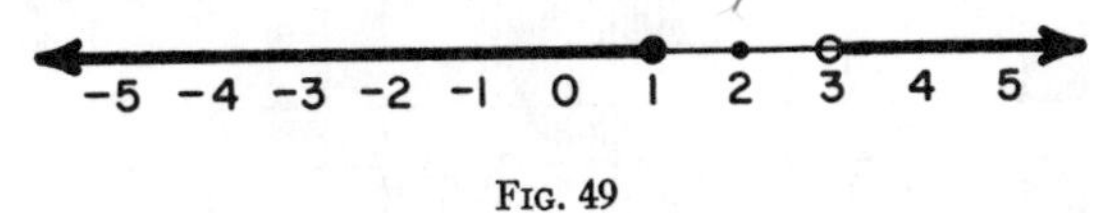

FIG. 49

12. Graph: $\{x \mid x > -2\} \cup \{x \mid x \leq -4\}$.

-7 -6 -5 -4 -3 -2 -1 0 1 2 3

FIG. 50 (Prob. 12)

13. Graph: $\{x \mid 2x - 1 \leq 3\} \cup \{x \mid 1 - x \geq 4\}$.

$$\{x \mid 2x - 1 \leq 3\} = \{x \mid 2x \leq 4\}$$
$$= \{x \mid x \leq 2\}$$

and

$$\{x \mid 1 - x \geq 4\} = \{x \mid -x \geq 3\}$$
$$= \{x \mid x \leq -3\}.$$

Thus, we have

$$\{x \mid x \leq 2\} \cup \{x \mid x \leq -3\}.$$

-5 -4 -3 -2 -1 0 1 2 3 4 5

FIG. 51 (Prob. 13)

14. Graph: $\{x \mid 0 < x \leq 5\} \cup \{x \mid -2 < x < 3\}$.

-4 -3 -2 -1 0 1 2 3 4 5 6

FIG. 52 (Prob. 14)

15. Graph: $\{x \mid -3 < x \leq 1\} \cup \{x \mid 3 < x < 5\}$.

-4 -3 -2 -1 0 1 2 3 4 5 6

FIG. 53 (Prob. 15)

Graphs of Relations Involving Inequalities. Relations expressed in terms of x and y (which were developed in Chapter 11) may be inequalities.

Consider the following problem.

Graph: $\{(x,y) \mid y > x + 1\}$, U is the set of all real numbers.

First we graph the expression as if it were an equality, in which case the graph would be a straight line. The line is drawn as a broken-line, however, to show that the graph does not include the line itself.

Three ordered pairs that satisfy the relation $y = x + 1$ are: $(-3,-2)$,$(0,1)$ and $(4,5)$. Locate these points in the coordinate plane as shown in Fig. 54. Connect these points with a broken-line, and place an arrow at each end of this line to show that it extends indefinitely in both directions.

Now begin to work with the inequality, $y > x + 1$. Select various values of x and determine whether the corresponding values of y fall above or below the broken line. As an example, if $x = 4$, then y is greater than 5. This means that the corresponding value of y is above the broken-line, a fact that will be verified for any value of x. Thus, the graph includes every point above the line $y = x + 1$. The graph of $y > x + 1$, then, is the shaded area in Fig. 54.

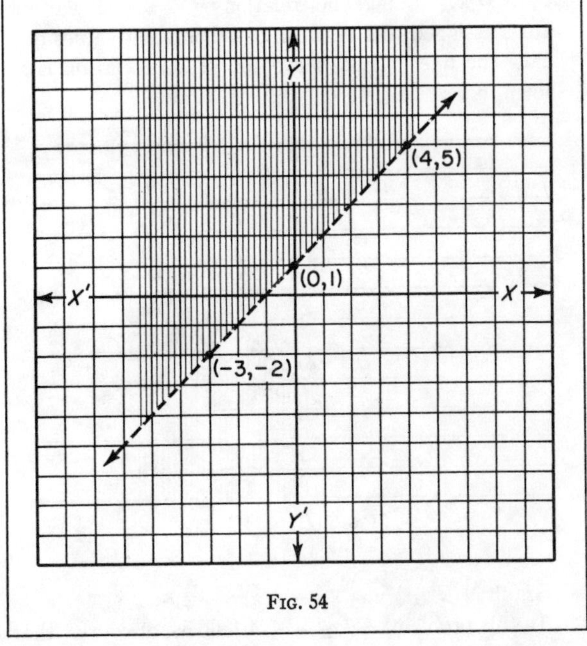

FIG. 54

16. Graph: $\{(x,y) \mid y \geq 1 - x\}$, $U =$ the set of all real numbers.

In this case, the line will not be broken because y is equal to $1 - x$ as well as greater than $1 - x$.

If $y = 1 - x$, the following three ordered pairs satisfy that relation: $(0,1)$,$(-3,4)$, and $(4,-3)$. Assigning values to x in $y \geq 1 - x$, we find that the corresponding values of y fall on or above the line. Hence, the graph includes the line and the area above the line, as shown in Fig. 55.

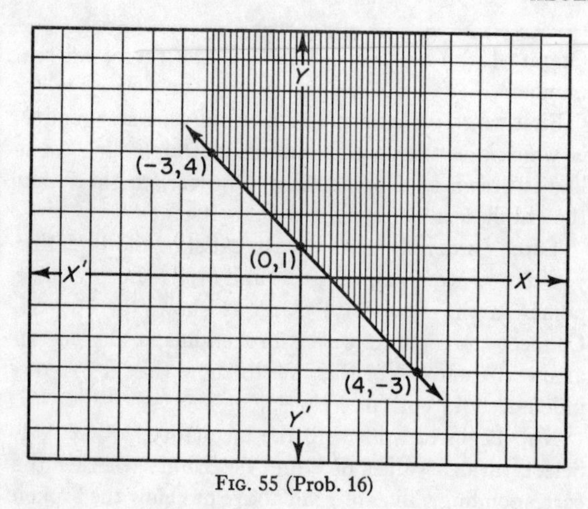

FIG. 55 (Prob. 16)

17. Graph: $\{(x,y) \mid y + 3 < x\}$, $U =$ the set of real numbers.

Graph the line $y + 3 = x$ or $y = x - 3$. Three ordered pairs that satisfy this relation are: (0,−3), (3,0), and (5,2). A broken line is used to indicate that $y + 3 \neq x$. Using the relation $y < x - 3$ various values assigned to x result in corresponding values of y below the line. Thus, the graph of this relation is as shown in Fig. 56.

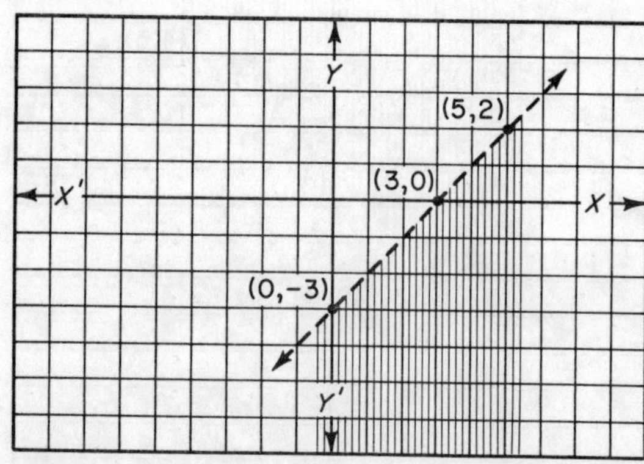

FIG. 56 (Prob. 17)

18. Graph: $\{(x,y) \mid y \geq 2x - 1 \text{ and } -2 \leq x \leq 4\}$.

In this problem, $-2 \leq x \leq 4$ tells us, of course, that x cannot be less than 2 nor greater than 4. The permissible values of y, then, are those that correspond to the limited values of x. Three ordered pairs associated with the relation, $y = 2x - 1$, are: (0,−1), (4,7), and (−2,−5). The graph of this is shown by the solid line segment in Fig. 57. To determine the shaded area we use the original relation, $y \geq 2x - 1$, and find that the values of y for each value of x lie on or above the dotted line. Heavy dots are placed at each end of the line segment since −2 and 4 were included in the permissible values of x.

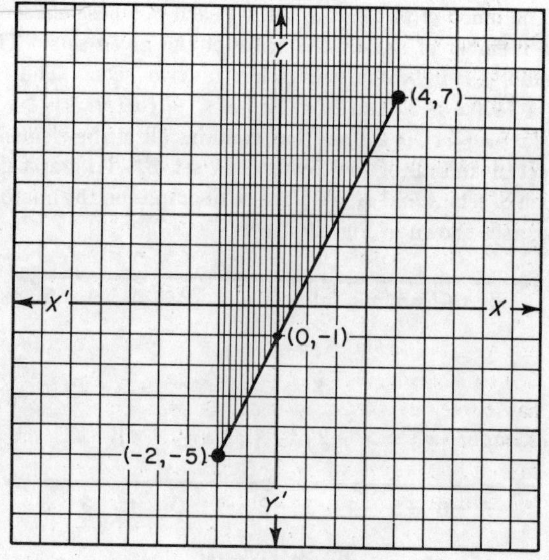

FIG. 57 (Prob. 18)

19. Graph: $\{(x,y) \mid y \geq 2 - x \text{ and } -3 < x \leq 3\}$.

Using the relation $y = 2 - x$, we find the following ordered pairs in the domain specified by $-3 < x \leq 3$: (−3,5), (0,2), and (3,−1). The line segment will not be broken on the graph of $y \geq 2 - x$ because y is equal to, as well as greater than, $2 - x$. The right endpoint of the line segment will be a heavy dot to indicate that x may equal 3. The left endpoint is a circle to show that x cannot equal −3. Each value assigned to x gives a corresponding value of y both on the line segment and above the line segment. Hence, the shaded area is above the line.

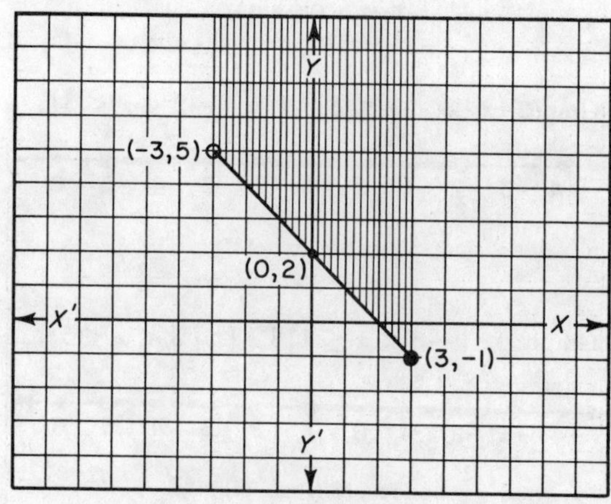

FIG. 58 (Prob. 19)

20. Graph: $\{(x,y) \mid y - 3 \geq x \text{ and } 2 < x < 5\}$.

Problems of this sort could be handled as graphs of intersections as well as in the way shown in Problems 43 and 44. Let us, then, rewrite Problem 45 in the way discussed on page 130.

$\{(x,y) \mid y - 3 \geq x\} + \{(x,y) \mid 2 < x < 5\}$.
Now we will graph each inequality separately on the same coordinate system.

First consider $\{(x,y) \mid y - 3 \geq x\}$. The line $y - 3 = x$, or $y = x + 3$, will not be broken because the set includes values of y which are equal to $x + 3$, as well as values which are greater than $x + 3$. Three ordered pairs which satisfy $y = x + 3$ are: (2,5), (3,6), and (5,8). If we assign values to x in $y = x + 3$, we find that the corresponding values of y are on or above the line. Hence, the shaded area ▥ is above the line $y = x + 3$.

Now consider $\{(x,y) \mid 2 < x < 5\}$. The inequality has three members. Hence we will graph the lines $2 = x$ and $x = 5$. Both lines will be broken since x cannot equal either 2 or 5. Since x is between 2 and 5, the shaded area ▤ is to the left of the broken line $x = 5$ and to the right of the broken line $x = 2$.

The graph of the intersection, then, is the area in which the graphs intersect. That is, the graph of the intersection is the area shaded ▦, and including the line segment from (2,5) to (5,8). Remember that the broken lines do not represent elements in the solution set. Therefore, circles have been used at (2,5) and (5,8).

In subsequent problems concerning the graphs of intersections, the original sets will not be graphed in their entirity; only the intersection will be completely graphed.

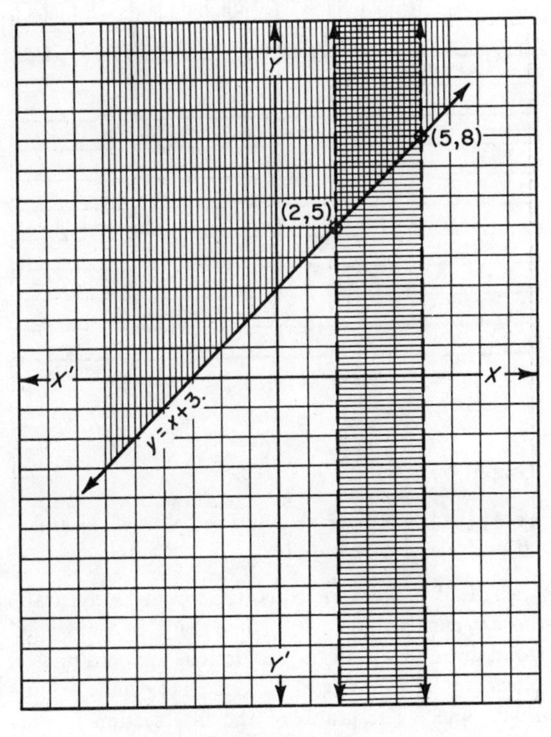

Fig. 59 (Prob. 20)

The Intersection of More Than Two Sets. The intersection of two or more sets may enclose an area. The following problems are representative of this concept.

21. Graph $A \cap B \cap C$ if $A = \{(x,y) \mid y > x - 3\}$, $B = \{(x,y) \mid x \geq 0\}$, and $C = \{(x,y) \mid y < 2 - x\}$. U is the set of all real numbers.

In set A, the values of y which correspond to each value of x are greater than those found on the broken line $y = x - 3$. Thus, the graph of this set consists of all regions above the line $y = x - 3$. In set B, we are restricted to positive values for x; that is, values to the right of the Y-axis. This, then, eliminates all regions to the left of the Y-axis. In set C, the values of y which correspond to assigned values of x are less than those found for the equation $y = 2 - x$. Thus, all area under the line $y = 2 - x$ would be shaded if set C were the only set to be graphed. It can be seen that each set eliminates certain parts of each other set from the graph of the intersection. The shaded area in Fig. 60 includes all ordered pairs that satisfy all three sets.

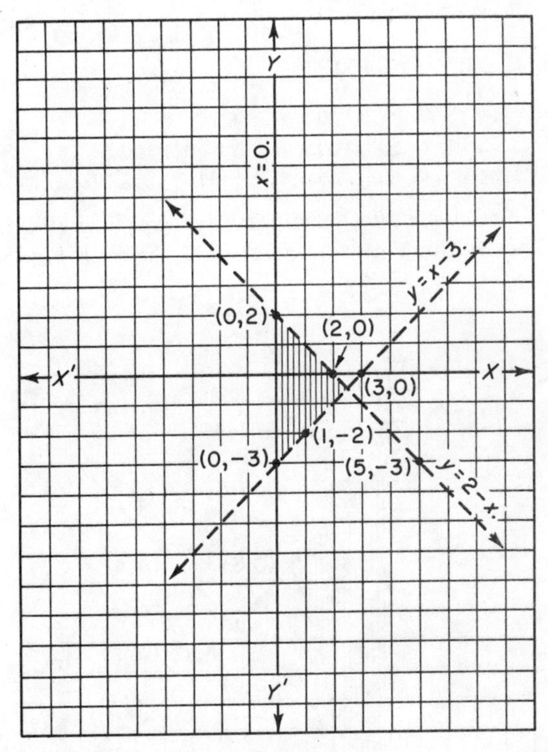

Fig. 60 (Prob. 21)

22. Graph $A \cap B \cap C$ if $A = \{(x,y) \mid y \leq x + 2\}$, $B = \{(x,y) \mid y < 2 - 3x\}$, and $C = \{(x,y) \mid y \geq 0\}$.

Sets A and C will require solid lines while set B will require a broken line. Graph these lines first, and then fill in the shaded area which represents elements of all three sets. This graph is shown in Fig. 61.

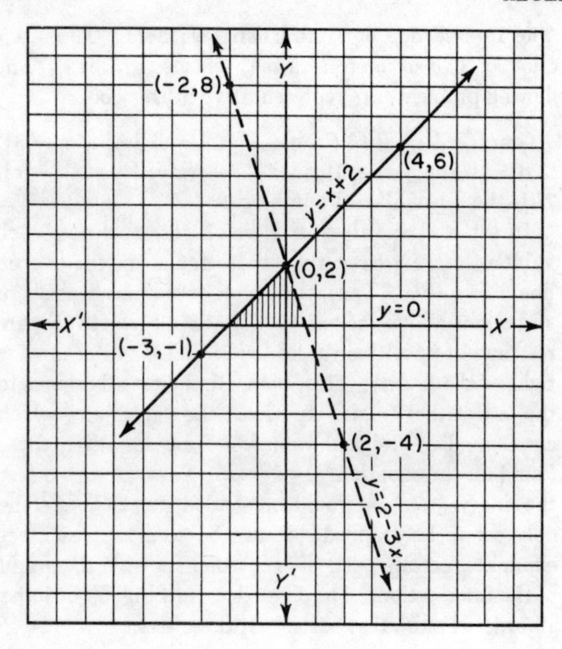

FIG. 61 (Prob. 22)

23. Graph $A \cap B \cap C$ if

$$A = \{(x,y) \mid y > x\},$$
$$B = \{(x,y) \mid x + y < 3\}, \text{ and}$$
$$C = \{(x,y) \mid x > -4\}.$$

Since these three sets are strictly inequalities all three lines will be dotted. The graph is shown in Fig. 62.

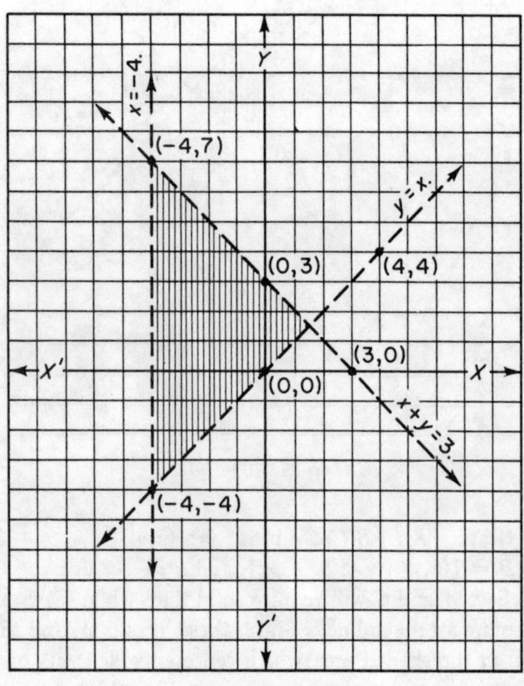

FIG. 62 (Prob. 23)

24. Graph $A \cap B \cap C$ if $A = \{(x,y) \mid x^2 + y^2 \leq 5\}$, $B = \{(x,y) \mid x^2 + y^2 > 2\}$, and $C = \{(x,y) \mid x < 5\}$.

The graph of $x^2 + y^2 = 5$ is a circle whose center is at the origin and whose radius is 5. The circumference of this circle will not be broken on the graph because set A is an equality as well as an inequality.

Set B, if it were an equality, would be a circle whose center is at the origin and whose radius is 2. Since it is not an equality its circumference will be represented by a broken line. Set B will include all points outside this circle. Set C will include all points to the left of the line $x = 5$. Hence, the graph of the intersection of A, B, and C will include the belt of area between the two circles and all the points, except (5, 0), on the circumference of the larger circle. The point (5, 0) is not included because set C does not include points on the line $x = 5$. The graph of the intersection is shown in Fig. 63.

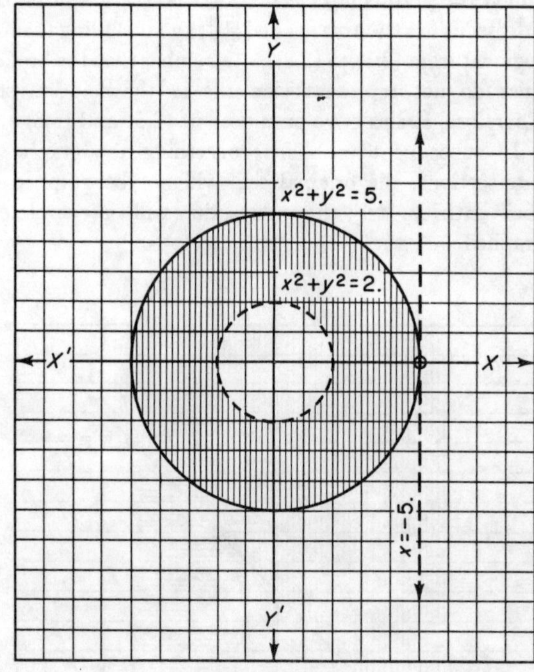

FIG. 63 (Prob. 24)

25. Graph $A \cap B$ if

$$A = \{(x,y) \mid y \leq x^2 - 1 \text{ and } -2 < x < 2\} \text{ and}$$
$$B = \{(x,y) \mid y > x + 1\}.$$

Set A, if it were an equality, would be a parabola whose vertex is at $y = -1$. The graph of set A will be a solid line and within the restrictions imposed upon the values of x by $-2 < x < 2$. Set B, of course, will be a broken line. The graph of the intersection is shown in Fig. 64.

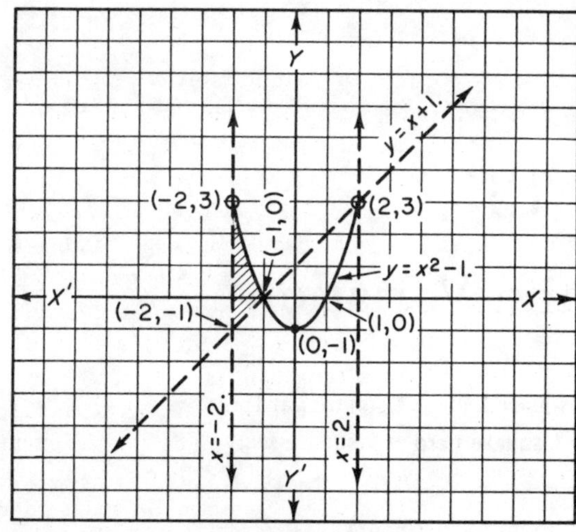

FIG. 64 (Prob. 25)

SUPPLEMENTARY PROBLEMS

Solve for x.

1. $2x + 1 > 5$.

2. $1 - \frac{2}{3}x > -\frac{5}{3}$.

3. $x + 1 < 3x - 2$.

4. $\frac{3}{4}x + 1 < \frac{2}{3}x + 2$.

5. $\frac{1}{3} - \frac{1}{3}x > \frac{1}{2} + \frac{2}{3}x$.

Find the solution set for the following inequalities.

6. $\{x \mid x - 2 < 5\}$, U = the set of natural numbers.

7. $\{x \mid 3 - 2x > x\}$, U = the set of natural numbers.

8. $\{x \mid \frac{1}{3}x > x - 5\}$, U = the set of natural numbers.

9. $\{x \mid 1 - \frac{1}{2}x < x + 3\}$, U = the set of one digit natural numbers.

10. $\{x \mid 3x - 1 > 3 - 5x\}$, $U = \{1,2,3,4,5,6\}$.

Solve for x. U = the set of real numbers.

11. $\left\{x \,\middle|\, \frac{2}{3x} > 1\right\}$.

12. $\left\{x \,\middle|\, \frac{3}{x-1} < 2\right\}$.

13. $|x + 3| \geq 2$.

14. $\left|\frac{x-1}{x+1}\right| \leq 4$.

15. $\left|\frac{3x}{1+x}\right| < 2$.

Chapter 13

Ratio, Proportion, and Variation

Ratios. A **ratio** is an expression indicating a relationship between two things. It is the quotient obtained by dividing the first item by the second item and is often expressed as a fraction. The ratio of 1 to 2 is $\frac{1}{2}$; 3 to 4 is $\frac{3}{4}$; and x to y is $x : y$. In this last form, the colon indicates division. Two things that are expressed as a ratio must be in terms of the same unit of measurement. In order to express the ratio of one pound to four ounces, the pounds must be converted into ounces. The ratio would be $\frac{16}{4}$ or $\frac{4}{1}$.

Express the following ratios as fractions and simplify.

1. $4 : 12$.

$$4 : 12 = \tfrac{4}{12} = \tfrac{1}{3}.$$

2. $18 : 24$.

$$18 : 24 = \tfrac{18}{24} = \tfrac{3}{4}.$$

3. $3x : 6x$.

$$3x : 6x = \frac{3x}{6x} = \tfrac{1}{2}.$$

4. $4x : 8$.

$$4x : 8 = \frac{4x}{8} = \frac{x}{2}.$$

5. 3 inches to 9 inches.

$$3 : 9 = \tfrac{3}{9} = \tfrac{1}{3}.$$

6. 4 feet to 16 inches.

4 feet = 48 inches.

$$48 : 16 = \tfrac{48}{16} = \tfrac{3}{1} = 3.$$

7. 1 nickel to 1 quarter.

Convert both to cents.

$$5 : 25 = \tfrac{5}{25} = \tfrac{1}{5}.$$

8. 1 square foot to 1 square yard.

1 square yard = 9 square feet.

$$1 : 9 = \tfrac{1}{9}.$$

9. 10 minutes to 2 hours.

Change hours to minutes.

$$10 : 120 = \tfrac{10}{120} = \tfrac{1}{12}.$$

10. 1 ton to 3,500 pounds.

Change tons to pounds.

$$2{,}000 : 3{,}500 = \tfrac{2000}{3500} = \tfrac{4}{7}.$$

Working with Proportions. A **proportion** is an expression which indicates that two ratios are equal. An example is $a : b = c : d$ or $\frac{a}{b} = \frac{c}{d}$.

In the expression $a : b = c : d$, the a is called the first term, b the second, c the third, and d the fourth term. The a and d are known as the **extremes** and c and b are the **means** of the proportion. In this case d is called the **fourth proportional** to a, b, and c.

In a proportion, if the second and third terms (b and c) are equal, then b is called the **mean proportional** between a and d ($a : b = b : d$), and d is known as the **third proportional** to a and b.

In solving many problems, three of the terms will be given and the other one will be unknown, but its value required. In such problems the following rule should be used: *The product of the means equals the product of the extremes.*

If $2 : 3 = 5 : x$, find the value of x.

$$2 : 3 = 5 : x.$$
$$2x = 15.$$
$$x = 7\tfrac{1}{2}.$$

Solve for x in each of the following problems and simplify the result:

11. $4 : x = 1 : 3$.

$$(x)(1) = (4)(3).$$
$$x = 12.$$

12. $x : 5 = 3 : 5.$

$$(x)(5) = (5)(3).$$
$$5x = 15.$$
$$x = 3.$$

13. $3 : 4 = 2 : x.$

$$(3)(x) = (4)(2).$$
$$3x = 8.$$
$$x = \tfrac{8}{3}.$$

14. $(x + 1) : 6 = 1 : 2.$

$$2(x + 1) = 6.$$
$$2x + 2 = 6.$$
$$2x = 4.$$
$$x = 2.$$

15. $(x + 2) : 3 = 1 : x.$

$$x(x + 2) = 3.$$
$$x^2 + 2x = 3.$$
$$x^2 + 2x - 3 = 0.$$

Factor the quadratic equation and find the roots.

$$(x - 1)(x + 3) = 0.$$
$$x - 1 = 0. \qquad x + 3 = 0.$$
$$x = 1. \qquad x = -3.$$

Hence, x could equal either 1 or -3.

16. $(x + 5) : (x + 3) = 4 : x.$

$$x(x + 5) = 4(x + 3).$$
$$x^2 + 5x = 4x + 12.$$
$$x^2 + x - 12 = 0.$$
$$(x + 4)(x - 3) = 0.$$
$$x + 4 = 0. \qquad x - 3 = 0.$$
$$x = -4. \qquad x = 3.$$

Hence, x can equal either -4 or 3.

17. $3 : x = x : 27.$

$$x^2 = 81.$$
$$x = \pm 9.$$

18. What is the fourth proportional to 5, 9, and 15?

$$5 : 9 = 15 : x.$$
$$5x = 135.$$
$$x = 27.$$

19. What is the third proportional to 3 and 5?

$$3 : 5 = 5 : x.$$
$$3x = 25.$$
$$x = 8\tfrac{1}{3}.$$

20. What is the mean proportional between 2 and 18?

$$2 : x = x : 18.$$
$$x^2 = 36.$$
$$x = \pm 6.$$

Variation. In physics and other sciences, many principles and laws are based upon a relationship between two quantities. As one of these quantities changes, there is a change in the other quantity. **Variation** refers to that relationship between the two quantities whereby a change in one of them results in a definite change in the other, based upon a specific ratio.

Direct Variation. A relationship between two unknowns such that as the first one increases or decreases, the second one will increase or decrease correspondingly at a given ratio is known as **direct variation.** An illustration of this concept might be the relationship between a man's income and the number of days he works. If his daily wage is a fixed amount, then his income is directly proportional to the number of days he works. The number of days and the income are the variables and the fixed daily wage is the constant ratio.

In the expression

$$y = kx,$$

y and x are the variables and k is the ratio or the **constant of proportionality.**

The expression $y = kx$ means that **y varies directly as x** or that **y is directly proportional to x.**

This expression may be written

$$\frac{y}{x} = k.$$

If we wish to find the value of k, it is necessary to know the values of x and y.

> If y is directly proportional to x, and $x = 3$ when $y = 12$, what is the constant of proportionality?
>
> $$y = kx.$$
> $$12 = k(3).$$
> $$3k = 12.$$
> $$k = 4.$$
>
> If y is directly proportional to x, and $x = 2$ when $y = 1$, what is x when $y = 7$?
>
> $$y = kx.$$
> $$1 = k(2).$$
> $$2k = 1.$$
> $$k = \tfrac{1}{2}.$$
>
> Then $$y = \tfrac{1}{2}x.$$
> $$\tfrac{1}{2}x = 7.$$
> $$x = 14.$$

21. If y varies directly as x, and $y = 3$ when $x = 15$, what is y when $x = 25$?

$$y = kx. \tag{1}$$

Substitute 3 for y and 15 for x, and solve for k.

$$3 = k(15).$$
$$15k = 3.$$
$$k = \tfrac{1}{5}. \quad (2)$$

Substitute this value of k back in the original equation.

$$y = kx.$$
$$y = \tfrac{1}{5}x. \quad (3)$$

Substitute the given value of $x = 25$ in equation (3).

$$y = \tfrac{1}{5}x \quad (3)$$
$$= \tfrac{1}{5}(25)$$
$$= 5.$$

Hence, when $x = 25$, $y = 5$.

22. If y varies directly as x, and $y = 27$ when $x = 3$, what is y when $x = 7$?

$$y = kx. \quad (1)$$
$$27 = k(3).$$
$$3k = 27.$$
$$k = 9. \quad (2)$$

Substitute the value of k from equation (2) in equation (1)

$$y = kx. \quad (1)$$
$$y = 9x. \quad (3)$$

Substitute the given value, $x = 7$, in equation (3).

$$y = 9x \quad (3)$$
$$= 9(7)$$
$$= 63.$$

Hence, when $x = 7$, $y = 63$.

23. If y is directly proportional to x^2, and $y = 36$ when $x = 3$, what is y when $x = 5$?

$$y = kx^2. \quad (1)$$
$$36 = k(3)^2.$$
$$36 = 9k.$$
$$9k = 36.$$
$$k = 4. \quad (2)$$

Substitute the value of k from equation (2) in equation (1).

$$y = kx^2. \quad (1)$$
$$y = 4x^2. \quad (3)$$

Substitute the given value, $x = 5$, in equation (3).

$$y = 4x^2 \quad (3)$$
$$= 4(5)^2$$
$$= 4(25)$$
$$= 100.$$

Hence, when $x = 5$, $y = 100$.

24. If x is directly proportional to y^3, and $x = 24$ when $y = 2$, what is x when $y = 4$?

$$x = ky^3. \quad (1)$$
$$24 = k(2)^3.$$
$$24 = 8k.$$
$$8k = 24.$$
$$k = 3. \quad (2)$$

Substitute the value of k from equation (2) in equation (1).

$$x = ky^3. \quad (1)$$
$$x = 3y^3. \quad (3)$$

Substitute the given value, $y = 4$, in equation (3).

$$x = 3(4)^3$$
$$= 3(64)$$
$$= 192.$$

Hence, $x = 192$ when $y = 4$.

25. If x is directly proportional to y^2, and $x = 6$ when $y = 2$, what is x when $y = 6$?

$$x = ky^2. \quad (1)$$
$$6 = k(2)^2.$$
$$6 = 4k.$$
$$4k = 6.$$
$$k = \tfrac{3}{2}. \quad (2)$$

Substitute the value of k from equation (2) in equation (1).

$$x = ky^2. \quad (1)$$
$$x = \tfrac{3}{2}y^2. \quad (3)$$

Substitute the given value, $y = 6$, in equation (3).

$$x = \tfrac{3}{2}(6)^2$$
$$= \tfrac{3}{2}(36)$$
$$= 54.$$

Hence, $x = 54$ when $y = 6$.

Inverse Variation. A relationship between two variables such that as the first one increases, the second one decreases (or as the first one decreases, the second one increases) is known as **inverse variation.** This concept is illustrated by the fact that the time required to travel a given distance increases as the rate is diminished. Likewise, if the time spent in traveling a given distance is decreased, the rate will be increased. The equation for inverse variation is

$$x = \frac{k}{y}.$$

By multiplying both members of the above equation by y, the equation becomes

$$xy = k.$$

Problems involving inverse variation are worked very much as those involving direct variation. A pair of values for the variables must be known to solve for k.

26. If x varies inversely as y, and $x = 3$ when $y = 6$, what is the value of x when $y = 12$?

$$x = \frac{k}{y}. \quad (1)$$
$$3 = \frac{k}{6}.$$
$$k = 18. \quad (2)$$

Substitute the value of k from equation (2) in equation (1).

$$x = \frac{18}{y}. \quad (3)$$

Substitute the value $y = 12$ in equation (3).

$$x = \tfrac{18}{12}$$
$$= \tfrac{3}{2}.$$

Hence, $x = \frac{3}{2}$ when $y = 12$.

27. If x is inversely proportional to y, and $x = 2$ when $y = 3$, what is x when $y = 4$?

$$x = \frac{k}{y}. \quad (1)$$
$$2 = \frac{k}{3}.$$
$$k = 6. \quad (2)$$

Substitute the value of k from equation (2) in equation (1).

$$x = \frac{6}{y}. \quad (3)$$

Substitute the value $y = 4$ in equation (3).

$$x = \tfrac{6}{4}$$
$$= \tfrac{3}{2}.$$

Hence, $x = \frac{3}{2}$ when $y = 4$.

28. If x varies inversely as y, and $x = \frac{1}{2}$ when $y = \frac{1}{3}$, what is the value of x when $y = 2$?

$$x = \frac{k}{y}. \quad (1)$$
$$\frac{1}{2} = \frac{k}{\frac{1}{3}}.$$
$$k = \tfrac{1}{6}. \quad (2)$$

Substitute the value of k from equation (2) in equation (1).

$$x = \frac{\frac{1}{6}}{y}.$$
$$x = \frac{1}{6y}. \quad (3)$$

Substitute the value $y = 2$ in equation (3).

$$x = \frac{1}{6(2)}$$
$$= \tfrac{1}{12}.$$

Hence, $x = \frac{1}{12}$ when $y = 2$.

29. If x varies inversely as y^2, and $x = 2$ when $y = 1$, what is the value of x when $y = 3$?

$$x = \frac{k}{y^2}. \quad (1)$$
$$2 = \frac{k}{1^2}.$$
$$k = 2. \quad (2)$$

Substitute the value of k from equation (2) in equation (1).

$$x = \frac{2}{y^2}. \quad (3)$$

Substitute the value $y = 3$ in equation (3).

$$x = \frac{2}{3^2}$$
$$= \tfrac{2}{9}.$$

Hence, $x = \frac{2}{9}$ when $y = 3$.

30. If y varies inversely as x^3, and $y = 2$ when $x = 2$, what is the value of y when $x = \frac{3}{2}$?

$$y = \frac{k}{x^3}. \quad (1)$$
$$2 = \frac{k}{2^3}.$$
$$k = 16. \quad (2)$$

Substitute the value of k from equation (2) in equation (1).

$$y = \frac{16}{x^3}. \quad (3)$$

Substitute the value $x = \frac{3}{2}$ in equation (3).

$$y = \frac{16}{(\frac{3}{2})^3}$$
$$= \frac{16}{\frac{27}{8}}$$
$$= 16 \cdot \tfrac{8}{27}$$
$$= \tfrac{128}{27}.$$

Hence, $y = \frac{128}{27}$ when $x = \frac{3}{2}$.

Joint Variation. If one quantity varies as the product of a constant and two or more other quantities, it **varies jointly** as the other variables. An example of this type of variation is the formula for the area of a triangle, $A = \frac{1}{2}bh$.

31. If x varies jointly as y and z, and $x = 18$ when $y = 2$ and $z = 3$, what is the value of x when $y = 3$ and $z = 1$?

$$x = kyz. \quad (1)$$
$$18 = k(2)(3).$$
$$6k = 18.$$
$$k = 3. \quad (2)$$

Substitute the value of k from equation (2) in equation (1)

$$x = 3yz. \quad (3)$$

Substitute the given values, $y = 3$ and $z = 1$, in equation (3).

$$x = 3(3)(1)$$
$$= 9.$$

Hence, when $y = 3$ and $z = 1$, $x = 9$.

32. If x varies jointly as y and z, and $x = 4$ when $y = 3$ and $z = 4$, what is the value of x when $y = 6$ and $z = 5$?

$$x = kyz. \quad (1)$$
$$4 = k(3)(4).$$
$$4 = 12k.$$
$$12k = 4.$$
$$k = \tfrac{1}{3}. \quad (2)$$

Substitute the value of k from equation (2) in equation (1).

$$x = \tfrac{1}{3}yz. \quad (3)$$

Substitute the given values, $y = 6$ and $z = 5$, in equation (3).

$$x = \tfrac{1}{3}(6)(5)$$
$$= 10.$$

Hence, $x = 10$ when $y = 6$ and $z = 5$.

33. If m varies jointly as n and p, and $m = 12$ when $n = 4$ and $p = 2$, what is the value of m when $n = 5$ and $p = 7$?

$$m = knp. \quad (1)$$
$$12 = k(4)(2).$$
$$8k = 12.$$
$$k = \tfrac{3}{2}. \quad (2)$$

Substitute the value of k from equation (2) in equation (1).

$$m = \tfrac{3}{2}np. \quad (3)$$

Substitute the given values, $n = 5$ and $p = 7$, in equation (3).

$$m = \tfrac{3}{2}(5)(7)$$
$$= \tfrac{105}{2}.$$

Hence, $m = \tfrac{105}{2}$ when $n = 5$ and $p = 7$.

Combined Variation. A combination of two or more of the foregoing types of variation in the same equation is called **combined variation.** This is the most general form of variation, as all other forms can be considered special cases of this one.

34. If y varies directly as x and inversely as z, and $y = 3$ when $x = 3$ and $z = 2$, what is the value of y when $x = 1$ and $z = 4$?

$$y = \frac{kx}{z}. \quad (1)$$
$$3 = \frac{k(3)}{2}.$$
$$3k = 6.$$
$$k = 2. \quad (2)$$

Substitute the value of k from equation (2) in equation (1).

$$y = \frac{2x}{z}. \quad (3)$$

Substitute the given values, $x = 1$ and $z = 4$, in equation (3).

$$y = \frac{2(1)}{4}.$$
$$4y = 2.$$
$$y = \tfrac{1}{2}.$$

Hence, $y = \tfrac{1}{2}$ when $x = 1$ and $z = 4$.

35. If x varies jointly as y and z, and inversely as w^2; and $x = 4$, $y = 2$, and $z = 4$ when $w = 2$; what is the value of x when $y = 3$, $z = 5$, and $w = 3$?

$$x = \frac{kyz}{w^2}. \quad (1)$$
$$4 = \frac{k(2)(4)}{2^2}.$$
$$8k = 16.$$
$$k = 2. \quad (2)$$

Substitute the value of k from equation (2) in equation (1).

$$x = \frac{2yz}{w^2}. \quad (3)$$

Substitute the given values, $y = 3$, $z = 5$, and $w = 3$, in equation (3).

$$x = \frac{2(3)(5)}{9}$$
$$= \tfrac{30}{9}$$
$$= \tfrac{10}{3}.$$

Hence, $x = \tfrac{10}{3}$ when $y = 3$, $z = 5$, and $w = 3$.

Charles' Law. It is a known fact that when a gas is heated, it expands, and when it is cooled, it contracts. This means that heated gas will occupy a greater volume than the same amount of gas when it is cooled. Charles' Law for gases states that the first volume is to the second volume as the first temperature is to the second temperature. As a proportion it is stated as

$$\frac{V_1}{V_2} = \frac{T_1}{T_2}.$$

36. By use of Charles' Law, determine how much volume a quantity of gas will occupy at 352° K. if it occupies 100 cubic centimeters at 320° K.? (K. refers to the Kelvin scale of temperature.)

$$\frac{V_1}{V_2} = \frac{T_1}{T_2}.$$

$$\frac{100}{V_2} = \frac{320}{352}.$$

$$320\ V_2 = 35{,}200.$$

$$V_2 = 110 \text{ cubic centimeters.}$$

37. If a certain gas occupies 1,000 cubic centimeters at a temperature of 360° K., at what temperature will it occupy 900 cubic centimeters?

$$\frac{V_1}{V_2} = \frac{T_1}{T_2}.$$

$$\frac{1{,}000}{900} = \frac{360}{T_2}.$$

$$1{,}000T_2 = 324{,}000.$$

$$T_2 = 324°\text{K.}$$

Boyle's Law. Boyle's Law for gases involves volume and pressure. It states that the first volume is to the second volume as the second pressure is to the first pressure. This principle involves the fact that the greater the pressure on a given quantity of gas, the smaller the volume it occupies. As a proportion, the law becomes

$$\frac{V_1}{V_2} = \frac{P_2}{P_1}.$$

38. If a certain gas occupies 600 cubic centimeters under a pressure of 40 pounds, what will its volume be if the pressure is raised to 70 pounds?

$$\frac{V_1}{V_2} = \frac{P_2}{P_1}.$$

$$\frac{600}{V_2} = \frac{70}{40}.$$

$$70V_2 = 24{,}000.$$

$V_2 = 342\frac{6}{7}$ cubic centimeters;

or, approximately 343 cubic centimeters.

39. If a certain gas occupies 270 cubic centimeters under a pressure of 30 pounds, how much pressure will need to be exerted to reduce the volume to 180 cubic centimeters?

$$\frac{V_1}{V_2} = \frac{P_2}{P_1}.$$

$$\frac{270}{180} = \frac{P_2}{30}.$$

$$180P_2 = 8{,}100.$$

$P_2 = 45$ pounds.

40. The weight of a body in space varies inversely as the square of its distance from the center of the earth. If a man weighs 150 pounds at the earth's surface, what would be his weight 150 miles in space, assuming 4,000 miles is the distance from the center of the earth to its surface? (See Fig. 65.)

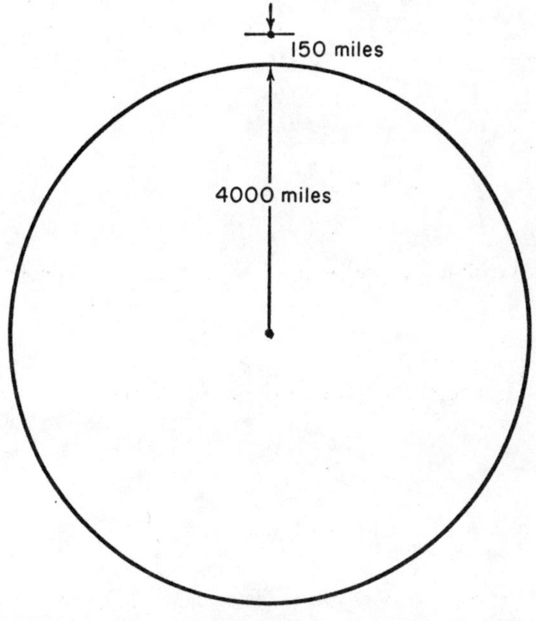

Fig. 65

$$\frac{W_1}{W_2} = \frac{(d_2)^2}{(d_1)^2}.$$

$$\frac{150}{W_2} = \frac{(4{,}150)^2}{(4{,}000)^2}.$$

$$\frac{150}{W_2} = \frac{17{,}222{,}500}{16{,}000{,}000}.$$

$$\frac{150}{W_2} = \frac{172{,}223}{160{,}000}.$$

$$172{,}223W_2 = 24{,}000{,}000.$$

$W_2 = 139.35$ pounds.

$$172{,}223W_2 = 24{,}000{,}000.$$

W_2 is approximately 139.35 pounds.

SUPPLEMENTARY PROBLEMS

Find the value of x in the following proportions.

1. $3 : x = 4 : 7.$
2. $x : 2 = 3 : 8.$
3. $2 : x = x : 6.$
4. $5 : 3 = 10 : x.$
5. $\frac{2}{3} : x = \frac{3}{4} : \frac{1}{2}.$
6. What is the mean proportional between 2 and 12?
7. What is the fourth proportional to x, $2x$, and $3x$?

8. What is the mean proportional between 4 and $(x + 1)$?
9. What is the third proportional to $(x + 1)$, $(x + 1)^2$, and $(x + 1)^2$?
10. If y varies directly as x and $y = 4$ when $x = 6$, what is y when $x = 7$?
11. If y varies directly as x and $y = 3$ when $x = 4\frac{1}{2}$, what is y when $x = 7\frac{1}{2}$?
12. If y varies inversely as x and $y = 3$ when $x = 8$, what is y when $x = 15$?
13. If y varies jointly as x and z and $y = 8$ when $x = 2$ and $z = 3$, what is y when $x = 3$ and $z = 4$?
14. If y varies directly as the cube of x and inversely as the square of z, and $y = 8$ when $x = 4$ and $z = 2$, what is y when $x = 3$ and $z = 3$?
15. If y varies jointly as x and the square of z and inversely as the cube of w, and $y = 16$ when $x = 4$, $z = 4$, and $w = 2$, what is y when $x = 2$, $z = 3$, and $w = 4$?

Chapter 14

Logarithms

Definitions. A **logarithm** is an exponent used to indicate the power to which a **base** must be raised in order to obtain a given number. In the expression $2^4 = 16$, 2 is the base, 4 is the power, and 16 is the given number. This is written $\log_2 16 = 4$. In the expression $4^3 = 64$, 4 is the base, 3 is the power or logarithm, and 64 is the given number. This is written $\log_4 64 = 3$.

For purposes of computation by the use of logarithms, only the base 10 will be used in this chapter. The logarithm table in the Appendix is calculated on the base 10, and such a table is often referred to as a table of **common** logarithms. In the expression $10^2 = 100$, 10 is the base, 100 is the number, and 2 is the logarithm. Since the base is 10 unless otherwise noted, the expression can be written $\log 100 = 2$.

The Base 10. The following list of values will be useful in working with logarithms to the base 10.

$$10^4 = 10{,}000.$$
$$10^3 = 1{,}000.$$
$$10^2 = 100.$$
$$10^1 = 10.$$
$$10^0 = 1.$$
$$10^{-1} = 0.1.$$
$$10^{-2} = 0.01.$$
$$10^{-3} = 0.001.$$
$$10^{-4} = 0.0001.$$

This list could be continued indefinitely in both a positive and a negative direction.

Note that in each of the above cases the power of 10 is an integer. Often the number with which we are working will not be an integral power of 10. For example, the power of 10 that would produce 76 must be between 1 and 2 since 76 is greater than 10^1 and less than 10^2. Therefore, in the expression $10^x = 76$, the exponent x is between 1 and 2. Logarithms usually represent that power correct to several decimal places. Four-place tables are used in this book, but tables of five, six, or more places are available to make more accurate computations. For most simple calculations, four-place tables are usually accurate enough. However, it is important to remember that when four-place tables are used, accurate results in the solution cannot possibly be expected past the fourth digit (starting with the first non-zero digit).

Determining the Characteristic. From the above discussion it can be seen that a logarithm consists of two parts: the integral or whole number part, and the decimal or fractional part. The integral part is called the **characteristic** and the decimal part is called the **mantissa.** In the expression $10^{3.4782}$, 3 is the characteristic and .4782 is the mantissa. The characteristic, if it is zero or a positive number, can be determined from the upper part of the above list. If we want the characteristic of 784, we observe that 784 is greater than 100 and less than 1,000. Hence, from the preceding chart, we see that the logarithm is greater than 2 and less than 3, so the characteristic is 2.

Determine the characteristics of the logarithms of the following numbers.

1. 47

$$47 = 10^{1+0.----}.$$

The four dashes are placed after the characteristic to indicate that a mantissa is necessary to complete the logarithm. This procedure will be followed until the method for determining the mantissa has been developed.

2. 1,736

$$1{,}736 = 10^{3+0.----}.$$

3. 273

$$273 = 10^{2+0.----}.$$

4. 7

$$7 = 10^{0+0.----}.$$

The number 7 is greater than 10^0, which equals 1, but less than 10^1.

5. 19

$$19 = 10^{1+0.----}.$$

Negative Characteristics. When the characteristic is a negative number, the list under Base 10 above can be used or the following rule can be observed: *The absolute value of the characteristic is one more than the number of zeros directly following the decimal point in the number.*

6. 0.037

$$0.037 = 10^{-2+0.----}.$$

7. 0.183

$$0.183 = 10^{-1+0.----}.$$

8. 0.00057

$$0.00057 = 10^{-4+0.----}.$$

9. 0.00478

$$0.00478 = 10^{-3+0.----}.$$

10. 0.0000076

$$0.0000076 = 10^{-6+0.----}.$$

Convention for Negative Characteristics. Computations involving negative characteristics are simplified considerably by the use of another method for denoting negative characteristics. If we add 9 and −10, the result is −1. The usual method for denoting a characteristic of −1 is 9 + 0.---- − 10, or simply 9.---- − 10. Again, the four dashes represent the mantissa. The expression 7.---- − 10 represents a characteristic of −3, and 4.---- − 10 indicates a characteristic of −6.

Finding the Mantissa. The mantissa is found from the Four-Place Common Logarithm Table in the Appendix. The number, if it contains two digits, is found by following down the *N* column. For example, to get the mantissa for the number 57, find 57 in the *N* column and look across to the 0 column. The mantissa is .7559. The decimal point is always placed immediately preceding the mantissa in the table. The characteristic of 57 is 1 since 57 falls between 10^1 and 10^2. Hence, the logarithm of 57 is 1.7559, or $57 = 10^{1.7559}$. The mantissa for 570 is the same as the mantissa for 57. The characteristic of 570 is 2 since 570 lies between 10^2 and 10^3. Therefore the logarithm of 570 is 2.7559, or $570 = 10^{2.7559}$.

The mantissa for 571 would be found in the 1 column to the right of 57 under the *N* column. This mantissa is .7566. Therefore, the logarithm of 571 is 2.7566, or $571 = 10^{2.7566}$. The logarithm of 57.1 has the same mantissa as the logarithm of 571. The characteristic is 1 since 57.1 lies between 10^1 and 10^2. The mantissa for 5.71 is also .7566, and the characteristic is 0 since 5.71 lies between 10^0 and 10^1. Therefore, the logarithm of 5.71 is 0.7566, or $5.71 = 10^{0.7566}$. The mantissa for 0.00571 is again .7566, and the characteristic is −3 or 7 − 10. Hence, the logarithm of 0.00571 is 7.7566 − 10, or $0.00571 = 10^{7.7566-10}$.

State each of the following numbers as a power of 10 and write the logarithm of each.

11. 31

$$31 = 10^{1.4914}.$$
$$\log 31 = 1.4914.$$

12. 377

$$377 = 10^{2.5763}.$$
$$\log 377 = 2.5763.$$

13. 409

$$409 = 10^{2.6117}.$$
$$\log 409 = 2.6117.$$

14. 0.0372

$$0.0372 = 10^{8.5705-10}.$$
$$\log 0.0372 = 8.5705 - 10.$$

15. 1.09

$$1.09 = 10^{0.0374}.$$
$$\log 1.09 = 0.0374.$$

Interpolation. When the logarithm of a number of four or more digits is to be found, an estimating technique is used. This technique is known as **interpolation,** and it assumes that small changes in the number are proportional to corresponding changes in the mantissa.

The number whose mantissa we are seeking is a fraction of the way up from the lower known number to the higher known number. This fraction can be expressed as the ratio of the *difference* between the number in question and the lower known number to the *tabular difference* between the two known numbers.

Similarly, the mantissa we are seeking will be a fraction of the way up from the lower known mantissa to the higher known mantissa. This fraction can be expressed as the ratio of the *difference* between the unknown mantissa and the lower known mantissa to the *tabular difference* between the two known mantissas.

Assuming that corresponding changes in each case are proportional, the two ratios are equal. From this relationship we can determine the difference between the unknown mantissa and the lower known mantissa. Adding this difference to the lower known mantissa will give us the value of the mantissa in question. The following example will illustrate this procedure.

Find the logarithm of 4733.
The characteristic is 3.

tab. diff. = 10 [log 4740 = 3.6758] tab. diff. = .0009

diff. = 3 [log 4733 = ?] diff. = x

log 4730 = 3.6749

Since we assume that corresponding changes are proportional,

$$\frac{3}{10} = \frac{x}{.0009}.$$
$$10x = .0027.$$
$$x = .00027 \quad \text{or} \quad .0003.$$

Adding the difference x to 3.6749 gives us the value of log 4733.

$$\log 4733 = 3.6749 + .0003$$
$$= 3.6752.$$

16. Find the logarithm of 30.63.

The characteristic is 1.

$$\begin{array}{lllll} & & \log 30.70 = 1.4871 & & \\ .10 & .03 & \log 30.63 = \ ? & x & .0014 \\ & & \log 30.60 = 1.4857 & & \end{array}$$

$$\frac{.03}{.10} = \frac{x}{.0014}.$$

$$.3 = \frac{x}{.0014}.$$

$$x = .00042 \quad \text{or} \quad .0004.$$

$$\log 30.63 = 1.4857 + .0004 = 1.4861.$$

17. Find the logarithm of 9,726.

The characteristic is 3.

$$\begin{array}{lllll} & & \log 9730 = 3.9881 & & \\ 10 & 6 & \log 9726 = \ ? & x & .0004 \\ & & \log 9720 = 3.9877 & & \end{array}$$

$$\frac{6}{10} = \frac{x}{.0004}.$$

$$.6 = \frac{x}{.0004}.$$

$$x = .00024 \quad \text{or} \quad .0002.$$

$$\log 9726 = 3.9877 + .0002 = 3.9879.$$

18. Find the logarithm of 1.878.

The characteristic is 0.

$$\begin{array}{lllll} & & \log 1.880 = 0.2742 & & \\ .010 & .008 & \log 1.878 = \ ? & x & .0024 \\ & & \log 1.870 = 0.2718 & & \end{array}$$

$$\frac{.008}{.010} = \frac{x}{.0024}.$$

$$.8 = \frac{x}{.0024}.$$

$$x = .00192 \quad \text{or} \quad .0019.$$

$$\log 1.878 = 0.2718 + 0.0019 = 0.2737.$$

19. Find the logarithm of 0.07483.

The characteristic is -2.

$$\begin{array}{lllll} & & \log 0.07490 = 8.8745 - 10 & & \\ .00010 & .00003 & \log 0.07483 = \ ? & x & .0006 \\ & & \log 0.07480 = 8.8739 - 10 & & \end{array}$$

$$\frac{.00003}{.00010} = \frac{x}{.0006}.$$

$$.3 = \frac{x}{.0006}.$$

$$x = .00018 \quad \text{or} \quad .0002.$$

$$\log 0.07483 = 8.8739 - 10 + .0002 = 8.8741 - 10.$$

20. Find the logarithm of 0.0007467.

The characteristic is -4.

$$\begin{array}{lllll} & & \log 0.0007470 = 6.8733 - 10 & & \\ .0000010 & .0000007 & \log 0.0007467 = \ ? & x & .0006 \\ & & \log 0.0007460 = 6.8727 - 10 & & \end{array}$$

$$\frac{.0000007}{.0000010} = \frac{x}{.0006}.$$

$$.7 = \frac{x}{.0006}.$$

$$x = .00042 \quad \text{or} \quad .0004.$$

$$\log 0.0007467 = 6.8727 - 10 + .0004 = 6.8731 - 10.$$

Five-Digit Numbers. The logarithms of five-digit numbers are found in the same manner as those of the four-digit numbers above, except that it will be necessary to add two zeros instead of one to the known three-digit numbers.

21. Find the logarithm of 28,834.

The characteristic is 4.

$$\begin{array}{lllll} & & \log 28900 = 4.4609 & & \\ 100 & 34 & \log 28834 = \ ? & x & .0015 \\ & & \log 28800 = 4.4594 & & \end{array}$$

$$\frac{34}{100} = \frac{x}{.0015}.$$

$$.34 = \frac{x}{.0015}.$$

$$x = .00051 \quad \text{or} \quad .0005.$$

$$\log 28{,}834 = 4.4594 + .0005 = 4.4599.$$

22. Find the logarithm of 72.963.

The characteristic is 1.

$$\begin{array}{lllll} & & \log 73.000 = 1.8633 & & \\ .100 & .063 & \log 72.963 = \ ? & x & .0006 \\ & & \log 72.900 = 1.8627 & & \end{array}$$

$$\frac{.063}{.100} = \frac{x}{.0006}.$$

$$.63 = \frac{x}{.0006}.$$

$$x = .000378 \quad \text{or} \quad .0004.$$

$$\log 72.963 = 1.8627 + .0004 = 1.8631.$$

23. Find the logarithm of 100.75.

The characteristic is 2.

$$\begin{array}{lllll} & & \log 101.00 = 2.0043 & & \\ 1.00 & .75 & \log 100.75 = \ ? & x & .0043 \\ & & \log 100.00 = 2.0000 & & \end{array}$$

$$\frac{.75}{1.00} = \frac{x}{.0043}.$$

$$.75 = \frac{x}{.0043}.$$

$$x = .003225 \quad \text{or} \quad .0032.$$

$$\log 100.75 = 2.0000 + .0032 = 2.0032.$$

Antilogarithms. When working problems by the use of logarithms, it is often necessary to find the number corresponding to a given logarithm. This process is the inverse of finding the logarithm of a number. The number is called the **antilogarithm** (abbreviated **antilog**) of the number.

In finding the antilog of a logarithm, two distinct operations are necessary. First the given logarithm is found in the table (interpolation is often necessary), and the required number is the one which corresponds to that logarithm. Then the position of the decimal point is determined from the characteristic of the given logarithm. To determine where the decimal point is placed, refer to the table on page 143 or use the following rules.

Rule 1. If the characteristic is zero or a positive number, add 1 to the characteristic and that sum will be the number of digits placed before the decimal point.

Rule 2. If the characteristic is a negative number, subtract 1 from the absolute value of that negative characteristic and the result will be the number of zeros that must follow the decimal point.

When the zeros have been recorded, follow them by the digits found when computing the antilog.

24. Find the number whose log is 1.8069.

$N =$ the number.
$\log N = 1.8069.$

The .8069 is found to the right of 64 (under column N) and in the 1 column. The decimal point is placed after the 4 since a characteristic of 1 indicates that the number is between 10 and 100. $N = 64.1.$

The decimal point could also have been located by using Rule 1, above.

25. Find the number whose log is 3.7372.

$N =$ the number.
$\log N = 3.7372.$
$N = 5{,}460.$

The zero is placed after the 546 since the characteristic 3 indicates that the number lies between 1,000 and 10,000.

26. Find the number whose log is $8.6212 - 10$.

$N =$ the number.
$\log N = 8.6212 - 10.$
$N = 0.0418.$

The zero is placed before the 4, as the characteristic is -2. By Rule 2, subtract 1 from the absolute value of the negative characteristic and follow the decimal point with that many zeros.

27. Find the number whose log is $7.9917 - 10$.

$N =$ the number.
$\log N = 7.9917.$
$N = 0.00981.$

Antilog Interpolation. If the mantissa of a given log does not appear in the table, interpolation will be required. The method of interpolation required is quite similar to that discussed on page 144 and carried out in the problems which followed.

Find the number whose logarithm is 1.2984.

$N =$ the number.
$\log N = 1.2984.$

tab. diff. = .10; diff. = x

$\log 19.90 = 1.2989$
$\log N = 1.2984$
$\log 19.80 = 1.2967$

diff. = .0017; tab. diff. = .0022

Since we assume that the corresponding changes are proportional,

$$\frac{x}{.10} = \frac{.0017}{.0022}.$$
$$\frac{x}{.10} = \frac{17}{22}.$$
$$22x = 1.7.$$
$$x = .07727 \text{ or } .08.$$

Adding the difference x to 19.80 gives us the value of N.

$N = 19.88.$

28. Find the number whose log is 2.8417.

$N =$ the number.
$\log N = 2.8417.$

1.0; x

$\log 695.0 = 2.8420$
$\log N = 2.8417$
$\log 694.0 = 2.8414$

.0003; .0006

$$\frac{x}{1.0} = \frac{.0003}{.0006}.$$
$$\frac{x}{1} = \frac{3}{6}.$$
$$6x = 3.$$
$$x = 0.5.$$
$$N = 694.5.$$

29. Find the number whose log is 3.0096.

$N =$ the number.
$\log N = 3.0096.$

10.0; x

$\log 1030.0 = 3.0128$
$\log N = 3.0096$
$\log 1020.0 = 3.0086$

.0010; .0042

$$\frac{x}{10.0} = \frac{.0010}{.0042}.$$
$$\frac{x}{10} = \frac{10}{42}.$$
$$42x = 100.$$
$$x = \text{(approximately) } 2.4 \text{ or } 2.$$
$$N = 1{,}022.$$

30. Find the number whose log is 7.0672 − 10.

$$N = \text{the number.}$$
$$\log N = 7.0672 - 10.$$

.000010   log .001170 = 7.0682 − 10   .0037
x log N = 7.0672 − 10   .0027
log .001160 = 7.0645 − 10

$$\frac{x}{.000010} = \frac{.0027}{.0037}.$$
$$\frac{x}{.00001} = \frac{27}{37}.$$
$$37x = .00027.$$
$$x = \text{(approximately) } .0000073 \quad \text{or} \quad .000007.$$
$$N = .001167.$$

Cologarithms. The **cologarithm** (abbreviated **colog**) of a number N is the log of $\frac{1}{N}$. The colog is sometimes convenient to use in solving problems by use of logarithms, as will be shown later. It should be remembered that logarithms are exponents, so the log of $\frac{1}{N}$ is the log of 1 minus the log of N (see Rule 2 under Computation with Logarithms below).

31. If the logarithm of N is 2.3711, what is the colog of N?

log 1 = 0.0000.
log N = 2.3711.

The log of 1, which is equal to zero, can be written as 10.0000 − 10. This is more convenient to work with.

log 1 = 10.0000 − 10
log N = 2.3711
colog N = 7.6289 − 10

32. If the logarithm of N is 4.9518, what is the colog of N?

log 1 = 10.0000 − 10
log N = 4.9518
colog N = 5.0482 − 10

33. If the logarithm of N is 0.1732, what is the colog of N?

log 1 = 10.0000 − 10
log N = 0.1732
colog N = 9.8268 − 10

34. If log N = 9.4738 − 10, what is colog N?

log 1 = 10.0000 − 10
log N = 9.4738 − 10
colog N = 0.5262

35. If log N = 7.4711 − 10, what is colog N?

log 1 = 10.0000 − 10
log N = 7.4711 − 10
colog N = 2.5289

Computation with Logarithms. The work of computation involved in finding products and quotients, raising to powers, and extracting roots can usually be reduced considerably by the use of logarithms. The rules of logarithm manipulation, together with the laws of exponents from which they are derived (Laws 2, 3, and 5 of Chap. 8), appear below.

$$x^m \cdot x^n = x^{m+n}, \qquad \log_x (mn) = \log_x m + \log_x n. \quad (1)$$
$$\frac{x^m}{x^n} = x^{m-n}, \qquad \log_x \left(\frac{m}{n}\right) = \log_x m - \log_x n. \quad (2)$$
$$(x^m)^n = x^{mn}, \qquad \log_x (m^n) = n \log m. \quad (3)$$

Solve the following problems by logarithms:

36. (471)(2.72).

$$N = (471)(2.72).$$
$$\log N = \log 471 + \log 2.72.$$

log 471 = 2.6730
log 2.72 = 0.4346
log N = 3.1076

10   log 1290 = 3.1106   .0034
x log N = 3.1076   .0004
log 1280 = 3.1072

$$\frac{x}{10} = \frac{.0004}{.0034}.$$
$$\frac{x}{10} = \frac{4}{34}.$$
$$34x = 40.$$
$$x = \text{(approximately) } 1.2 \quad \text{or} \quad 1.$$
$$N = 1{,}281.$$

37. (4.06)(118).

$$N = (4.06)(118).$$
$$\log N = \log 4.06 + \log 118.$$

log 4.06 = 0.6085
log 118 = 2.0719
log N = 2.6804

1.0   log 480.0 = 2.6812   .0009
x log N = 2.6804   .0001
log 479.0 = 2.6803

$$\frac{x}{1.0} = \frac{.0001}{.0009}.$$
$$\frac{x}{1} = \frac{1}{9}.$$
$$9x = 1.$$
$$x = .11 \quad \text{or} \quad .1.$$
$$N = 479.1.$$

38. (486)(20.7)(1.159).

$$N = (486)(20.7)(1.159).$$
$$\log N = \log 486 + \log 20.7 + \log 1.159.$$

log 486 = 2.6866
log 20.7 = 1.3160
log 1.159 = 0.0641
log N = 4.0667

$$
100\left[\begin{array}{l} \log 11{,}700 = 4.0682 \\ x\left[\begin{array}{l}\log N \quad\;\; = 4.0667 \\ \log 11{,}600 = 4.0645\end{array}\right] .0022 \end{array}\right] .0037
$$

$$\frac{x}{100} = \frac{.0022}{.0037}.$$

$$\frac{x}{100} = \frac{22}{37}.$$

$$37x = 2{,}200.$$

$$x = 59.4594 \quad \text{or} \quad 59.$$

$$\text{N} = 11{,}659.$$

39. (0.0321)(7.163)(1.584).

$$N = (0.0321)(7.163)(1.584).$$

$$\log N = \log 0.0321 + \log 7.163 + \log 1.584.$$

$$\begin{array}{l} \log 0.0321 = 8.5065 - 10 \\ \log 7.163 \;\;= 0.8551 \\ \log 1.584 \;\;= 0.1998 \\ \hline \log N = 9.5614 - 10 \end{array}$$

$$
.0010\left[\begin{array}{l} \log 0.3650 = 9.5623 - 10 \\ x\left[\begin{array}{l}\log \quad N \;\; = 9.5614 - 10 \\ \log 0.3640 = 9.5611 - 10\end{array}\right] .0003 \end{array}\right] .0012
$$

$$\frac{x}{.0010} = \frac{.0003}{.0012}.$$

$$\frac{x}{.001} = \frac{3}{12}.$$

$$12x = .003.$$

$$x = .00025 \quad \text{or} \quad .0003.$$

$$N = .3643.$$

40. $\frac{368}{171}$.

$$N = \frac{368}{171}.$$

$$\log N = \log 368 - \log 171.$$

$$\begin{array}{l} \log 368 = 2.5658 \\ \log 171 = 2.2330 \\ \hline \log N \;\;= 0.3328 \end{array}$$

$$
.010\left[\begin{array}{l} \log 2.160 = 0.3345 \\ x\left[\begin{array}{l}\log \quad N \;\; = 0.3328 \\ \log 2.150 = 0.3324\end{array}\right] .0004 \end{array}\right] .0021
$$

$$\frac{x}{.010} = \frac{.0004}{.0021}.$$

$$\frac{x}{.01} = \frac{4}{21}.$$

$$21x = .04.$$

$$x = \text{(approximately) } .0019 \quad \text{or} \quad .002.$$

$$N = 2.152.$$

41. $\frac{43.8}{761}$.

$$N = \frac{43.8}{761}.$$

$$\log N = \log 43.8 - \log 761.$$

$$\log 43.8 = 1.6415.$$

$$\log 761 \;= 2.8814.$$

Since the log of the denominator is larger than the log of the numerator, it is convenient to add 10 and subtract 10 from the characteristic of the numerator.

$$\begin{array}{l} \log 43.8 = 11.6415 - 10 \\ \log 761 \;\;= \;\;2.8814 \\ \hline \log N = \;\;8.7601 - 10 \end{array}$$

$$
.00010\left[\begin{array}{l} \log .05760 = 8.7604 - 10 \\ x\left[\begin{array}{l}\log \quad N \;\; = 8.7601 - 10 \\ \log .05750 = 8.7597 - 10\end{array}\right] .0004 \end{array}\right] .0007
$$

$$\frac{x}{.00010} = \frac{.0004}{.0007}.$$

$$\frac{x}{.0001} = \frac{4}{7}.$$

$$7x = .0004.$$

$$x = \text{(approximately) } .000057 \quad \text{or} \quad .00006.$$

$$N = .05756.$$

42. $\frac{(47.6)(203.6)}{(158)}$.

$$N = \frac{(47.6)(203.6)}{(158)}.$$

$$\log N = \log 47.6 + \log 203.6 - \log 158.$$

The logarithms of the two factors in the numerator are added, since those factors are multiplied together. Then log 158 is subtracted from their sum, since the numerator is divided by the denominator.

$$\begin{array}{r} \log\;\; 47.6 = 1.6776 \\ \log 203.6 = 2.3088 \\ \hline \log 47.6 + \log 203.6 = 3.9864 \end{array}$$

$$\begin{array}{r} \log 47.6 + \log 203.6 = 3.9864 \\ \log 158 \quad = 2.1987 \\ \hline \log N = 1.7877 \end{array}$$

Log N can also be found by use of a colog. Instead of subtracting the log of 158 from the sum of log 47.6 and log 203.6, the colog of 158 is added to that sum. This simplifies the computation, since it can be accomplished in a single operation.

$$\begin{array}{r} \log\;\; 47.6 = 1.6776 \\ \log 203.6 = 2.3088 \\ \text{colog } 158 \quad = 7.8013 - 10 \\ \hline \log N = 11.7877 - 10 \\ \log N = 1.7877. \end{array}$$

$$
.10\left[\begin{array}{l} \log 61.40 = 1.7882 \\ x\left[\begin{array}{l}\log \quad N \;\; = 1.7877 \\ \log 61.30 = 1.7875\end{array}\right] .0002 \end{array}\right] .0007
$$

$$\frac{x}{.10} = \frac{.0002}{.0007}.$$

$$\frac{x}{.1} = \frac{2}{7}.$$

$$7x = .2.$$

$$x = \text{(approximately) } .029 \quad \text{or} \quad .03.$$

$$N = 61.33.$$

43. $\dfrac{(298)}{(47.3)(2.617)}$.

$$\log N = \log 298 - (\log 47.3 + \log 2.617).$$

In this problem the logarithms in the denominator may be added together and their sum subtracted from the logarithm of the numerator. However, it is more convenient to add the colog of each factor in the denominator to the log of the numerator.

$$\log N = \log 298 + \text{colog } 47.3 + \text{colog } 2.617.$$

$$\begin{aligned}
\log 298 &= 2.4742 \\
\text{colog } 47.3 &= 8.3251 - 10 \\
\text{colog } 2.617 &= 9.5822 - 10 \\
\hline
\log N &= 20.3815 - 20 \\
\log N &= 0.3815.
\end{aligned}$$

$$.010 \left[\begin{array}{l} \log 2.410 = 0.3820 \\ x \left[\begin{array}{l} \log\ N = 0.3815 \\ \log 2.400 = 0.3802 \end{array} \right] .0013 \end{array} \right] .0018$$

$$\begin{aligned}
\frac{x}{.010} &= \frac{.0013}{.0018}. \\
\frac{x}{.010} &= \frac{13}{18}. \\
18x &= .13. \\
x &= \text{(approximately) } .0072 \quad \text{or} \quad .007. \\
N &= 2.407.
\end{aligned}$$

44. $\dfrac{(28.6)(7.056)(151.7)}{(26.19)(48.7)}$.

$$N = \frac{(28.6)(7.056)(151.7)}{(26.19)(48.7)}.$$

$$\log N = \log 28.6 + \log 7.056 + \log 151.7 + \text{colog } 26.19 + \text{colog } 48.7.$$

$$\begin{aligned}
\log 28.6 &= 1.4564 \\
\log 7.056 &= 0.8486 \\
\log 151.7 &= 2.1810 \\
\text{colog } 26.19 &= 8.5819 - 10 \\
\text{colog } 48.7 &= 8.3125 - 10 \\
\hline
\log N &= 21.3804 - 20 \\
\log N &= 1.3804.
\end{aligned}$$

$$.10 \left[\begin{array}{l} \log 24.10 = 1.3820 \\ x \left[\begin{array}{l} \log\ N = 1.3804 \\ \log 24.00 = 1.3802 \end{array} \right] .0002 \end{array} \right] .0018$$

$$\begin{aligned}
\frac{x}{.10} &= \frac{.0002}{.0018}. \\
\frac{x}{.1} &= \frac{1}{9}. \\
9x &= .1. \\
x &= \text{(approximately) } .011 \quad \text{or} \quad .01. \\
N &= 24.01.
\end{aligned}$$

45. $\dfrac{(409.6)(32.16)}{(2.57)(29.72)(47.18)}$.

$$N = \frac{(409.6)(32.16)}{(2.57)(29.72)(47.18)}.$$

$$\log N = \log 409.6 + \log 32.16 + \text{colog } 2.57 + \text{colog } 29.72 + \text{colog } 47.18.$$

$$\begin{aligned}
\log 409.6 &= 2.6124 \\
\log 32.16 &= 1.5073 \\
\text{colog } 2.57 &= 9.5901 - 10 \\
\text{colog } 29.72 &= 8.5269 - 10 \\
\text{colog } 47.18 &= 8.3263 - 10 \\
\hline
\log N &= 30.5630 - 30 \\
\log N &= 0.5630.
\end{aligned}$$

$$.010 \left[\begin{array}{l} \log 3.660 = 0.5635 \\ x \left[\begin{array}{l} \log\ N = 0.5630 \\ \log 3.650 = 0.5623 \end{array} \right] .0007 \end{array} \right] .0012$$

$$\begin{aligned}
\frac{x}{.010} &= \frac{.0007}{.0012}. \\
\frac{x}{.01} &= \frac{7}{12} \\
12x &= .07 \\
x &= .0058\overline{3} \quad \text{or} \quad .006 \\
N &= 3.656
\end{aligned}$$

46. $(231)^2$.

Remember, to get the logarithm of a number with an exponent, find the logarithm of the number and multiply it by the exponent (Rule 3, p. 147).

$$\begin{aligned}
N &= (231)^2. \\
\log N &= 2 \log 231 \\
&= 2(2.3636) \\
&= 4.7272.
\end{aligned}$$

$$100 \left[\begin{array}{l} \log 53{,}400 = 4.7275 \\ x \left[\begin{array}{l} \log\ N = 4.7272 \\ \log 53{,}300 = 4.7267 \end{array} \right] .0005 \end{array} \right] .0008$$

$$\begin{aligned}
\frac{x}{100} &= \frac{.0005}{.0008}. \\
\frac{x}{100} &= \frac{5}{8}. \\
8x &= 500. \\
x &= 62.5 \quad \text{or} \quad 63. \\
N &= 53{,}363.
\end{aligned}$$

The actual value of $(231)^2$ is 53,361. The logarithmic evaluation is slightly inaccurate because we used only four-place logs. If we had used five-place logs, we would have obtained the more accurate result 53,360.

47. $(23)^2(15)^3$.

$$\begin{aligned}
N &= (23)^2(15)^3. \\
\log N &= 2 \log 23 + 3 \log 15 \\
&= 2(1.3617) + 3(1.1761) \\
&= 2.7234 + 3.5283 \\
&= 6.2517.
\end{aligned}$$

$$10{,}000 \left[\begin{array}{l} \log 1{,}790{,}000 = 6.2529 \\ x \left[\begin{array}{l} \log\ N = 6.2517 \\ \log 1{,}780{,}000 = 6.2504 \end{array} \right] .0013 \end{array} \right] .0025$$

$$\frac{x}{10{,}000} = \frac{.0013}{.0025}.$$

$$\frac{x}{10{,}000} = \frac{13}{25}.$$

$$25x = 130{,}000.$$

$$x = 5{,}200 \quad \text{or} \quad 5{,}000.$$

$$N = 1{,}785{,}000.$$

48. $\dfrac{47}{(1.09)^3}$.

$$N = \frac{47}{(1.09)^3}.$$

$$\begin{aligned}
\log N &= \log 47 + 3 \operatorname{colog} 1.09 \\
&= 1.6721 + 3(9.9626 - 10) \\
&= 1.6721 + 29.8878 - 30 \\
&= 1.6721 + (9.8878 - 10) \\
&= 11.5599 - 10 \\
&= 1.5599.
\end{aligned}$$

$$N = 36.3.$$

49. $\dfrac{(75)^4}{(21)^3}$.

$$N = \frac{(75)^4}{(21)^3}.$$

$$\begin{aligned}
\log N &= 4 \log 75 - 3 \log 21. \\
&= 4(1.8751) - 3(1.3222) \\
&= 7.5004 - 3.9666 \\
&= 3.5338.
\end{aligned}$$

$$10 \left[\begin{array}{l} \log 3{,}420 = 3.5340 \\ x \left[\begin{array}{l} \log \;\; N = 3.5338 \\ \log 3{,}410 = 3.5328 \end{array}\right] .0010 \end{array}\right] .0012$$

$$\frac{x}{10} = \frac{.0010}{.0012}.$$

$$\frac{x}{10} = \frac{5}{6}.$$

$$6x = 50.$$

$$x = 8.\overline{3} \quad \text{or} \quad 8.$$

$$N = 3{,}418.$$

50. $\dfrac{(2.48)^2(76.71)^2}{(29.2)^3}$.

$$N = \frac{(2.48)^2(76.71)^2}{(29.2)^3}.$$

$$\log N = 2 \log 2.48 + 2 \log 76.71 + 3 \operatorname{colog} 29.2.$$

$$\begin{array}{rcl}
2 \log 2.48 &=& 0.7890 \\
2 \log 76.71 &=& 3.7698 \\
3 \operatorname{colog} 29.2 &=& 5.6038 - 10 \\
\hline
\log N &=& 10.1626 - 10 \\
\log N &=& 0.1626.
\end{array}$$

$$.010 \left[\begin{array}{l} \log 1.460 = 0.1644 \\ x \left[\begin{array}{l} \log \;\; N = 0.1626 \\ \log 1.450 = 0.1614 \end{array}\right] .0012 \end{array}\right] .0030$$

$$\frac{x}{.010} = \frac{.0012}{.0030}.$$

$$\frac{x}{.01} = \frac{2}{5}.$$

$$5x = .02.$$

$$x = .004.$$

$$N = 1.454.$$

51. $\sqrt{478}$.

$$\begin{aligned}
N &= \sqrt{478} \\
&= (478)^{\frac{1}{2}}. \\
\log N &= \tfrac{1}{2} \log 478 \\
&= \tfrac{1}{2}(2.6794) \\
&= 1.3397.
\end{aligned}$$

$$.10 \left[\begin{array}{l} \log 21.90 = 1.3404 \\ x \left[\begin{array}{l} \log \;\; N = 1.3397 \\ \log 21.80 = 1.3385 \end{array}\right] .0012 \end{array}\right] .0019$$

$$\frac{x}{.10} = \frac{.0012}{.0019}.$$

$$\frac{x}{.1} = \frac{12}{19}.$$

$$19x = 1.2.$$

$$x = (\text{approximately})\ .063 \quad \text{or} \quad .06.$$

$$N = 21.86.$$

52. $\sqrt[3]{864.2}$.

$$\begin{aligned}
N &= \sqrt[3]{864.2} \\
&= (864.2)^{\frac{1}{3}}. \\
\log N &= \tfrac{1}{3} \log 864.2 \\
&= \tfrac{1}{3}(2.9366) \\
&= 0.9788\overline{6} \quad \text{or} \quad 0.9789.
\end{aligned}$$

$$.010 \left[\begin{array}{l} \log 9.530 = 0.9791 \\ x \left[\begin{array}{l} \log \;\; N = 0.9789 \\ \log 9.520 = 0.9786 \end{array}\right] .0003 \end{array}\right] .0005$$

$$\frac{x}{.010} = \frac{.0003}{.0005}.$$

$$\frac{x}{.01} = \frac{3}{5}.$$

$$5x = .03.$$

$$x = .006.$$

$$N = 9.526.$$

53. $\sqrt{0.00786}$.

$$\begin{aligned}
N &= \sqrt{0.00786} \\
&= (0.00786)^{\frac{1}{2}}. \\
\log N &= \tfrac{1}{2} \log 0.00786 \\
&= \tfrac{1}{2}(7.8954 - 10) \\
&= 3.9477 - 5 \\
&= 8.9477 - 10.
\end{aligned}$$

$$0.00010 \left[\begin{array}{l} \log 0.08870 = 8.9479 - 10 \\ x \left[\begin{array}{l} \log \;\;\; N = 8.9477 - 10 \\ \log 0.08860 = 8.9474 - 10 \end{array}\right] .0003 \end{array}\right] .0005$$

$$\frac{x}{0.00010} = \frac{.0003}{.0005}.$$
$$\frac{x}{0.0001} = \frac{3}{5}.$$
$$5x = .0003.$$
$$x = .00006.$$
$$N = .08866.$$

54. $\dfrac{\sqrt{73.22}}{\sqrt[3]{96.56}}.$

$$\begin{aligned} N &= \frac{\sqrt{73.22}}{\sqrt[3]{96.56}}. \\ &= \frac{(73.22)^{\frac{1}{2}}}{(96.56)^{\frac{1}{3}}}. \\ \log N &= \tfrac{1}{2}\log 73.22 - \tfrac{1}{3}\log 96.56 \\ &= \tfrac{1}{2}(1.8646) - \tfrac{1}{3}(1.9848) \\ &= 0.9323 - 0.6616 \\ &= 0.2707. \end{aligned}$$

$$.010\ \boxed{\begin{array}{l}\log 1.870 = 0.2718 \\ x\left[\begin{array}{l}\log\ N\ = 0.2707 \\ \log 1.860 = 0.2695\end{array}\right] .0012\end{array}}\ .0023$$

$$\frac{x}{.010} = \frac{.0012}{.0023}.$$
$$\frac{x}{.01} = \frac{12}{23}.$$
$$23x = .12.$$
$$x = \text{(approximately) } .005.$$
$$N = 1.865.$$

55. $71\sqrt{8.752}.$

$$\begin{aligned} N &= 71\sqrt{8.752} \\ &= 71(8.752)^{\frac{1}{2}}. \\ \log N &= \log 71 + \tfrac{1}{2}\log 8.752 \\ &= 1.8513 + \tfrac{1}{2}(0.9421) \\ &= 1.8513 + 0.47105 \\ &= 2.32235 \quad \text{or} \quad 2.3224. \end{aligned}$$

$$1.0\ \boxed{\begin{array}{l}\log 211.0 = 2.3243 \\ x\left[\begin{array}{l}\log\ N\ = 2.3224 \\ \log 210.0 = 2.3222\end{array}\right] .0002\end{array}}\ .0021$$

$$\frac{x}{1.0} = \frac{.0002}{.0021}.$$
$$\frac{x}{1} = \frac{2}{21}.$$
$$x = \text{(approximately) } .095 \quad \text{or} \quad .1.$$
$$N = 210.1.$$

56. $\sqrt[3]{(275)^2}.$

$$\begin{aligned} N &= \sqrt[3]{(275)^2} \\ &= [(275)^2]^{\frac{1}{3}} \\ &= (275)^{\frac{2}{3}}. \\ \log N &= \tfrac{2}{3}\log 275 \\ &= \tfrac{2}{3}(2.4393) \\ &= 1.6262. \end{aligned}$$

$$.10\ \boxed{\begin{array}{l}\log 42.30 = 1.6263 \\ x\left[\begin{array}{l}\log\ N\ = 1.6262 \\ \log 42.20 = 1.6253\end{array}\right] .0009\end{array}}\ .0010$$

$$\frac{x}{.10} = \frac{.0009}{.0010}.$$
$$\frac{x}{.1} = \frac{9}{10}.$$
$$10x = .9.$$
$$x = .09.$$
$$N = 42.29.$$

57. $\sqrt[4]{(26.3)(4.07)^2}.$

$$\begin{aligned} N &= \sqrt[4]{(26.3)(4.07)^2} \\ &= [(26.3)(4.07)^2]^{\frac{1}{4}} \\ &= (26.3)^{\frac{1}{4}}(4.07)^{\frac{1}{2}}. \\ \log N &= \tfrac{1}{4}\log 26.3 + \tfrac{1}{2}\log 4.07 \\ &= \tfrac{1}{4}(1.4200) + \tfrac{1}{2}(0.6096) \\ &= 0.3550 + 0.3048 \\ &= 0.6598. \end{aligned}$$

$$.010\ \boxed{\begin{array}{l}\log 4.570 = 0.6599 \\ x\left[\begin{array}{l}\log\ N\ = 0.6598 \\ \log 4.560 = 0.6590\end{array}\right] .0008\end{array}}\ .0009$$

$$\frac{x}{.010} = \frac{.0008}{.0009}.$$
$$\frac{x}{.01} = \frac{8}{9}.$$
$$9x = .08.$$
$$x = .0089 \quad \text{or} \quad .009.$$
$$N = 4.569.$$

58. $\sqrt{\sqrt{479}}.$

$$\begin{aligned} N &= \sqrt{\sqrt{479}} \\ &= [(479)^{\frac{1}{2}}]^{\frac{1}{2}} \\ &= (479)^{\frac{1}{4}}. \\ \log N &= \tfrac{1}{4}\log 479 \\ &= \tfrac{1}{4}(2.6803) \\ &= 0.6701. \end{aligned}$$

$$.010\ \boxed{\begin{array}{l}\log 4.680 = 0.6702 \\ x\left[\begin{array}{l}\log\ N\ = 0.6701 \\ \log 4.670 = 0.6693\end{array}\right] .0008\end{array}}\ .0009$$

$$\frac{x}{.010} = \frac{.0008}{.0009}.$$
$$\frac{x}{.01} = \frac{8}{9}.$$
$$9x = .08.$$
$$x = .008\bar{8} \quad \text{or} \quad .009.$$
$$N = 4.679.$$

59. $\sqrt[3]{\sqrt{4711}}.$

$$\begin{aligned} N &= \sqrt[3]{\sqrt{4711}} \\ &= [(4711)^{\frac{1}{2}}]^{\frac{1}{3}} \\ &= (4711)^{\frac{1}{6}}. \end{aligned}$$

$$\log N = \tfrac{1}{6}\log 4711$$
$$= \tfrac{1}{6}(3.6731)$$
$$= 0.61218\bar{3} \quad \text{or} \quad 0.6122.$$

$$.010 \left[\begin{array}{l} \log 4.100 = 0.6128 \\ x \left[\begin{array}{l} \log\ N \ \ = 0.6122 \\ \log 4.090 = 0.6117 \end{array} \right] .0005 \end{array} \right] .0011$$

$$\frac{x}{.010} = \frac{.0005}{.0011}.$$
$$\frac{x}{.01} = \frac{5}{11}.$$
$$11x = .05.$$
$$x = .0045 \quad \text{or} \quad .005.$$
$$N = 4.095.$$

60. $\sqrt[3]{\sqrt{(6.03)^3}}$.

$$N = \sqrt[3]{\sqrt{(6.03)^3}}$$
$$= \{[(6.03)^3]^{\frac{1}{2}}\}^{\frac{1}{3}}$$
$$= [(6.03)^{\frac{3}{2}}]^{\frac{1}{3}}$$
$$= (6.03)^{\frac{1}{2}}.$$
$$\log N = \tfrac{1}{2}\log 6.03$$
$$= \tfrac{1}{2}(0.7803)$$
$$= 0.39015 \quad \text{or} \quad 0.3902.$$

$$.010 \left[\begin{array}{l} \log 2.460 = 0.3909 \\ x \left[\begin{array}{l} \log\ N \ \ = 0.3902 \\ \log 2.450 = 0.3892 \end{array} \right] .0010 \end{array} \right] .0017$$

$$\frac{x}{.010} = \frac{.0010}{.0017}.$$
$$\frac{x}{.01} = \frac{10}{17}.$$
$$17x = .1$$
$$x = \text{(approximately) } .0059 \quad \text{or} \quad .006.$$
$$N = 2.456.$$

SUPPLEMENTARY PROBLEMS

Find the common logarithm of the following numbers.

1. 31.71
2. 401.8
3. 0.7846
4. 0.009671
5. 0.00008345
6. 9.877

Find the number whose logarithms are given below.

7. 3.0471
8. 1.6783
9. 0.3862
10. 7.4071 − 10
11. 9.9354 − 10
12. 8.0669 − 10

Find the value of the following by the use of logarithms.

13. $(37)^2$
14. $(17)^3$
15. $(401)^2$
16. $\sqrt{671}$
17. $\sqrt[5]{47}$
18. $\sqrt[3]{40.31}$
19. $(2.038)(37.45)$
20. $(46.08)(2.706)$
21. $(47)(8.61)(0.07382)$
22. $\dfrac{231}{196}$
23. $\dfrac{87.04}{123}$
24. $\dfrac{831.7}{0.0963}$
25. $\dfrac{(23.4)(2.782)}{4.091}$
26. $\dfrac{47.83}{(21.67)(0.00831)}$
27. $\dfrac{(409.1)(2.86)}{(0.173)(67.39)}$
28. $\sqrt{\dfrac{275.8}{40.07}}$
29. $\sqrt{43\sqrt[3]{(67.8)^2}}$
30. $\sqrt{\sqrt[3]{\sqrt{2319}}}$

Chapter 15

Progressions

Arithmetic Progressions. A sequence of numbers each of which can be obtained from the one that precedes it by adding a fixed quantity is known as an **arithmetic progression** (abbreviated **A.P.**). The fixed quantity is known as the **common difference** between any two terms in the sequence.

In this discussion, the following letters will be used to denote certain parts of the progression.

a = the first term;
l = the last term;
n = the number of terms in the sequence;
d = the common difference.

If a is the first term, the second one will be obtained by adding the common difference d to it.

a = the first term;
$a + d$ = the second term;
$a + 2d$ = the third term;
$a + 3d$ = the fourth term.

It can be seen that the coefficient of the d in any term is one less than the number of that term.

Therefore, the formula for the last (or nth) term in an arithmetic progression would be

$$l = a + (n - 1)d.$$

1. What is the last term in a sequence of 8 terms that starts 3, 6, 9, . . .?

$$a = 3; \quad n = 8; \quad d = 3.$$

$$\begin{aligned} l &= a + (n - 1)d \\ &= 3 + (8 - 1)3 \\ &= 3 + (7)3 \\ &= 3 + 21 \\ &= 24. \end{aligned}$$

2. What is the 12th term in the arithmetic progression 2, 6, 10, . . .?

$$a = 2; \quad n = 12; \quad d = 4.$$

$$\begin{aligned} l &= a + (n - 1)d \\ &= 2 + (12 - 1)4 \\ &= 2 + (11)4 \\ &= 2 + 44 \\ &= 46. \end{aligned}$$

3. What is the 9th term in the A.P. 8, 6, 4, . . .?

$$a = 8; \quad n = 9; \quad d = -2.$$

$$\begin{aligned} l &= a + (n - 1)d \\ &= 8 + (9 - 1)(-2) \\ &= 8 + (8)(-2) \\ &= 8 - 16 \\ &= -8. \end{aligned}$$

4. If the 8th term in an A.P. is 19 and the 17th term is 37, what are the first six terms of the progression?

Use the formula $l = a + (n - 1)d$, assuming 8 to be the number of the last term and 19 the value of that term.

$$l = a + (n - 1)d.$$
$$19 = a + (8 - 1)d.$$
$$19 = a + 7d. \quad (1)$$

Use the formula again, assuming 37 to be the value of the last or 17th term.

$$l = a + (n - 1)d.$$
$$37 = a + (17 - 1)d.$$
$$37 = a + 16d. \quad (2)$$

Solve equations (1) and (2) simultaneously.

$$\begin{aligned} 37 &= a + 16d \quad (2) \\ 19 &= a + 7d \quad (1) \\ \hline 18 &= 9d \end{aligned}$$

$$9d = 18.$$
$$d = 2. \quad (3)$$

Substitute the value of d from (3) in (1).

$$19 = a + 7d. \quad (1)$$
$$19 = a + (7)(2).$$
$$19 = a + 14.$$
$$a = 5.$$

Since the first term is 5 and the common difference is 2, the first six terms are 5, 7, 9, 11, 13, and 15.

5. In the A.P. 2, $3\frac{1}{2}$, 5, . . ., what term has a value of 20?

$$l = a + (n - 1)d.$$
$$20 = 2 + (n - 1)\tfrac{3}{2}.$$
$$20 = 2 + \frac{3(n-1)}{2}.$$

Clear the equation of fractions and solve for n.

$$40 = 4 + 3n - 3.$$
$$-3n = -40 + 4 - 3.$$
$$-3n = -39.$$
$$n = 13.$$

Hence, 20 is the 13th term.

Sum of an Arithmetic Progression. In certain types of problems it is required that we find the sum of all the terms in an arithmetic progression. The following formula is used to find that sum.

$$S = \frac{n}{2}(a + l).$$

If the last term is unknown, the above equation is not convenient to use. It can be converted into a more convenient one by substituting the value of l in it.

$$S = \frac{n}{2}[2a + (n - 1)d].$$

6. What is the sum of the first 31 terms in the A.P. 1, 4, 7, . . .?

$$S = \frac{n}{2}[2a + (n - 1)d]$$
$$= \tfrac{31}{2}[2(1) + (31 - 1)3]$$
$$= \tfrac{31}{2}[2 + (30)(3)]$$
$$= \tfrac{31}{2}[2 + 90]$$
$$= \tfrac{31}{2}[92]$$
$$= 1{,}426.$$

7. What is the sum of the first 14 terms in the A.P. $3\frac{1}{2}$, $2\frac{1}{2}$, $1\frac{1}{2}$, . . .?

$$S = \frac{n}{2}[2a + (n - 1)d]$$
$$= \tfrac{14}{2}[(2)(3\tfrac{1}{2}) + (14 - 1)(-1)]$$
$$= 7[7 + 13(-1)]$$
$$= 7[7 - 13]$$
$$= 7[-6]$$
$$= -42.$$

8. Find the 11th term in the A.P. 11, 16, 21, . . . and the sum of the first 11 terms.

$$l = a + (n - 1)d$$
$$= 11 + (11 - 1)5$$
$$= 11 + (10)5$$
$$= 11 + 50$$
$$= 61.$$

$$S = \frac{n}{2}(a + l)$$
$$= \tfrac{11}{2}(11 + 61)$$
$$= \tfrac{11}{2}(72)$$
$$= 396.$$

9. Find the 16th term in the A.P. 17, 11, 5, . . . and the sum of the first 16 terms.

$$l = a + (n - 1)d$$
$$= 17 + (16 - 1)(-6)$$
$$= 17 + (15)(-6)$$
$$= 17 - 90$$
$$= -73.$$

$$S = \frac{n}{2}(a + l)$$
$$= \tfrac{16}{2}[17 + (-73)]$$
$$= 8[17 - 73]$$
$$= 8[-56]$$
$$= -448.$$

10. Find the 22nd term in the A.P. $7\frac{1}{4}$, $6\frac{1}{2}$, $5\frac{3}{4}$, . . . and the sum of the first 22 terms.

The problem is easier to solve if the given terms in the progression are changed to fractions.

$$7\tfrac{1}{4}, 6\tfrac{1}{2}, 5\tfrac{3}{4}, \ldots = \tfrac{29}{4}, \tfrac{26}{4}, \tfrac{23}{4}, \ldots.$$
$$l = a + (n - 1)d$$
$$= \tfrac{29}{4} + (22 - 1)(-\tfrac{3}{4})$$
$$= \tfrac{29}{4} + (21)(-\tfrac{3}{4})$$
$$= \tfrac{29}{4} - \tfrac{63}{4}$$
$$= -\tfrac{34}{4}$$
$$= -\tfrac{17}{2}.$$

$$S = \frac{n}{2}(a + l)$$
$$= \tfrac{22}{2}[\tfrac{29}{4} + (-\tfrac{17}{2})]$$
$$= 11[\tfrac{29}{4} - \tfrac{34}{4}]$$
$$= 11[-\tfrac{5}{4}]$$
$$= -\tfrac{55}{4}$$
$$= -13\tfrac{3}{4}.$$

Arithmetic Means. The first and last terms of an A.P. are called the **extremes** of the progression. All the terms between the first and last are known as the **arithmetic means.** The number of arithmetic means in a progression will obviously be two less than the number of terms. (These extremes and means should not be confused with those of a proportion.)

11. Insert five arithmetic means between 3 and 15.

$$a = 3;$$
$$n = 7;$$
$$l = 15.$$
$$l = a + (n - 1)d.$$
$$15 = 3 + (7 - 1)d.$$
$$15 = 3 + 6d.$$
$$-6d = -15 + 3.$$
$$-6d = -12.$$
$$d = 2.$$

Hence, the five arithmetic means are 5, 7, 9, 11, and 13.

12. Insert four arithmetic means between 17 and 57.

$$a = 17;$$
$$n = 6;$$
$$l = 57.$$
$$l = a + (n - 1)d.$$
$$57 = 17 + (6 - 1)d.$$
$$57 = 17 + 5d.$$
$$-5d = -57 + 17.$$
$$-5d = -40.$$
$$d = 8.$$

Hence, the four arithmetic means are 25, 33, 41, and 49.

13. Insert six arithmetic means between 21 and −21.

$$a = 21;$$
$$n = 8;$$
$$l = -21.$$
$$l = a + (n - 1)d.$$
$$-21 = 21 + (8 - 1)d.$$
$$-21 = 21 + 7d.$$
$$-7d = 21 + 21.$$
$$-7d = 42.$$
$$d = -6.$$

Hence, the six arithmetic means are 15, 9, 3, −3, −9, and −15.

14. A man bought an automobile on which the mortgage was $1,500.00, to be paid off at the rate of $50 per month for 30 months. If he paid the interest at the end of the month with each payment and the interest rate was 1% per month on the unpaid balance, what was the total amount he paid the holder of the mortgage?

The payments form an arithmetic progression of 30 terms.

$$\text{amount of first payment} = \$50 + 1\% \text{ of } \$1500 = \$65.$$
$$\text{amount of last payment} = \$50 + 1\% \text{ of } \$50 = \$50.50.$$

The sum of all the payments can be found by the use of the formula for the sum of an arithmetic progression.

$$S = \frac{n}{2}(a + l)$$
$$= \tfrac{30}{2}(\$65 + \$50.50)$$
$$= 15(\$115.50)$$
$$= \$1,732.50.$$

15. A man accepts a position at an annual starting salary of $3,600, with the provision that he will receive an increase of $300 per year for 15 years if his services are satisfactory. What will be his annual salary in the 15th year, and how much will he have been paid at the end of 15 years?

$$a = \$3,600;$$
$$n = 15;$$
$$d = \$300.$$
$$l = a + (n - 1)d$$
$$= \$3,600 + (15 - 1)(\$300)$$
$$= \$3,600 + 14\ (\$300)$$
$$= \$3,600 + \$4,200$$
$$= \$7,800.$$

$$S = \frac{n}{2}(a + l)$$
$$= \tfrac{15}{2}(\$3,600 + \$7,800)$$
$$= 7.5(\$11,400)$$
$$= \$85,500.$$

Geometric Progressions. A sequence of numbers so related that each one after the first is obtained by multiplying the preceding one by a fixed constant is known as a **geometric progression** (abbreviated **G.P.**). The fixed constant is called the **common ratio.**

As with the arithmetic progression, the first and last terms of a geometric progression are called the **extremes** and the terms between the extremes are known as the **geometric means.**

The following letters are used in the formulas dealing with geometric progressions.

a = the first term;
l = the last term;
n = the number of terms;
r = the common ratio;
S = the sum of the terms.

By definition, each term after the first is found by multiplying the preceding term by the common ratio. Hence, the first five terms would be

a = the first term;
ar = the second term;
ar^2 = the third term;
ar^3 = the fourth term;
ar^4 = the fifth term.

It can be seen from the above that the power of r for any term is one less than the number of the term. Hence, the last term would be represented by the formula

$$l = ar^{n-1}.$$

16. Find the 7th term of the G.P. 4, 8, 16, ...

$$n = 7;$$
$$a = 4;$$
$$r = 2.$$
$$l = ar^{n-1}$$
$$= 4(2)^6$$
$$= 4(64)$$
$$= 256.$$

17. Find the 9th term of the G.P. 16, 8, 4, ...

$$a = 16;\quad r = \tfrac{1}{2};\quad n = 9.$$

$$l = ar^{n-1} = 16(\tfrac{1}{2})^8 = (16)(\tfrac{1}{256}) = \tfrac{16}{256} = \tfrac{1}{16}.$$

18. Find the 7th term in the G.P. 10, −20, 40, ...

$$a = 10;\quad r = -2;\quad n = 7.$$

$$l = ar^{n-1} = 10(-2)^6 = 10(64) = 640.$$

19. Find the 5th term of the G.P. $\frac{1}{4}, \frac{1}{2}, 1, \ldots$.

$$a = \tfrac{1}{4};\quad r = 2;\quad n = 5.$$

$$l = ar^{n-1} = (\tfrac{1}{4})(2)^4 = (\tfrac{1}{4})(16) = 4.$$

20. Find the 8th term of the G.P. $-3\frac{1}{2}, 1\frac{3}{4}, -\frac{7}{8}, \ldots$.

Change each term to its fractional equivalent, and the G.P. will read: $-\frac{7}{2}, \frac{7}{4}, -\frac{7}{8}, \ldots$.

$$a = -\tfrac{7}{2};\quad r = -\tfrac{1}{2};\quad n = 8.$$

$$l = ar^{n-1} = (-\tfrac{7}{2})(-\tfrac{1}{2})^7 = (-\tfrac{7}{2})(-\tfrac{1}{128}) = \tfrac{7}{256}.$$

Sum of a Geometric Progression. The formula for finding the sum of the terms in a geometric progression is

$$S = \frac{ar^n - a}{r - 1}$$

or

$$S = \frac{a(r^n - 1)}{r - 1}.$$

21. Find the sum of the first 9 terms in the G.P. 2, 6, 18, ...

$$a = 2;\quad r = 3;\quad n = 9.$$

$$S = \frac{a(r^n - 1)}{r - 1} = \frac{2[(3)^9 - 1]}{3 - 1} = \frac{2[19{,}683 - 1]}{2} = 19{,}682.$$

22. Find the sum of the first 7 terms in the G.P. 7, −14, 28, ...

$$a = 7;\quad r = -2;\quad n = 7.$$

$$S = \frac{a(r^n - 1)}{r - 1} = \frac{7[(-2)^7 - 1]}{-2 - 1} = \frac{7[-128 - 1]}{-3} = \frac{7[-129]}{-3} = \frac{-903}{-3} = 301.$$

23. Find the sum of the first 8 terms in the G.P. $\frac{1}{4}, -\frac{1}{8}, \frac{1}{16}, \ldots$.

$$a = \tfrac{1}{4};\quad r = -\tfrac{1}{2};\quad n = 8.$$

$$S = \frac{a(r^n - 1)}{r - 1} = \frac{\frac{1}{4}[(-\frac{1}{2})^8 - 1]}{-\frac{1}{2} - 1} = \frac{\frac{1}{4}[\frac{1}{256} - 1]}{-\frac{3}{2}} = \tfrac{1}{4}(-\tfrac{2}{3})(\tfrac{1}{256} - \tfrac{256}{256}) = -\tfrac{1}{6}(-\tfrac{255}{256}) = \frac{85}{2(256)} = \tfrac{85}{512}.$$

24. Insert two geometric means between 3 and 24.

$$a = 3;\quad n = 4;\quad l = 24.$$

$$l = ar^{n-1}.\quad 24 = (3)(r)^3.\quad r^3 = 8.\quad r = 2.$$

Hence, the two geometric means are 6 and 12.

25. Insert four geometric means between 4 and -128.

$$a = 4;$$
$$n = 6;$$
$$l = -128.$$
$$l = ar^{n-1}.$$
$$-128 = (4)(r)^5.$$
$$r^5 = -32.$$
$$r = -2.$$

Hence, the four geometric means are -8, 16, -32, and 64.

26. Insert three geometric means between $\frac{5}{2}$ and $\frac{40}{81}$.

$$a = \tfrac{5}{2};$$
$$n = 5;$$
$$l = \tfrac{40}{81}.$$
$$l = ar^{n-1}.$$
$$\tfrac{40}{81} = \tfrac{5}{2}r^4.$$
$$r^4 = \tfrac{16}{81}.$$
$$r = \pm\tfrac{2}{3}.$$

Hence, the three geometric means are $\frac{5}{3}$, $\frac{10}{9}$, and $\frac{20}{27}$ or $-\frac{5}{3}$, $\frac{10}{9}$, and $-\frac{20}{27}$.

27. Insert four geometric means between $-\frac{3}{4}$ and $\frac{3}{128}$.

$$a = -\tfrac{3}{4};$$
$$n = 6;$$
$$l = \tfrac{3}{128}.$$
$$l = ar^{n-1}.$$
$$\tfrac{3}{128} = -\tfrac{3}{4}r^5.$$
$$r^5 = -\tfrac{4}{128}$$
$$= -\tfrac{1}{32}.$$
$$r = -\tfrac{1}{2}.$$

Hence, the four geometric means are $\frac{3}{8}$, $-\frac{3}{16}$, $\frac{3}{32}$, and $-\frac{3}{64}$.

28. Find the number of terms in the G.P. 2, 4, 8, . . . whose sum is 4,094.

$$a = 2;$$
$$r = 2;$$
$$S = 4{,}094.$$
$$S = \frac{a(r^n - 1)}{r - 1}.$$
$$4{,}094 = \frac{2(2^n - 1)}{2 - 1}.$$
$$4{,}094 = 2(2^n) - 2.$$
$$2(2^n) = 4{,}096.$$
$$2^n = 2{,}048$$
$$= 2^{11}.$$

Hence, $n = 11$ terms.

29. A city school district board of trustees has been told by statisticians that they can expect the school population to triple every five years for the next fifteen years. If the present school population is 12,000 students, how many students must they plan facilities for at the end of fifteen years?

a = present number of students;
$n - 1$ = number of 5-year intervals;
r = ratio of increase for each interval;
l = number of students after 15 years.

$$a = 12{,}000;$$
$$n - 1 = 3;$$
$$r = 3.$$
$$l = ar^{n-1}$$
$$= (12{,}000)(3)^3$$
$$= (12{,}000)(27)$$
$$= 324{,}000 \text{ students.}$$

30. An amateur gardener bought three expensive bulbs of his favorite rare flower. He was assured that at the end of the season each bulb would produce two new ones that could be separated from the original and thus have three flowering plants the next year from each original bulb. If he continued the process for five years, how many bulbs could he expect to have to start the sixth season?

$$a = 3;$$
$$r = 3;$$
$$n - 1 = 5.$$
$$l = ar^{n-1}$$
$$= 3 \cdot 3^5$$
$$= 3(243)$$
$$= 729.$$

31. How many ancestors (parents, grandparents, great-grandparents, etc.) would an individual have in ten generations?

$a = 2$ (the parents);
$n = 10$;
$r = 2$ (two parents for each parent).

$$S = \frac{a(r^n - 1)}{r - 1}$$
$$= \frac{2(2^{10} - 1)}{2 - 1}$$
$$= \frac{2(1024 - 1)}{1}$$
$$= 2(1023)$$
$$= 2{,}046.$$

32. A man decided to save money for fifteen days. The first day he put one cent into his savings account, the second day two cents, the third day four cents, and each day thereafter double the amount he had deposited the preceding day. How much had he saved at the end of fifteen days?

$a = 1$ cent;
$n = 15$;
$r = 2$.

$$S = \frac{a(r^n - 1)}{r - 1}$$
$$= \frac{1(2^{15} - 1)}{2 - 1}$$
$$= \frac{2^{15} - 1}{1}$$
$$= 2^{15} - 1$$
$$= 32{,}768 - 1$$
$$= 32{,}767 \text{ cents}$$
$$= \$327.67.$$

SUPPLEMENTARY PROBLEMS

Solve the following problems.

1. Find the 9th term of the following: 3, 7, 11, ...

2. Find the 17th term of the following: 16, 13, 10, ...

3. Find the 6th term of the following: $-2, -6, -18, \ldots$.

4. Find the 7th term of the following: 2, $\frac{4}{3}$, $\frac{8}{9}$, ...

5. Insert three geometric means between 24 and $\frac{3}{2}$.

6. Insert four geometric means between 48 and $6\frac{26}{81}$.

7. Insert six arithmetic means between 5 and 26.

8. Insert six arithmetic means between 1 and 2.

9. Find the sum of the first 30 odd integers.

10. Find the sum of the first eight terms in the following: 22, 27, 32, ...

11. Find the sum of the first five terms in the following: 81, $40\frac{1}{2}$, $20\frac{1}{4}$, ...

12. In the first term in an arithmetic progression is -6, the last term is 10 and the common differences is 2, how many terms are there in the progression?

13. If -2 is the first term in an arithmetic progression, 26 is the last term and there are 9 terms, what is the common difference?

14. If, in an arithmetic progression, $a = 26$, $l = -10$, $d = -3$, what is n?

15. If, in an arithmetic progression, $S = 99$, $l = 29$, $a = 4$, and $d = 5$, what is n?

16. If, in an arithmetic progression the 4th term is 5 and the 7th term is 9, what is d?

17. If, in a geometric progression, $a = \frac{2}{3}$, $n = 5$ and the last term is $\frac{32}{243}$, what is r?

18. If, in a geometric progression, $r = -\frac{2}{3}$, $l = \frac{4}{27}$ and $n = 5$, what is a?

19. If, in a geometric progression, the third term is $5\frac{3}{4}$ and the 5th term is $1\frac{7}{16}$, what is a?

20. If, in a geometric progression, $a = -\frac{3}{4}$, $r = -\frac{3}{4}$, and $l = -\frac{243}{1024}$, what is n?

Chapter 16

The Binomial Theorem

Expansions. The binomial theorem is used in raising a binomial to any power. Consider the following expansions:

$$(x + y)^2 = x^2 + 2xy + y^2.$$
$$(x + y)^3 = x^3 + 3x^2y + 3xy^2 + y^3.$$
$$(x + y)^4 = x^4 + 4x^3y + 6x^2y^2 + 4xy^3 + y^4.$$

Let n be any power to which the binomial $x + y$ is raised. The following conditions exist for the expansion of $(x + y)^n$.

1. The number of terms is equal to $n + 1$.
2. The first term is x^n; the last term is y^n.
3. The exponent of x decreases by 1 from term to term; the exponent of y increases by 1 from term to term.
4. The sum of the exponents of x and y in any term is n.
5. The coefficient of the first term is 1; the coefficient of the second term is n; the coefficient of any term after the first is the coefficient of the preceding term multiplied by the exponent of the x-factor of the preceding term divided by the number of the preceding term. Thus, for example, the coefficient of the third term of
$$(a + b)^{10} = a^{10} + 10a^9b + \ldots + b^{10}$$
could be found from the second term as follows.
$$\frac{(10)(9)}{2} = \frac{90}{2} = 45.$$
6. The coefficients of terms equidistant from the ends of the expansion are the same.

By making use of these conditions, any expression of the form $(x + y)^n$ can be expanded. Note that in the expansion of an expression of the form $(x - y)^n$, the signs will alternate, starting with a positive value for the first term.

1. Expand $(a + c)^3$.

$$(a + c)^3 = a^3 + 3a^2c + 3ac^2 + c^3.$$

2. Expand $(m - 2)^3$.

$$(m - 2)^3 = m^3 - 3m^2(2) + 3m(2)^2 - (2)^3$$
$$= m^3 - 6m^2 + 12m - 8.$$

3. Expand $(x^2 + y^2)^3$.

$$(x^2 + y^2)^3 = (x^2)^3 + 3(x^2)^2(y^2) + 3(x^2)(y^2)^2 + (y^2)^3$$
$$= x^6 + 3x^4y^2 + 3x^2y^4 + y^6.$$

4. Expand $(3 + p)^4$.

$$(3 + p)^4 = (3)^4 + 4(3)^3p + 6(3)^2(p)^2 + 4(3)(p)^3 + (p)^4$$
$$= 81 + 108p + 54p^2 + 12p^3 + p^4.$$

5. Expand $(x + 1)^6$.

$$(x + 1)^6 = x^6 + 6x^5(1) + 15x^4(1)^2 + 20x^3(1)^3 + 15x^2(1)^4 + 6x(1)^5 + (1)^6.$$
$$= x^6 + 6x^5 + 15x^4 + 20x^3 + 15x^2 + 6x + 1.$$

6. Expand $(2x + 5)^4$.

$$(2x + 5)^4 = (2x)^4 + 4(2x)^3(5) + 6(2x)^2(5)^2 + 4(2x)(5)^3 + (5)^4$$

Note that we do *not* take coefficients of the terms of the unexpanded expression into consideration when determining coefficients of the terms of the expanded expression. Only after the coefficients of the expansion are determined are they multiplied by the coefficients of the terms of the unexpanded expression.

$$= 16x^4 + 4(8x^3)(5) + 6(4x^2)(25) + 8x(125) + 625$$
$$= 16x^4 + 160x^3 + 600x^2 + 1000x + 625.$$

7. Expand $(x + 3)^5$.

$$(x + 3)^5 = x^5 + 5x^4(3) + 10x^3(3)^2 + 10x^2(3)^3 + 5x(3)^4 + (3)^5$$
$$= x^5 + 5x^4(3) + 10x^3(9) + 10x^2(27) + 5x(81) + 243$$
$$= x^5 + 15x^4 + 90x^3 + 270x^2 + 405x + 243.$$

8. Expand $(3x - 2)^4$.

$$(3x - 2)^4 = (3x)^4 - 4(3x)^3(2) + 6(3x)^2(2)^2 - 4(3x)(2)^3 + (2)^4$$
$$= 81x^4 - 4(27x^3)(2) + 6(9x^2)(4) - 12x(8) + 16$$
$$= 81x^4 - 216x^3 + 216x^2 - 96x + 16.$$

9. Expand $(x + y + z)^2$.

By regrouping, and thinking of the $x + y$ as the first term, this problem can be solved by use of the binomial theorem as shown below.

$$[(x+y)+z]^2 = (x+y)^2 + 2(x+y)(z) + z^2$$
$$= x^2 + 2xy + y^2 + 2xz + 2yz + z^2.$$

10. Expand $(m - n + p)^2$.

Regroup this as shown below and solve by use of the Binomial Theorem.

$$[(m-n)+p]^2 = (m-n)^2 + 2(m-n)(p) + p^2$$
$$= m^2 - 2mn + n^2 + 2mp - 2np + p^2.$$

11. Expand $(2x - y - 2z)^3$.

Regroup as shown below.

$$[(2x-y)-2z]^3 = (2x-y)^3 - 3(2x-y)^2(2z)$$
$$+ 3(2x-y)(2z)^2 - (2z)^3$$

Use the binomial theorem to expand the first term and the second term.

$$= [(2x)^3 - 3(2x)^2(y) + 3(2x)(y^2) - y^3]$$
$$- 6z[(2x)^2 - 2(2x)(y) + y^2]$$
$$+ (2x-y)(12z^2) - 8z^3$$
$$= 8x^3 - 12x^2y + 6xy^2 - y^3 - 24x^2z$$
$$+ 24xyz - 6y^2z + 24xz^2 - 12yz^2 - 8z^3.$$

12. Expand $(a - b + c - d)^2$.

$$(a-b+c-d)^2 = [(a-b)+(c-d)]^2$$
$$= (a-b)^2 + 2(a-b)(c-d) + (c-d)^2$$
$$= a^2 - 2ab + b^2 + 2ac - 2ad - 2bc$$
$$+ 2bd + c^2 - 2cd + d^2.$$

13. Expand $(x - y + a + b)^2$.

$$(x-y+a+b)^2 = [(x-y)+(a+b)]^2$$
$$= (x-y)^2 + 2(x-y)(a+b) + (a+b)^2$$
$$= x^2 - 2xy + y^2 + 2ax - 2ay + 2bx$$
$$- 2by + a^2 + 2ab + b^2.$$

14. Find the value of $(2.03)^6$ by use of the binomial theorem.

$$(2.03)^6 = (2 + 0.03)^6$$
$$= 2^6 + 6(2)^5(0.03) + 15(2)^4(0.03)^2 + 20(2)^3(0.03)^3$$
$$+ 15(2)^2(0.03)^4 + 6(2)(0.03)^5 + (0.03)^6$$
$$= 64 + 6(32)(0.03) + 15(16)(0.0009)$$
$$+ 20(8)(0.000027) + 15(4)(0.00000081)$$
$$+ 6(2)(0.0000000243) + (0.000000000729).$$

Because they have no effect on the figure in the second decimal place, the last three terms will be discarded.

$$(2.03)^6 = 64 + 5.76 + 0.216 + 0.00432$$
$$= 69.98032 \quad \text{or} \quad 69.98.$$

15. Find the square root of 96 by use of the binomial theorem.

$$\sqrt{96} = (96)^{\frac{1}{2}}$$
$$= (100 - 4)^{\frac{1}{2}}$$
$$= (10^2 - 2^2)^{\frac{1}{2}}$$
$$= (10^2)^{\frac{1}{2}} - \tfrac{1}{2}(10^2)^{-\frac{1}{2}}(2^2)^1 + \tfrac{1}{8}(10^2)^{-\frac{3}{2}}(2^2)^2$$
$$- \tfrac{1}{16}(10^2)^{-\frac{5}{2}}(2^2)^3 + \cdots.$$

After four terms the amount added or subtracted has no effect on the figure in the second decimal place. Hence, there is no need to carry the expansion any further.

$$\sqrt{96} = (10^2)^{\frac{1}{2}} - \tfrac{1}{2}(10^2)^{-\frac{1}{2}}(2^2) + \tfrac{1}{8}(10^2)^{-\frac{3}{2}}(2^2)^2$$
$$- \tfrac{1}{16}(10^2)^{-\frac{5}{2}}(2^2)^3$$
$$= 10 - \tfrac{1}{2}(10^{-1})(2^2) + \tfrac{1}{8}(10^{-3})(2^4) - \tfrac{1}{16}(10^{-5})(2^6)$$
$$= 10 - \frac{1}{2}\left(\frac{1}{10}\right)(4) + \frac{1}{8}\left(\frac{1}{10^3}\right)(16) - \frac{1}{16}\left(\frac{1}{10^5}\right)(64)$$
$$= 10 - 0.2 + 0.002 - 0.00004$$
$$= 9.80196 \quad \text{or} \quad 9.80.$$

***N* Factorial.** The expression n factorial (written $n!$) means the product of n and all positive integers less than n. Examples of this concept are: $6! = 6 \cdot 5 \cdot 4 \cdot 3 \cdot 2 \cdot 1$; $3! = 3 \cdot 2 \cdot 1$; $n! = n(n-1)(n-2) \cdots$ until the last factor is 1.

The expansions of binomials are sometimes very long and complicated. There are times when only a certain term in the expansion is needed. The following formula may be used to find any term in the expansion of a binomial without knowing the preceding term:

$$r\text{th term} = \frac{n(n-1)(n-2)\cdots(n-r+2)}{(r-1)!} x^{n-r+1}y^{r-1}.$$

Remember that even-number terms of expansions of the form $(x - y)^n$ will be negative.

16. Find the fifth term in the expansion of $(a + x)^7$.

$$r = 5;$$
$$n = 7.$$

The last term in the numerator will be $(n - r + 2)$ or $7 - 5 + 2$ or 4. Then

$$5\text{th term} = \frac{7 \cdot 6 \cdot 5 \cdot 4}{4 \cdot 3 \cdot 2 \cdot 1} a^3x^4$$
$$= 35a^3x^4.$$

17. Find the seventh term in the expansion of $(m^2 - 1)^9$.

$$r = 7;$$
$$n = 9.$$
$$7\text{th term} = \frac{9 \cdot 8 \cdot 7 \cdot 6 \cdot 5 \cdot 4}{6 \cdot 5 \cdot 4 \cdot 3 \cdot 2 \cdot 1}(m^2)^3(1)^6$$
$$= 84m^6.$$

18. Find the fourth term in the expansion of $(2x - 3y^2)^8$.

$$r = 4;$$
$$n = 8.$$
$$4\text{th term} = -\frac{8 \cdot 7 \cdot 6}{3 \cdot 2 \cdot 1}(2x)^5(3y^2)^3$$
$$= -56(32x^5)(27y^6)$$
$$= -48{,}384x^5y^6.$$

SUPPLEMENTARY PROBLEMS

Expand each of the following.

1. $(a + x)^5$
2. $(m - t)^7$
3. $(2m + 3p)^5$
4. $(3d - 1)^4$
5. $\left(\frac{d}{3} + \frac{c}{2}\right)^5$
6. $\left(\frac{2m}{3} - \frac{3p}{2}\right)^4$
7. $\left(\frac{1}{a - b}\right)^4$
8. $\left(4 - \frac{x}{3}\right)^5$
9. What is the 5th term of $(a - 3b)^9$?
10. What is the 4th term of $(x - y)^6$?
11. Expand $(x - y + z)^3$ by use of the Binomial Theorem.
12. Expand $(2a - b + c)^3$ by use of the Binomial Theorem.
13. Find the value of $(1.03)^5$, to the nearest thousandth, by use of the Binomial Theorem.
14. Find the value of $(0.96)^4$, to the nearest hundredth, by use of the Binomial Theorem.
15. Find the square of 104 by use of the Binomial Theorem.

Chapter 17

Permutations and Combinations

The Fundamental Principle. *If one thing can be done in* h *ways and a second thing can be done in* k *ways, then the two things can be done in the stated order* h · k *ways.*

Any act we do in a given number of ways is thought of as a positive number of ways. Therefore, in this discussion, we shall use only the positive integers.

As an illustration of the above Principle let us assume an employer has two positions to fill in an office—one for a file clerk and the other for a typist. He has three applicants for the position as file clerk and four for the position as typist. In how many different ways can the pair of vacancies be filled? For each of the file clerks who have applied there are four possible typists from which to make a choice. Thus, if *FC* represents the applicants for the position of file clerk and *T* represents the applicants for the position of typist, then the employer may hire any one of the following pairs.

1st *FC*—1st *T*	2nd *FC*—1st *T*	3rd *FC*—1st *T*
1st *FC*—2nd *T*	2nd *FC*—2nd *T*	3rd *FC*—2nd *T*
1st *FC*—3rd *T*	2nd *FC*—3rd *T*	3rd *FC*—3rd *T*
1st *FC*—4th *T*	2nd *FC*—4th *T*	3rd *FC*—4th *T*

Since there are three applicants for file clerk there would be $3 \cdot 4$ or 12 possible pairings from the list of applicants for the two positions.

The above Principle holds true for more than two different variables. If something can be done in h ways, a second thing can be done in k different ways and a third in m different ways, then the number of possiblities of doing the three things, in stated order, would be $h \cdot k \cdot m$ ways.

If a first thing can be done in h ways and a second thing can be done in k ways, but both cannot take place at the same time, we say they are **mutually exclusive.** In such cases, one of the first thing in one of h ways, or one of the other thing in one of k ways, can take place at a given time, but *both cannot occur simultaneously.* Thus, if a person goes into a cafe for a meal and decides that he will take either a salad or a dessert, but not both, with his meal, then the salad and the dessert are mutually exclusive items. If, in this case, there are three salads and five desserts on the menu, then the possibilities, as far as salad or dessert are concerned, would be $3 + 5 = 8$.

A **permutation** of n different things taken r at a time is a unique ordered arrangement of r of those n things. It must be kept in mind that the order of things is important when we are dealing with permutations. The symbolism ${}_nP_r$ designates the number of permutations of n things taken r at a time. The number of permutations of the letters a, b, and c taken two at a time is 6. They are: *ab*, *ac*, *ba*, *bc*, *ca*, and *cb*. Note that each of these is unique if order is considered; *ab*, for example, is not the same as *ba*.

If there are no restrictions whatever placed on the items in a given situation, the formula for finding the number of permutations is

$${}_nP_r = n(n-1)(n-2)\ldots(n-r+1).$$

Certain problems, however, involve restrictions on the things under consideration. As these arise in the problems that follow, attention will be directed to them and how they may be incorporated into the solution. These restrictions vary considerably and, as a result, there is no uniformity in virtue of which a rule or formula can apply to all such restrictions.

How many three-digit numerals can be made from the digits 1, 2, 3, 4, and 5?

In this problem, $n = 5$ and $r = 3$. Thus,

$${}_nP_r = {}_5P_3 = 5 \cdot 4 \cdot \ldots (5 - 3 + 1).$$

This means that the factors in the answer start with 5 and continue, with each succeeding factor one less than the one that precedes it, such that the last factor is $(n - r + 1)$ or, in this case, 3. Then

$$\begin{aligned} {}_5P_3 &= 5 \cdot 4 \cdot 3 \\ &= 60. \end{aligned}$$

Hence, with the five digits given in the problem, and using three at a time, we could form sixty unique three-digit numerals.

Any five one-digit numerals chosen (for example, 1, 3, 5, 7, 9) would result in the same answer—the five numerals taken three at a time would have sixty permutations; though they would be different three-digit numerals. In this problem, we are not concerned with the values of the numbers named, we are concerned only with how many of them we can formulate.

1. If seven chairs are placed in a row and seven persons are to be seated in them, how many different seating arrangements can be made?

Since there are seven chairs, $n = 7$.

Since there are seven persons, $r = 7$.

Thus, $(n - r + 1)$, in this case, is equal to 1; and, therefore, the last factor in the answer is 1. We have

$$\begin{aligned} {}_7P_7 &= 7 \cdot 6 \cdot 5 \cdot 4 \cdot 3 \cdot 2 \cdot 1 \\ &= 7! \\ &= 5{,}040. \end{aligned}$$

This problem is illustrative of a special case of the permutation formula given on page 162. If $n = r$, then

$${}_nP_n = n!$$

2. If there are fifteen members in a club and four offices are to be filled, in how many ways could the group of four be selected?

$$\begin{aligned} n &= 15. \\ r &= 4. \\ n - r + 1 &= 12. \\ {}_{15}P_4 &= 15 \cdot 14 \cdot 13 \cdot 12 \\ &= 32{,}760. \end{aligned}$$

3. How many four-digit numerals could be formed using the digits 4, 5, 6, 7, 8, and 9 if no digit is repeated?

The restriction, "if no digit is repeated" means that a number such as 4,456 in which the 4 appears twice would not be one of the permutations of the six digits taken four at a time. Recall the discussion on page 162 in which the given formula does not allow for such repetitions.

$$\begin{aligned} n &= 6. \\ r &= 4. \\ n - r + 1 &= 3. \\ {}_6P_4 &= 6 \cdot 5 \cdot 4 \cdot 3 \\ &= 360. \end{aligned}$$

4. In how many ways may six persons be seated in a row of eight chairs?

In this problem, assign chairs to persons: eight chairs are to be taken six at a time. Then

$$\begin{aligned} n &= 8. \\ r &= 6. \\ n - r + 1 &= 3. \\ {}_8P_6 &= 8 \cdot 7 \cdot 6 \cdot 5 \cdot 4 \cdot 3 \\ &= 20{,}160. \end{aligned}$$

5. How many three-letter symbols could be made from the letters in the word *doctrine* if no letter is repeated?

$$\begin{aligned} n &= 8. \\ r &= 3. \\ n - r + 1 &= 6. \\ {}_8P_3 &= 8 \cdot 7 \cdot 6 \\ &= 336. \end{aligned}$$

6. How many different automobile license plates of six digits could be made from the numerals 1, 2, 3, 4, 5, 6, 7, 8, and 9 if no digit is repeated?

$$\begin{aligned} n &= 9. \\ r &= 6. \\ n - r + 1 &= 4. \\ {}_9P_6 &= 9 \cdot 8 \cdot 7 \cdot 6 \cdot 5 \cdot 4 \\ &= 60{,}480 \end{aligned}$$

7. An instructor has six students whom he suspects of cheating. He decides to place them in the front row of twelve chairs during the final examination. If they are seated such that there is a vacant chair between each two of them, how many ways could they be seated?

Either the first or second chair from one end could be selected for the first student; and, in either case, the six students could be seated in alternate chairs. Each case is the same as a situation in which there are only six chairs and six students to be seated. Thus

$$\begin{aligned} n &= 6. \\ r &= 6. \\ n - r + 1 &= 1. \\ {}_6P_6 &= 6 \cdot 5 \cdot 4 \cdot 3 \cdot 2 \cdot 1 \\ 2({}_6P_6) &= 2(6 \cdot 5 \cdot 4 \cdot 3 \cdot 2 \cdot 1) \\ &= 2(720). \\ &= 1{,}440. \end{aligned}$$

8. From the digits 5, 6, 7, 8, and 9, how many odd and how many even three-digit numerals could be formed if *digits can be repeated*?

This problem involves an exception that cannot be incorporated into the formula—it allows for a repetition of digits. Also, it asks for two answers, the number of permutations that will be even and the number of permutations that will be odd. We will consider each case individually. In either case, the last digit has to be considered first because the last digit determines whether the numeral will name an odd or even number.

Case 1. To find how many names for odd numbers can be formed. Let us start with the three blank spaces shown below, and fill the last one first. From the given set of numerals, three are odd. Since ${}_3P_3 = 3$, the last space could be filled in three different ways; and we would start as shown below. (The dots are used to indicate multiplication, for by the fundamental principle (page 162), the required number of permutations will be the product of the number of permutations for the first digit times the number of permutations for the second digit times the number of permutations for the third digit.)

$$\_ \cdot \_ \cdot \underline{3}$$

After that place is filled, we still have five numerals from which to choose for the second place because digits can be repeated. Therefore, the second factor will be ${}_5P_1$ or 5. Then we have

$$\_ \cdot \underline{5} \cdot \underline{3}.$$

Again, since digits can be repeated, we have five choices for filling the remaining place and our problem becomes

$$\underline{5} \cdot \underline{5} \cdot \underline{3} = 75.$$

Case 2. To find how many names for even numbers can be formed. The procedure is similar to that of Case 1, except that there are only two ways to fill the last blank. Thus, we have

$$\underline{5} \cdot \underline{5} \cdot \underline{2} = 50.$$

This problem can be checked readily as follows. We have three blanks to fill and, since digits can be repated, there are five choices each time. Hence, to find the total number of three-digit numerals that can be formed from the five given digits, we use the fundamental principle.

$$\underline{5} \cdot \underline{5} \cdot \underline{5} = 125.$$

In the original problem we found that seventy-five odd numbers and fifty even numbers would be named by the permutations. The total number of three-digit numerals that can be formed, therefore, is the sum of 75 and 50; and this sum corresponds with the result obtained in checking the problem.

9. In Problem 8, how many of each could be formed if no digit is repeated?

Case 1. To determine how many names for odd numbers could be formed, we would start as we did in Problem 8.

$$\_ \cdot \_ \cdot \underline{3}.$$

Now since we have used one digit, there are only four from which to select the next one. Then we would have

$$\_ \cdot \underline{4} \cdot \underline{3}.$$

We have three possible choices left from which to fill the remaining place. The problem becomes

$$\underline{3} \cdot \underline{4} \cdot \underline{3} = 36.$$

Case 2. In a similar manner we can determine the number of names for even numbers.

$$\underline{3} \cdot \underline{4} \cdot \underline{2} = 24.$$

This problem can be checked in a manner similar to the procedure used in Problem 8. The sum of the two answers obtained above is sixty. This means, of course, that there is a total of sixty three-digit numerals that can be formed from the given five numerals if no digit can be repeated. Since digits are not repeated, we can check this result by use of the formula for ${}_nP_r$.

$$n = 5.$$
$$r = 3.$$
$$n - r + 1 = 5 \cdot 4 \cdot 3$$
$$= 60.$$

10. How many different six-digit automobile license plates could be formed from the nine one-digit numerals for natural numbers? (Note that this problem is similar to Problem 6—the difference being that the digits may be repeated).

Since there are nine digits from which to choose, each blank space will be filled by the numeral 9. Thus

$$\underline{9} \cdot \underline{9} \cdot \underline{9} \cdot \underline{9} \cdot \underline{9} \cdot \underline{9} = 9^6 = 531{,}441$$

In general, if n things are taken r at a time and each n can be repeated up to r times, then the number of permutations is n^r.

11. How many different automobile license plates could be made by using three letters from the English alphabet followed by three digits selected from the one-digit numerals from our number system?

There are twenty-six letters to choose from so the first three blanks would contain three factors each of which is 26. There are ten one-digit numerals (including zero) so each of the last three blanks would contain the factor 10. Thus

$$\underline{26} \cdot \underline{26} \cdot \underline{26} \cdot \underline{10} \cdot \underline{10} \cdot \underline{10} = 26^3 \cdot 10^3$$
$$= 17{,}576 \cdot 1{,}000$$
$$= 17{,}576{,}000.$$

In practical use, this figure is too high. States which use this system for license-plate symbols have made it a practice to omit permutations containing certain three letter words that might be objectionable to the general public.

12. In how many different ways could six books be stacked in a row between book ends?

This problem has no limitations placed upon it, so the formula for ${}_nP_r$ may be used.

$$n = 6.$$
$$r = 6.$$
$$n - r + 1 = 1.$$
$${}_6P_6 = 6 \cdot 5 \cdot 4 \cdot 3 \cdot 2 \cdot 1$$
$$= 720.$$

13. If the books referred to in Problem 12 consisted of three textbooks and three novels, how many different ways could they be arranged such that all three textbooks were side by side and all novels were together?

This consists of two separate permutations whose values must be multiplied together. This product then would be multiplied by 2 since there are two ways of grouping those permutations—by having the textbooks in first position and then regrouping them with the novels first. Our problem, then, is set up as

$$2({}_3P_3 \cdot {}_3P_3) = 2(3 \cdot 2 \cdot 1 \cdot 3 \cdot 2 \cdot 1)$$
$$= 2(36)$$
$$= 72.$$

14. In how many different ways can a group of books consisting of three English textbooks, five from the science field, and six mathematics books be arranged on a shelf if all books of the same subject matter are grouped together?

This is similar to Problem 13. The main difference is the factor by which the products of the permutations is multiplied. In this case there are three groups which form a permutation of three things used three at a time. This produces a factor of 6. Then our problem becomes

$$6({}_3P_3 \cdot {}_5P_5 \cdot {}_6P_6) = 6(3 \cdot 2 \cdot 1 \cdot 5 \cdot 4 \cdot 3 \cdot 2 \cdot 1 \cdot 6 \cdot 5 \cdot 4 \cdot 3 \cdot 2 \cdot 1)$$
$$= 6(518{,}400)$$
$$= 3{,}110{,}400.$$

15. In how many different ways may a group of seven girls be seated in a row of seven chairs if three of the girls insist on sitting side by side?

In the first place there will be five positions the three girls may occupy together. This is shown by the following diagram

[— — —] — — — —
— [— — —] — — —
— — [— — —] — —
— — — [— — —] —
— — — — [— — —]

The three girls themselves form a permutation of three things using three at a time.

The remaining four girls, once the three have been seated, form a permutation of four things using four at a time. Our problem, then, will be set up as

$$5({}_3P_3 \cdot {}_4P_4) = 5(3 \cdot 2 \cdot 1 \cdot 4 \cdot 3 \cdot 2 \cdot 1)$$
$$= 5(144)$$
$$= 720.$$

16. If six girls are to be seated in a row of six chairs and two of them refuse to sit next to each other, in how many different ways may they be arranged?

The two girls who refuse to sit side by side may be arranged

in four different arrangements if there is one chair between them;

in three different arrangements if there are two chairs between them;

in two different arrangements if there are three chairs between them; or

in only one arrangement if there are four chairs between them.

This provides ten possibilities in which the two girls are in the same relative position—the first girl on the left of the second one. If they change relative positions this number is doubled. The remaining four girls form a permutation of four things using four at a time. Thus, we have

$$20({}_4P_4) = 20(4 \cdot 3 \cdot 2 \cdot 1)$$
$$= 20(24)$$
$$= 480.$$

17. How many three-letter symbols can be formed from the letters *a*, *b*, *c*, *d*, and *e* if each one ends in a vowel and no letter is repeated in the same symbol?

In this problem the last blank must be filled first in order to meet the restriction imposed. Since there are two vowels from which to choose we begin with

$$\_ \cdot \_ \cdot \underline{2}.$$

Now we have used one of the five possible letters so the next letter can be chosen in four ways. The final letter can be selected from any of the three remaining ones. Thus, we have

$$\underline{3} \cdot \underline{4} \cdot \underline{2} = 24.$$

18. How many four-letter symbols can be formed from the letters *c*, *d*, *e*, *f*, *g*, *h* and *i* if the first one must be a vowel, the last one a consonant and no letter can be repeated in any symbol?

We have two choices for the first blank and five for the last one. Therefore we begin with

$$\underline{2} \cdot \_ \cdot \_ \cdot \underline{5}.$$

We started with seven possible choices and thus far have used two of them, so the next blank can be filled from any one of the remaining five letters. Then there are four choices left for the last one. Hence, our problem becomes

$$\underline{2} \cdot \underline{5} \cdot \underline{4} \cdot \underline{5} = 200.$$

19. How many names for natural numbers can be formed from *some or all* of the digits 1, 2, 3, and 4 if digits can be repeated in the same numeral?

The solution to this problem consists of four groups of permutations—four things using four at a time with repetitions permitted, four things using three at a time with repetitions permitted, four things using two at a time with repetitions permitted, and four things using one at a time. In other words, some of the numerals have four digits, some have three digits, some have two digits, and some have only one digit. The total number of numerals will be the sum of the number of permutations in each of the four groups, for the groups are mutually exclusive. Our problem is set up as

$$(\underline{4} \cdot \underline{4} \cdot \underline{4} \cdot \underline{4}) + (\underline{4} \cdot \underline{4} \cdot \underline{4}) + (\underline{4} \cdot \underline{4}) + (\underline{4})$$
$$= 256 + 64 + 16 + 4$$
$$= 340.$$

20. How many numerals for real numbers greater than 3,000 can be formed from 0, 1, 2, 3, and 4 if no digits can be repeated in the same numeral?

The permutations asked for in this problem may be either four-digit or five-digit numerals.

Case 1. For the four-digit numerals, the first digit must be either 3 or 4. Then for our first digit we have two choices.

$$\underline{2} \cdot \_ \cdot \_ \cdot \_.$$

The next blank may be chosen from any of the four digits left after the first one is selected. The third one may be chosen from the remaining three digits, and the last one may be selected in one of two possible ways. Thus if the numerals contain four digits our problem becomes

$$\underline{2} \cdot \underline{4} \cdot \underline{3} \cdot \underline{2} = 48.$$

Case 2. If the numerals are composed of five digits, we may use in the first blank any one of those given, except 0. Thus, we start with

$$\underline{4} \cdot \underline{\ } \cdot \underline{\ } \cdot \underline{\ } \cdot \underline{\ }.$$

For the second digit we may use the zero, so there are still four possible choices for that digit. There are three choices for the third digit; two for the fourth digit; and only one choice left for the fifth digit. For all of those numerals containing five digits, the problem becomes

$$\underline{4} \cdot \underline{4} \cdot \underline{3} \cdot \underline{2} \cdot \underline{1} = 96.$$

Thus, the total number of permutations that can be formed from the given numerals such that the numbers thereby named are greater than 3,000 is

$$48 + 96 = 144,$$

the sum of the above two solutions.

Permutations in Which Some Things Are Alike. The permutations found in Problems 1-20 each had distinct letters, numerals, or objects. In some cases, however, two or more of the things to be permuted are indistinguishable one from the other. Consider, for example, the words *cheats* and *teller*. If no repetitions are allowed, the number of symbols that can be formed from the letters of the word *cheats* is $6 \cdot 5 \cdot 4 \cdot 3 \cdot 2 \cdot 1$ or 720. Now, if we wish to find the number of symbols which can be formed from the word *teller*, there would be, again, 720 such symbols. However, since *teller* contains two letters *e* and two letters *l*, many of those 720 symbols would be identical. In such cases, all the possible choices are different ones, but some cannot be distinguished one from the other.

Before introducing the formula by which such problems about permutations are solved, we must recall the symbol $n!$ which was discussed on page 160. This symbol is called n-factorial, and is defined as the product of n and all of the positive intergers less than n. Thus, $n!$ is equal to $n\ (n-1)\ (n-2)\ \ldots\ (n-n+1)$. Observe that $(n-n+1)$ is equal to 1. Thus, $4! = 4 \cdot 3 \cdot 2 \cdot 1 = 24$, and $6! = 6 \cdot 5 \cdot 4 \cdot 3 \cdot 2 \cdot 1 = 720$.

The formula by which to determine the number of permutations of n objects taken all at a time, some of which are indistinguishable, is

$$\frac{n!}{m!\ p!\ q! \ldots}$$

where n represents the total number of objects and m, p, and q represent indistinguishable objects.

21. How many different arrangements can be formed from the letters in the word *parallel*?

In this problem $n = 8$, the total number of letters in the given word. The *a* is used twice so one of the factors in the denominator of the formula will be $2!$ · The *l* is used 3 times in the word so the second factor in the denominator will be $3!$ · Thus, the number of permutations will be

$$\frac{8!}{2!\ 3!} = \frac{\overset{2}{8} \cdot 7 \cdot 6 \cdot 5 \cdot 4 \cdot 3 \cdot 2 \cdot 1}{2 \cdot 1 \cdot 3 \cdot 2 \cdot 1} = 3{,}360.$$

22. How many different six-digit numerals can be formed from the digits 2, 2, 3, 3, 4, 4?

$$\frac{6!}{2!\ 2!\ 2!} = \frac{6 \cdot 5 \cdot 4 \cdot 3 \cdot 2 \cdot 1}{2 \cdot 1 \cdot 2 \cdot 1 \cdot 2 \cdot 1} = \frac{720}{8} = 90.$$

23. A man has a collection of coins consisting of 5 nickels, 8 dimes, and 2 quarters. In the street near his home, 15 boys are playing; and, since he has 15 coins in his collection, he decides to distribute them among the boys such that each receives one coin. In how many distinct ways could the coins be distributed?

$$\frac{15!}{5!\ 8!\ 2!} = \frac{15 \cdot 14 \cdot 13 \cdot 12 \cdot 11 \cdot 10 \cdot 9 \cdot 8!}{5 \cdot 4 \cdot 3 \cdot 2 \cdot 1 \cdot 8! \cdot 2 \cdot 1} = \frac{15 \cdot 14 \cdot 13 \cdot 12 \cdot 11 \cdot 10 \cdot 9}{5 \cdot 4 \cdot 3 \cdot 2 \cdot 1 \cdot 2 \cdot 1} = \frac{\overset{7}{14} \cdot 13 \cdot \overset{3}{12} \cdot 11 \cdot \overset{5}{10} \cdot 9}{4 \cdot 2 \cdot 1 \cdot 2 \cdot 1} = 135{,}135.$$

Circular Permutations. When objects are arranged in circular form only the relative order is considered. If four persons are sitting around a circular table and each moves one place to the right (or left) the order has not changed—for each individual, the one on his right and the one on his left remains the same though everyone has changed chairs. The formula for finding the number of arrangements in circular formation is $(n-1)!$.

24. In how many ways may six individuals be seated around a circular table?

$$n = 6.$$
$$n - 1 = 5.$$
$$(n-1)! = 5! = 5 \cdot 4 \cdot 3 \cdot 2 \cdot 1 = 120.$$

25. In how many ways can four men and four women be seated at a circular table if the men and women alternate positions?

In this case, we first seat the women in alternate chairs. We merely think of four places and $n - 1 = 3$. Once the women have been seated, the positional (as opposed to relative) order must be taken into consideration for the men. Therefore, we use a regular permutation for the men. Thus, our problem becomes

$$(n-1)! \cdot {}_4P_4 = (3 \cdot 2 \cdot 1)(4 \cdot 3 \cdot 2 \cdot 1) = 6 \cdot 24 = 144.$$

26. In how many different ways may a group of six women sit at a circular table if two of them insist on sitting side by side?

We consider the two as occupying one position in the ring. Thus, we have 5 positions and $n - 1 = 4$. However, the two women may exchange relative positions which will double the circular permutations. Then our problem is set up as

$$2(n-1)! = 2(4 \cdot 3 \cdot 2 \cdot 1) = 2(24) = 48.$$

Combinations. A combination is defined as a selection of n things taken r at a time with no consideration as to their arrangement. Thus, the number of combinations of the letters, x, y, and z taken two at a time is 3. The combinations are: xy, xz, and yz. If order were being considered, there would be six permutations: xy, yx, xz, zx, yz, and zy. Thus, xy and yx, for example, are different permutations, but are the same combination.

The formula for finding the number of combinations of n things taken r at a time is

$${}_nC_r = \frac{n!}{r!(n-r)!}.$$

27. Find the number of combinations of ten things taken six at a time.

$$n = 10.$$
$$r = 6.$$
$${}_{10}C_6 = \frac{10!}{6!\,(10-6)!} = \frac{10!}{6!\,4!} = \frac{\overset{3}{10} \cdot 9 \cdot 8 \cdot 7 \cdot 6!}{4 \cdot 3 \cdot 2 \cdot 1 \cdot 6!} = 210.$$

28. In a certain office there are seven women employees and six men employees. A committee of five is to be selected to formulate plans for the annual Christmas party. In how many ways may such a committee be formed if it is agreed that there shall be exactly three women on the committee?

The committee, then, will consist of three women and two men. The three women can be selected in ${}_7C_3$ ways. Thus

$${}_7C_3 = \frac{7!}{3!(7-3)!} = \frac{7!}{3!4!} = \frac{7 \cdot 6 \cdot 5 \cdot 4!}{3 \cdot 2 \cdot 1 \cdot 4!} = 35.$$

The two men can be selected in ${}_6C_2$ ways. Thus

$${}_6C_2 = \frac{6!}{2!(6-2)!} = \frac{6!}{2!4!} = \frac{\overset{3}{6} \cdot 5 \cdot 4!}{2 \cdot 1 \cdot 4!} = 15.$$

The total number of ways in which the entire committee can be selected is

$${}_7C_3 \cdot {}_6C_2 = 35 \cdot 15 = 525.$$

29. Work Problem 28 if the agreement is that there shall be at least one man on the committee. In this problem, there are five possible cases: one man and four women, three men and two women, four men and one woman, or five men.

Case 1. In the case of one man and four women, the number of combinations would be

$${}_6C_1 \cdot {}_7C_4 = \frac{6!}{1!5!} \cdot \frac{7!}{4!3!} = \frac{6 \cdot 5!}{1 \cdot 5!} \cdot \frac{7 \cdot 6 \cdot 5 \cdot 4!}{3 \cdot 2 \cdot 1 \cdot 4!} = 6 \cdot 35 = 210.$$

Case 2. If the committee is composed of two men and three women we use the answer from Problem 28 which is

525.

Case 3. If the committee is composed of three men and two women the number would be

$${}_6C_3 \cdot {}_7C_2 = \frac{6!}{3!3!} \cdot \frac{7!}{2!5!} = \frac{6 \cdot 5 \cdot 4 \cdot 3!}{3 \cdot 2 \cdot 1 \cdot 3!} \cdot \frac{7 \cdot \overset{3}{6} \cdot 5!}{2 \cdot 1 \cdot 5!} = 20 \cdot 21 = 420.$$

Case 4. If the committee is composed of four men and one women the number would be

$${}_6C_4 \cdot {}_7C_1 = \frac{6!}{4!2!} \cdot 7 = \frac{\overset{3}{6} \cdot 5 \cdot 4!}{2 \cdot 1 \cdot 4!} \cdot 7 = 15 \cdot 7 = 105.$$

Case 5. If the committee is composed of five men the number would be

$$
\begin{aligned}
{}_6C_5 &= \frac{6!}{5!1!} \\
&= \frac{6 \cdot 5!}{1 \cdot 5!} \\
&= 6.
\end{aligned}
$$

The total, then, is the sum of all of the above answers which is

$$210 + 525 + 420 + 105 + 6 = 1{,}266.$$

30. Four persons are sitting at a card table with the aces, kings, queens, and jacks from a deck of regular playing cards. In how many ways may one person draw a hand of four cards?

$$
\begin{aligned}
{}_{16}C_4 &= \frac{16!}{4!(16-4)!} \\
&= \frac{16!}{4!12!} \\
&= \frac{\overset{4}{16} \cdot \overset{5}{15} \cdot \overset{7}{14} \cdot 13 \cdot 12!}{4 \cdot 3 \cdot 2 \cdot 1 \cdot 12!} \\
&= 1{,}820.
\end{aligned}
$$

SUPPLEMENTARY PROBLEMS

1. In how many ways may nine persons be seated in a row of nine chairs?

2. In how many ways may those nine persons in Problem 1 be seated if the chairs form a circle?

3. In how many ways may those nine persons in Problem 1 be seated if four of them insist on sitting next to each other in a group of four?

4. A men's physical education class, whose specialty is tennis, consists of twenty-six class members. In how many ways can a team of two members be chosen to send to the singles division in a regional tournament?

5. In Problem 4, in how many ways could a team of four members for the doubles division be chosen?

6. How many distinguishable symbols can be arranged from the letters in the word *pessimist*?

7. In Problem 2 how many ways would there be if three of the persons insisted on sitting next to each other?

8. A club of fifteen members desires to form a social committee composed of six persons. The president names the committee and refrains from naming herself to the committee. How many different committees could be formed?

9. In how many distinct arrangements may the letters in the word *problem* be set up if each one is to begin and end with a vowel?

10. A college baseball squad is composed of twelve men. Only two are qualified as catcher and of the remaining ten only three can serve as pitcher. After one pitcher and one catcher are chosen, the remaining members of the squad can play any of the other positions. How many different nine-man batting line-ups could be arranged if each line-up contains one pitcher and one catcher?

Chapter 18

Higher-Degree Equations

In the previous discussions we have dealt with linear and quadratic equations, and we were especially interested in finding the roots of those equations.

In the present chapter, we shall be concerned with equations whose highest-power variable has an exponent greater than 2. A **cubic equation** is one in which the highest power of the variable is 3. Thus, $x^3 + 3x^2 - 2x + 1 = 0$ is a cubic equation. An equation whose highest power is 4 is called a **quartic equation,** while an equation whose highest power is 5 is known as a **quintic equation.**

Our primary concern with these higher-degree equations will be that of finding their roots.

FUNDAMENTAL CONCEPTS AND THEOREMS

An equation in the form

$$a_0x^n + a_1x^{n-1} + a_2x^{n-2} + \ldots + a_{n-1}x + a_n = 0$$

where n is a positive integer and each a with its subscript is a constant, is called a **polynomial equation in x** or a **rational integral equation in x.** We sometimes designate a polynomial equation by the symbolism $f(x) = 0$.

As with linear and quadratic equations, the **root** or roots of higher-degree equations are those values that satisfy the equations. As an example, if $f(x) = x^3 + 6x^2 + 3x - 10 = 0$, then any number for which $f(x)$ equals zero when that number is substituted for x, is a root of the equation. Let us try $r = 1$ for the above equation.

$$f(x) = x^3 + 6x^2 + 3x - 10.$$
$$f(1) = (1)^3 + 6(1)^2 + 3(1) - 10$$
$$= 1 + 6 + 3 - 10 = 0.$$

Hence, 1 is a root of the equation.

The following theorems are fundamental to our approach to the study of higher-degree equations.

1. **The Fundamental Theorem of Algebra.** Every polynomial equation has a root among the complex numbers. (The set of **complex numbers** is the union of the set of real numbers and the set of imaginary numbers.)
2. **The Remainder Theorem.** If a polynomial $f(x)$ of degree 1 or greater is divided by a factor in the form $x - r$, the remainder is $R = f(r)$. As an example, let us again use the above polynomial and let $r = 2$. (This means that $x - 2$ is a divisor of the polynomial.)

$$f(x) = x^3 + 6x^2 + 3x - 10.$$
$$f(2) = (2)^3 + 6(2)^2 + 3(2) - 10$$
$$= 8 + 24 + 6 - 10$$
$$= 28.$$

By the remainder theorem, $R = f(r)$. The r corresponds to the 2 of our example, and R corresponds to the 28. This means that if we had used $x - 2$ as a divisor, of $x^3 + 6x^2 + 3x - 10$, we would have had a remainder of 28.

3. **The Factor Theorem.** If $f(r) = 0$, then $x - r$ is a factor of the polynomial $f(x)$. In other words, if the remainder is zero (if there is no remainder), then r is a root of the equation.

Find the remainder when $f(x) = x^3 - 2x^2 + 3$ is divided by $x - 1$.

$$f(x) = x^3 - 2x^2 + 3.$$
$$f(1) = 1 - 2 + 3$$
$$= 2.$$

Hence, 2 is the remainder.

Show that $f(x) = x^3 + 3x^2 - x - 3$ is divisible by $x + 3$.

$$f(x) = x^3 + 3x^2 - x - 3.$$
$$f(-3) = (-3)^3 + 3(-3)^2 - (-3) - 3$$
$$= -27 + 27 + 3 - 3$$
$$= 0.$$

Since there is no remainder, $f(x) = x^3 + 3x^2 - x - 3$ is divisible by $x + 3$. Hence, -3 is a root of the equation.

1. Find the remainder when $x^3 + 3x^2 - 2x - 3$ is divided by $x - 1$.

$$f(x) = x^3 + 3x^2 - 2x - 3.$$
$$f(1) = (1)^3 + 3(1)^2 - 2(1) - 3$$
$$= 1 + 3 - 2 - 3$$
$$= -1.$$

Hence, the remainder is -1.

2. Find the remainder when $x^4 - 3x^3 - 2x + 4$ is divided by $x - 2$.

$$f(x) = x^4 - 3x^3 - 2x + 4.$$

$$\begin{aligned} f(2) &= (2)^4 - 3(2)^3 - 2(2) + 4 \\ &= 16 - 24 - 4 + 4 \\ &= -8. \end{aligned}$$

Hence, the remainder is -8.

3. Find the remainder when $x^4 + 6x^2 - 3x - 6$ is divided by $x + 1$.

$$\begin{aligned} f(x) &= x^4 + 6x^2 - 3x - 6. \\ f(-1) &= (-1)^4 + 6(-1)^2 - 3(-1) - 6 \\ &= 1 + 6 + 3 - 6 \\ &= 4. \end{aligned}$$

Hence, the remainder is 4.

4. Find the remainder when $x^5 - 3x^3 + 6x^2 - x + 4$ is divided by $x - 2$.

$$\begin{aligned} f(x) &= x^5 - 3x^3 + 6x^2 - x + 4. \\ f(2) &= (2)^5 - 3(2)^3 + 6(2)^2 - (2) + 4 \\ &= 32 - 24 + 24 - 2 + 4 \\ &= 34. \end{aligned}$$

Hence, the remainder is 34.

5. Show that $x^3 + 4x^2 - x - 4$ is exactly divisible by $x + 1$.

$$\begin{aligned} f(x) &= x^3 + 4x^2 - x - 4. \\ f(-1) &= (-1)^3 + 4(-1)^2 - (-1) - 4 \\ &= -1 + 4 + 1 - 4 \\ &= 0. \end{aligned}$$

Since there is no remainder, $x^3 + 4x^2 - x - 4$ is exactly divisible by $x + 1$; and -1 is a root of the equation.

6. Show that $x^3 - x^2 - 4x + 4$ is exactly divisible by $x + 2$.

$$\begin{aligned} f(x) &= x^3 - x^2 - 4x + 4. \\ f(-2) &= (-2)^3 - (-2)^2 - 4(-2) + 4 \\ &= -8 - 4 + 8 + 4 \\ &= 0. \end{aligned}$$

Since there is no remainder, the division is possible; and -2 is a root of the equation.

7. Show that $x^3 - 7x - 6$ is exactly divisible by $x - 3$.

$$\begin{aligned} f(x) &= x^3 - 7x - 6. \\ f(3) &= (3)^3 - 7(3) - 6 \\ &= 27 - 21 - 6 \\ &= 0. \end{aligned}$$

Since the remainder is 0, the division is possible; and 3 is a root of the equation.

8. Show that $x^3 - x^2 - 8x + 12$ is divisible by $x + 3$.

$$\begin{aligned} f(x) &= x^3 - x^2 - 8x + 12. \\ f(-3) &= (-3)^3 - (-3)^2 - 8(-3) + 12 \\ &= -27 - 9 + 24 + 12 = 0. \end{aligned}$$

Since the remainder is 0 the division is possible; and -3 is a root of the equation.

9. Is $x^3 - 6x^2 + 11x - 6$ exactly divisible by $x - 3$?

$$\begin{aligned} f(x) &= x^3 - 6x^2 + 11x - 6. \\ f(3) &= (3)^3 - 6(3)^2 + 11(3) - 6 \\ &= 27 - 54 + 33 - 6 = 0. \end{aligned}$$

Since there is no remainder $x^3 - 6x^2 + 11x - 6$ *is* divisible by $x - 3$. Also, 3 is a root of the equation.

10. Is $x^4 + x^3 - 16x^2 - 4x + 48$ exactly divisible by $x + 3$?

$$\begin{aligned} f(x) &= x^4 + x^3 - 16x^2 - 4x + 48. \\ f(-3) &= (-3)^4 + (-3)^3 - 16(-3)^2 - 4(-3) + 48 \\ &= 81 - 27 - 144 + 12 + 48 \\ &= -30. \end{aligned}$$

Since there is a remainder, $x^4 + x^3 - 16x^2 - 4x + 48$ is *not* exactly divisible by $x + 3$.

Synthetic Division. Consider the following problem in long division.

$$\begin{array}{r|l} & x^2 - x - 6 \\ \hline x + 3 & x^3 + 2x^2 - 9x - 18 \\ & \underline{x^3 + 3x^2} \\ & -x^2 - 9x \\ & \underline{-x^2 - 3x} \\ & -6x - 18 \\ & \underline{-6x - 18} \end{array}$$

Since this division has no remainder, $x^3 + 2x^2 - 9x - 18$ is exactly divisible by $x + 3$ and the quotient is $x^2 - x - 6$.

Synthetic division is a second and more concise method of making the division. When algebraic expressions are concerned, synthetic division can be completed with less computational effort and somewhat more rapidly than ordinary long division. To divide synthetically, take the dividend of the given problem and use only the numerical coefficients of each term, each preceded by the same sign as was the case in the given problem. To the right of this sequence of numerals, place the **multiplier**—the numerical term from the binomial divisor.

$$1 + 2 - 9 - 18 \,\lfloor\, 3$$

Then under the 1 and below the horizontal line place a 1. (The first term in the sequence of numerical coefficients is rewritten directly under itself and under the horizontal line.) Multiply that 1 by the multiplier 3, and place the product under the 2 and above the horizontal line. Then subtract as we do in regular long division.

$$\begin{array}{l} 1 + 2 - 9 - 18 \,\lfloor\, 3 \\ \underline{\quad\;\; + 3 \qquad\qquad} \\ 1 - 1 \end{array}$$

Now multiply that -1 by the multiplier, place this product under the -9, and subtract.

$$\begin{array}{l} 1 + 2 - 9 - 18 \,\lfloor\, 3 \\ \underline{\quad\;\; + 3 - 3 \qquad} \\ 1 - 1 - 6 \end{array}$$

Finally multiply the -6 by the 3, place the product under the -18, and subtract.

$$\begin{array}{l} 1 + 2 - 9 - 18 \,\lfloor\, 3 \\ \underline{\quad\;\; + 3 - 3 - 18} \\ 1 - 1 - 6 + \;0 \end{array}$$

Observe that the first three numbers in the row below the horizontal line are the coefficients of the quotient obtained in the original solution done by regular long division.

The first term in the answer of a long division problem, if the variable in our divisor is of the first power, will be one lower power than the first term in the dividend. Thus, in our answer obtained by synthetic division, the first term is to the second power and that term will be x^2. Each succeeding number in the answer is the numerical coefficient of the next term and the power of the variable is one less than the preceding one. The last term in the answer in a problem in synthetic division is the remainder. In the answer to the above problem, the last term was zero, so the dividend is exactly divisible by the divsior, and the quotient is $x^2 - x - 6$.

When using synthetic division we usually change the sign of the multiplier (the numerical part of the binomial divisor which was 3 in the example above). When this is done, we add instead of subtracting to compensate for the change in sign. Another advantage of making this change will become obvious in a subsequent discussion.

Let us solve the above problem, changing the sign of the multiplier.

$$\begin{array}{r|l} 1 + 2 - 9 - 18 & -3 \\ -3 + 3 + 18 & \\ \hline 1 - 1 - 6 + 0 & \end{array}$$

Our quotient, again, is $x^2 - x - 6$. If we set this quotient equal to zero, we have $x^2 - x - 6 = 0$. This equation is called the first **depressed equation** in our process of finding the roots of an equation. We get this name, depressed equation, from the fact that the degree is depressed or lowered by 1.

> Divide $x^3 - x^2 - 9x + 9$ by $x + 3$, using synthetic division.
>
> $$\begin{array}{r|l} 1 - 1 - 9 + 9 & -3 \\ -3 + 12 - 9 & \\ \hline 1 - 4 + 3 + 0 & \end{array}$$
>
> Thus, $x^3 - x^2 - 9x + 9$ is exactly divisible by $x + 3$ and the quotient is $x^2 - 4x + 3$.

11. Divide $x^3 - 21x + 20$ by $x - 4$.

Before we start this solution observe that the first term is x^3 and that there is no x^2-term. When one or more powers of the variable are missing, its numerical coefficient is zero. When solving by synthetic division, always replace such omissions with a zero and then proceed as we did above.

$$\begin{array}{r|l} 1 + 0 - 21 + 20 & 4 \\ +4 + 16 - 20 & \\ \hline 1 + 4 - 5 + 0 & \end{array}$$

Hence, the quotient is $x^2 + 4x - 5$.

12. Divide $2x^3 - x^2 - 27x + 36$ by $x + 4$.

$$\begin{array}{r|l} 2 - 1 - 27 + 36 & -4 \\ -8 + 36 - 36 & \\ \hline 2 - 9 + 9 + 0 & \end{array}$$

Thus, the quotient is $2x^2 - 9x + 9$.

13. Divide $x^4 - 7x^3 + 3x^2 + 63x - 108$ by $x - 4$.

$$\begin{array}{r|l} 1 - 7 + 3 + 63 - 108 & 4 \\ +4 - 12 - 36 + 108 & \\ \hline 1 - 3 - 9 + 27 + 0 & \end{array}$$

Thus, the quotient is $x^3 - 3x^2 - 9x + 27$.

14. Divide $x^3 - 19x - 18$ by $x - 5$.

$$\begin{array}{r|l} 1 + 0 - 19 - 18 & 5 \\ +5 + 25 + 30 & \\ \hline 1 + 5 + 6 + 12 & \end{array}$$

Hence, the quotient is $x^2 + 5x + 6$ with a remainder of 12. This answer, of course, could be written $x^2 + 5x + 6 + \dfrac{12}{x - 5}$.

15. Divide $x^5 - 32$ by $x - 2$.

$$\begin{array}{r|l} 1 + 0 + 0 + 0 + 0 - 32 & 2 \\ +2 + 4 + 8 + 16 + 32 & \\ \hline 1 + 2 + 4 + 8 + 16 + 0 & \end{array}$$

The quotient, then, is $x^4 + 2x^3 + 4x^2 + 8x + 16$.

INTEGRAL ROOTS OF HIGHER-DEGREE EQUATIONS

Integral roots are, of course, those that are integers. If a_0, the leading coefficient of $f(x)$, is 1, then any integral root of $f(x)$ must be a divisor of a_n, the last term or the constant in the function. Thus, in the equation

$$f(x) = x^3 - 7x + 6 = 0$$

if there are any integral roots of the equation, they must be exact divisors of 6. Hence, the only possible integral roots are ± 1, ± 2, ± 3, and ± 6. At this stage, our only recourse is to try these possibilities until we find one that leaves no remainder, or until we have shown that all of them leave remainders. The method of finding integral roots of an equation is shown in the following illustration.

> Find the integral roots of $f(x) = x^3 - 7x + 6 = 0$.
> If 2 is a root then $x - 2$ is a divisor of $f(x)$.
>
> $$\begin{array}{r|l} 1 + 0 - 7 + 6 & 2 \\ +2 + 4 - 6 & \\ \hline 1 + 2 - 3 + 0 & \end{array}$$
>
> Thus, 2 is a root since there is no remainder, and $x^2 + 2x - 3 = 0$ is the first depressed equation.
>
> Now take the depressed equation and try another of the possible roots.
>
> Let us try 1.
>
> $$\begin{array}{r|l} 1 + 2 - 3 & 1 \\ +1 + 3 & \\ \hline 1 + 3 + 0 & \end{array}$$
>
> Again, since we have no remainder, 1 is a root and $x + 3 = 0$ is our second depressed equation.
>
> By use of the subtraction axiom, if $x + 3 = 0$, then $x = -3$.

Then our three roots are 2, 1 and -3.

In an algebraic equation there are as many roots as the highest power of the variable if we count duplicate roots as two roots. But all these roots need not be integral. Some or all of the roots may be real and non-integral or imaginary.

16. Find all of the integral roots of the equation

$$f(x) = x^3 + x^2 - 10x + 8 = 0.$$

If there are integral roots they must be chosen from divisors of 8 which are: ±1, ±2, ±4, and ±8. Let us try 4.

$$\begin{array}{l|l} 1 + 1 - 10 + 8 & 4 \\ \quad + 4 + 20 + 40 & \\ \hline 1 + 5 + 10 + 48 & \end{array}$$

Since there is a remainder we must discard 4 as a possible root. Let us try -4.

$$\begin{array}{l|l} 1 + 1 - 10 + 8 & -4 \\ \quad - 4 + 12 - 8 & \\ \hline 1 - 3 + 2 + 0 & \end{array}$$

Thus -4 is a root and $x^2 - 3x + 2 = 0$ is our first depressed equation. The only possible integral roots of this depressed equation are divisors of 2 which are ±1 and ±2. Let us try 2.

$$\begin{array}{l|l} 1 - 3 + 2 & 2 \\ \quad + 2 - 2 & \\ \hline 1 - 1 + 0 & \end{array}$$

Then 2 is a root and our second depressed equation is $x - 1 = 0$. The root of this equation is 1.

Thus, the three roots are -4, 2, and 1.

If we can see into the possibilities and avoid too much trial and error, we may solve in a more compact manner as shown below.

$$\begin{array}{l|l} 1 + 1 - 10 + 8 & -4 \\ \quad - 4 + 12 - 8 & \\ \hline 1 - 3 + 2 + 0 & 2 \\ \quad + 2 - 2 & \\ \hline 1 - 1 + 0 & 1 \\ \quad + 1 & \\ \hline 1 + 0 & \end{array}$$

17. Find all of the integral roots of the equation

$$f(x) = x^3 - 5x^2 + 8x - 4 = 0.$$

The only possibilities are divisors of 4 which are: ±1, ±2, and ±4. Let us try 4.

$$\begin{array}{l|l} 1 - 5 + 8 - 4 & 4 \\ \quad + 4 - 4 + 16 & \\ \hline 1 - 1 + 4 + 12 & \end{array}$$

Since there is a remainder, 4 is not a root. Let us try -4.

$$\begin{array}{l|l} 1 - 5 + 8 - 4 & -4 \\ \quad - 4 + 36 - 176 & \\ \hline 1 - 9 + 44 - 180 & \end{array}$$

Neither is -4 a root. Let us try -2.

$$\begin{array}{l|l} 1 - 5 + 8 - 4 & -2 \\ \quad - 2 + 14 - 44 & \\ \hline 1 - 7 + 22 - 48 & \end{array}$$

Nor is -2 a root. Let us try 2.

$$\begin{array}{l|l} 1 - 5 + 8 - 4 & 2 \\ \quad + 2 - 6 + 4 & \\ \hline 1 - 3 + 2 + 0 & \end{array}$$

Since there is no remainder, 2 is a root. Let us now try 1, using the first depressed equation.

$$\begin{array}{l|l} 1 - 3 + 2 & 1 \\ \quad + 1 - 2 & \\ \hline 1 - 2 + 0 & \end{array}$$

Thus, 1 is a root and the second depressed equation is $x - 2 = 0$, and from that linear equation $x = 2$. Observe that 2 is a root twice in this problem. We therefore say that it is a **double root** or a **root of multiplicity two.**

Hence, our roots are 2, 2, and 1.

18. Find the integral roots of the equation

$$f(x) = x^3 - 3x^2 - 16x + 48 = 0.$$

There are twenty possible integral roots of this equation. Needless to say, the trial and error method we have been using could become a tremendously tedious task if one were not able to anticipate the correct roots early in the attempts at solution. Moreover, if these twenty possibilities failed to provide the three roots of the cubic equation, we would have to search further for non-integral real roots and/or for imaginary roots.

In order to reduce the amount of work involved in finding the real roots of an equation, we have two techniques by which we may find certain information about the roots before we attempt to find them. The first of these is to find the **upper and lower bounds** between which any real roots will be found. By the upper bound we mean a value above which a given equation can have no roots; and by the lower bound we mean a value below which there are no roots. By synthetic division, let us see if 6 is a root of the equation given in Problem 18.

$$\begin{array}{l|l} 1 \; -3 \; -16 \; +48 & 6 \\ \quad +6 \; +18 \; +12 & \\ \hline 1 \; +3 \; +2 \; +60 & \end{array}$$

Since there is a remainder, we know, of course, that 6 is not a root. Observe, however, that in the above quotient all of the signs are positive. When this happens, we know that the root we are trying is an upper bound. In such cases, no number larger than the one we are trying (in this case, 6) can be a root of the equation. Thus, we have eliminated six of the twenty possibilities. That is, 8, 12, 16, 24, and 48 cannot be roots of the given equation because each is greater than the upper bound; and 6 cannot be a root because $f(6)$ has a remainder.

To find the lower bound, first substitute $-x$ for x in the given equation. Then use that new equation to try for possible roots. Substituting $-x$ for x in the given equation gives us

$$f(-x) = (-x)^3 - 3(-x)^2 - 16(-x) + 48 = 0.$$
$$-x^3 - 3x^2 + 16x + 48 = 0.$$

Multiply both members of this last equation by -1 and we get

$$x^3 + 3x^2 - 16x - 48 = 0.$$

In this equation try 6. (The root thus being tested is actually -6, for we have substituted $-x$ for x in the original equation.)

$$\begin{array}{rrrr|l} 1 + 3 & - 16 & - 48 & & 6 \\ + 6 & + 54 & + 228 & & \\ \hline 1 + 9 & + 38 & + 180 & & \end{array}$$

Since all of the signs are positive in this quotient, -6 is a lower bound. Now we have eliminated six more possible roots: $-6, -8, -12, -16, -24$, and -48.

In these two trials, we have eliminated twelve of the twenty possible integral roots of the original equation; and have left as possibilities only the following: $\pm1, \pm2, \pm3$, and ±4. Let us try 4 in the original equation.

$$\begin{array}{rrrr|l} 1 - 3 & - 16 & + 48 & & 4 \\ + 4 & + 4 & - 48 & & \\ \hline 1 + 1 & - 12 & - 0 & & \end{array}$$

Thus 4 is a root and $x^2 + x - 12 = 0$ is our first depressed equation. Rather than use synthetic division to find the remaining roots, the student may wish to solve the depressed equation as a quadratic equation by the methods learned earlier. Thus,

$$x^2 + x - 20 = 0.$$
$$(x + 4)(x - 3) = 0.$$
$$x + 4 = 0. \qquad x - 3 = 0.$$
$$x = -4. \qquad x = 3.$$

Hence, the other two roots are -4 and 3, and the three roots of the original equation are 4, -4, and 3.

19. Find the integral roots of the equation

$$f(x) = x^3 + 12x^2 + 20x - 96 = 0.$$

Again we are faced with the task of finding, out of 24 possibilities, at most 3 integral roots. Let us try 8 as a possible root.

$$\begin{array}{rrrr|l} 1 + 12 & + 20 & - 96 & & 8 \\ + 8 & + 160 & + 1440 & & \\ \hline 1 + 20 & + 180 & + 1344 & & \end{array}$$

Observe the size of the remainder. This, with the signs all positive, indicates that in all probability, there is an upper bound much lower than 8. Let us try 2.

$$\begin{array}{rrrr|l} 1 + 12 & + 20 & - 96 & & 2 \\ + 2 & + 28 & + 96 & & \\ \hline 1 + 14 & + 48 & + 0 & & \end{array}$$

Thus, 2 is a root, and since all the signs are positive we have eliminated 10 of the 12 possible positive roots. There remains only 1, or duplications of 1 and 2, as other possible positive roots and twelve possible negative roots. However, before we attempt any of those left, let us introduce the second technique by which our labors in finding any real roots may be reduced.

There are *at most* as many positive roots as there are variations in sign in the equation. By a **variation in sign** we mean a change from + to − or from − to + as we progress from left to right across the given equation. Let us go back for a moment to the given equation $x^3 + 12x^2 + 20x - 96 = 0$. As we observe the terms in this equation, there is only one variation in sign—from $+ 20x$ to $- 96$. Therefore, there is only one positive root which we have found to be 2.

If we take the original equation and substitute $-x$ for x we get a different equation. Taking this new equation we, again, observe the variations in sign and this number of variations will tell us the maximum number of negative roots possible in the original equation. Thus,

$$f(-x) = (-x)^3 + 12(-x)^2 + 20(-x) - 96 = 0.$$
$$-x^3 + 12x^2 - 20x - 96 = 0.$$

We can observe two variations in sign. Thus, there are *at most* two negative roots of the original equation.

This procedure, known as **Descartes' Rule of Signs,** was derived by the famous French mathematician, Rene Descartes.

Going back to our solution we found 2 to be a root. Since there can be at most one positive root we need try no more positive values. Now let us find the lower bound for roots. In the above discussion, we substituted $-x$ for each x in the original equation, and obtained

$$-x^3 + 12x^2 - 20x - 96 = 0.$$

Multiplying both members of this equation by -1, we obtain

$$x^3 - 12x^2 + 20x + 96 = 0.$$

Let us try 12, keeping in mind that the value being tested is -12.

$$\begin{array}{rrrr|l} 1 - 12 & + 20 & + 96 & & 12 \\ + 12 & + 0 & + 240 & & \\ \hline 1 + 0 & + 20 & + 336 & & \end{array}$$

Since we have a remainder, -12 is not a root. But -12 is a lower bound because all signs in the quotient are positive.

Let us try -6.

$$\begin{array}{rrrr|l} 1 - 12 & + 20 & + 96 & & 6 \\ + 6 & - 36 & - 96 & & \\ \hline 1 - 6 & - 16 & + 0 & & \end{array}$$

Thus -6 *is* a root and our depressed equation is

$$x^2 - 6x - 16 = 0.$$

Factoring, we get

$$(x - 8)(x + 2) = 0.$$
$$x - 8 = 0. \qquad x + 2 = 0.$$
$$x = 8. \qquad x = -2.$$

which seems to indicate two more roots, -8 and $+2$. But there cannot be a double root of 2 because, by the rule of signs, we found that there was at most one positive root. Thus, our three roots are 2, -6 and -8.

20. Find the integral roots of the equation

$$x^5 + 2x^4 - 2x^2 - x = 0.$$

Since the highest power of the variable in this equation is 5, the equation has five roots. Let us begin our search for the integral roots with the observation that every term in the left member of the equation contains an x. Therefore, by a simple factorization (see page 14), we get

$$x(x^4 + 2x^3 - 2x - 1) = 0.$$

Now since this product is equal to zero, the equation will be satisfied if both factors are equal to zero. Thus, we obtain

$$x = 0.$$

$$x^4 + 2x^3 - 2x - 1 = 0.$$

The first of these equations, $x = 0$, tells us that 0 is a root of $x^5 + 2x^4 - 2x^2 - x = 0$. It remains for us to find the roots which make the left member of the depressed equation, $x^4 + 2x^3 - 2x - 1 = 0$, equal to zero. Since the constant term in the left member of this equation is -1, the only possible non-zero integral roots are ± 1. Therefore, there is no need to find upper and lower bounds. Moreover, since there is only one change in sign, there is at most one positive integral root.

For that one possible positive integral root let us try 1.

$$\begin{array}{l|l} 1 + 2 + 0 - 2 - 1 & 1 \\ + 1 + 3 + 3 + 1 & \\ \hline 1 + 3 + 3 + 1 + 0 & \end{array}$$

Hence, 1 is a root, and our first depressed equation is $x^3 + 3x^2 + 3x + 1 = 0$. With that equation let us try -1.

$$\begin{array}{l|l} 1 + 3 + 3 + 1 & -1 \\ - 1 - 2 - 1 & \\ \hline 1 + 2 + 1 + 0 & \end{array}$$

Then -1 is a root, and our second depressed equation is $x^2 + 2x + 1 = 0$. Solve this equation by factoring and we have

$$(x + 1)(x + 1) = 0.$$

$$x + 1 = 0. \qquad x + 1 = 0.$$

$$x = -1. \qquad x = -1.$$

Thus, there are two more roots of -1, and the five roots of the solution are: $-1, -1, -1, 1,$ and 0.

RATIONAL ROOTS OF HIGHER-DEGREE EQUATIONS

We defined a rational number, in an earlier discussion (page 2), as any number that can be expressed as $\frac{a}{b}$, where a and b are integers and $b \neq 0$.

The methods for finding the rational roots of an equation involve the same principles as apply to integers with one variation. The possible rational roots may be any rational number that can be formed by using for the numerator of that rational number any factor of a_n, the constant, and for the denominator any factor of a_0, the leading coefficient. Thus, the numerator of any root must be a factor of the numerator of $\frac{a_n}{a_0}$, and the denominator of any root must be a factor of the denominator of $\frac{a_n}{a_0}$.

1. Find all rational roots of the equation

$$f(x) = 2x^3 + x^2 - 2x - 1 = 0.$$

$$\frac{a_n}{a_0} = -\frac{1}{2}.$$

The only possible factors of the numerator are ± 1, and the only possible factors of the denominator are ± 1 and ± 2. Thus, the only possible rational roots are $\pm \frac{1}{2}$ and ± 1. Let us try 1.

$$\begin{array}{l|l} 2 + 1 - 2 - 1 & 1 \\ + 2 + 3 + 1 & \\ \hline 2 + 3 + 1 + 0 & \end{array}$$

Hence, 1 is a root, and our first depressed equation is $2x^2 + 3x + 1 = 0$. Using this equation, let us try -1.

$$\begin{array}{l|l} 2 + 3 + 1 & -1 \\ - 2 - 1 & \\ \hline 2 + 1 + 0 & \end{array}$$

Thus, -1 is also a root, and our second depressed equation is $2x + 1 = 0$. Solving this linear equation we get

$$2x = -1.$$

$$x = -\tfrac{1}{2}.$$

Thus, the three roots are $-\frac{1}{2}$, -1, and 1.

2. Find the rational roots of the equation

$$f(x) = 2x^3 - 3x^2 - 2x = 0.$$

$$2x^3 - 3x^2 - 2x = 0.$$

$$x(2x^2 - 3x - 2) = 0.$$

$$x = 0. \qquad 2x^2 - 3x - 2 = 0.$$

Hence, 0 is a root of the equation, and the first depressed equation is $2x^2 - 3x - 2 = 0$.

Dividing the constant by 2 gives us $\frac{a_n}{a_0} = -\frac{2}{2}$.

This rational number is not reduced before we determine the possibilities. Hence, the possibilities are $\pm \frac{1}{2}$, ± 1, and ± 2. Let us try 2.

$$\begin{array}{l|l} 2 - 3 - 2 & 2 \\ + 4 + 2 & \\ \hline 2 + 1 + 0 & \end{array}$$

Hence, 2 is a root, and our second depressed equation is $2x + 1 = 0$. Solve this linear equation, and we have $x = -\frac{1}{2}$.

Then our roots are $-\frac{1}{2}$, 0, and 2.

After having found the root $x = 0$, the depressed equation, $2x^2 - 3x - 2 = 0$, could have been solved by any of the methods used for solving quadratic equations rather than by the above method.

3. Find the rational roots of the equation

$$f(x) = 4x^3 + 12x^2 - x - 3 = 0.$$

$$\frac{a_n}{a_0} = -\frac{3}{4}$$

Then our possible roots would be ± 3, $\pm\frac{3}{2}$, $\pm\frac{3}{4}$, ± 1, $\pm\frac{1}{2}$, and $\pm\frac{1}{4}$. Let us try -3.

$$\begin{array}{l|l} 4 + 12 - 1 - 3 & -3 \\ \quad - 12 + 0 + 3 & \\ \hline 4 + \ \ 0 - 1 + 0 & \end{array}$$

Thus, -3 *is* a root and the first depressed equation is $4x^2 - 1 = 0$. Solving this quadratic equation we find the roots are $-\frac{1}{2}$ and $\frac{1}{2}$.

Hence, the three roots are -3, $-\frac{1}{2}$, and $\frac{1}{2}$.

4. Find the rational roots of the equation

$$f(x) = 6x^3 - 37x^2 + 58x - 24 = 0.$$

Since there are a great many possibilities, let us find the upper and lower bounds. Try 8.

$$\begin{array}{l|l} 6 - 37 + \ 58 - \ \ 24 & 8 \\ \quad + 48 + \ 88 + 1168 & \\ \hline 6 + 11 + 146 + 1144 & \end{array}$$

Certainly, 8 is an upper bound since all of the signs are positive; but, with such a large remainder, there may be a number smaller than 8 that is an upper bound. Let us drop down to 4 and try that as a possible upper bound.

$$\begin{array}{l|l} 6 - 37 + 58 - 24 & 4 \\ \quad + 24 - 52 + 24 & \\ \hline 6 - 13 + \ 6 + \ 0 & \end{array}$$

And this gives us a root of 4. (This sometimes happens—when attempting to find either the upper or lower bound we may be fortunate enough to find one of the roots.) The first depressed equation is $6x^2 - 13x + 6 = 0$. We need not look for a lower bound since we can proceed with the solution of this quadratic equation.

$$6x^2 - 13x + 6 = 0.$$

Then by factoring we have

$$(2x - 3)(3x - 2) = 0.$$

$$\begin{array}{rcl} 2x - 3 = 0. & \qquad & 3x - 2 = 0. \\ 2x = 3. & & 3x = 2. \\ x = \frac{3}{2}. & & x = \frac{2}{3}. \end{array}$$

Thus, our three roots are 4, $\frac{3}{2}$ and $\frac{2}{3}$.

Thus far, all the higher-degree equations we have solved have had all real roots. In some cases, however, an equation has imaginary roots. In most of these cases, we will be interested only in the real roots, and the imaginary roots will be disregarded when the answer is stated.

5. Find the rational roots of the equation

$$f(x) = x^4 - 1 = 0.$$

The only possibilities, of course, are ± 1. Let us try 1.

$$\begin{array}{l|l} 1 + 0 + 0 + 0 - 1 & 1 \\ \quad + 1 + 1 + 1 + 1 & \\ \hline 1 + 1 + 1 + 1 + 0 & \end{array}$$

Hence, 1 is a root, and our first depressed equation is $x^3 + x^2 + x + 1 = 0$. A fourth degree equation has four roots so it would seem that there might be a double root of 1. Let us try 1 with this first depressed equation.

$$\begin{array}{l|l} 1 + 1 + 1 + 1 & 1 \\ \quad + 1 + 2 + 3 & \\ \hline 1 + 2 + 3 + 4 & \end{array}$$

Then, there is not a second root of 1. Let us try -1 with the first depressed equation.

$$\begin{array}{l|l} 1 + 1 + 1 + 1 & -1 \\ \quad - 1 + 0 - 1 & \\ \hline 1 + 0 + 1 + 0 & \end{array}$$

Since the remainder is 0, -1 is a root, and our second depressed equation is $x^2 + 1 = 0$. Solving this quadratic equation we find the two roots i and $-i$ both of which are imaginary.

Then the only rational roots are -1 and 1.

It will be observed that we found two imaginary roots in the above solution. In standard form (see page 82), those roots are $0 + i$ and $0 - i$ which are conjugates of each other. *Imaginary roots of equations whose coefficients are real numbers always come in pairs which are conjugates of each other.* The same holds true for roots in the form of a binomial containing a quadratic surd, such as $a + \sqrt{b}$ and $a - \sqrt{b}$ where a and b are rational and $\sqrt{b}$ is irrational. It will be recalled from the work with quadratic equations, especially those solved by use of the quadratic formula (see page 89 ff.) that this was the case. In those solutions, if a radical appeared in one of the roots, there was always another root containing the radical. And these two roots were conjugates of each other.

6. Find the real roots of the equation $x^4 - 4x^2 + 3 = 0$.

This equation has four roots. Let us try 1.

$$\begin{array}{l|l} 1 + 0 - 4 + 0 + 3 & 1 \\ \quad + 1 + 1 - 3 - 3 & \\ \hline 1 + 1 - 3 - 3 + 0 & \end{array}$$

Thus, 1 is a root, and our first depressed equation is $x^3 + x^2 - 3x - 3 = 0$. Let us try -1 with this depressed equation.

$$\begin{array}{l|l} 1 + 1 - 3 - 3 & -1 \\ \quad - 1 + 0 + 3 & \\ \hline 1 + 0 - 3 + 0 & \end{array}$$

Hence, -1 is also a root, and the second depressed equation is $x^2 - 3 = 0$. Solving this equation we obtain

$$x^2 - 3 = 0$$
$$x^2 = 3$$
$$x = \pm\sqrt{3}$$

Hence, the four real roots are $-1, 1, -\sqrt{3}$, and $\sqrt{3}$. Only two of these, -1 and 1, are rational roots; $-\sqrt{3}$ and $\sqrt{3}$ are irrational.

7. Find the real roots of the equation

$$6x^4 - 7x^3 + 3x^2 - 7x - 3 = 0.$$

Let us try $\frac{3}{2}$ as a possible root.

$$\begin{array}{l} 6 - 7 + 3 - 7 - 3 \,\lfloor \tfrac{3}{2} \\ \quad + 9 + 3 + 9 + 3 \\ \hline 6 + 2 + 6 + 2 + 0 \end{array}$$

Thus, $\frac{3}{2}$ is a root, and our depressed equation is $6x^3 + 2x^2 + 6x + 2 = 0$. Now let us try $-\frac{1}{3}$ with this depressed equation.

$$\begin{array}{l} 6 + 2 + 6 + 2 \,\lfloor -\tfrac{1}{3} \\ \quad - 2 + 0 - 2 \\ \hline 6 + 0 + 6 + 0 \end{array}$$

Hence, $-\frac{1}{3}$ is also a root, and our second depressed equation is $6x^2 + 6 = 0$ or $x^2 + 1 = 0$. Solving this we find the two imaginary roots i and $-i$.

Hence, the only real roots are $\frac{3}{2}$ and $-\frac{1}{3}$.

8. Find the real roots of the equation

$$f(x) = x^5 + 5x^4 + 7x^3 + 17x^2 + 10x - 40 = 0.$$

There are sixteen possible integral roots. Let us find upper and lower bounds to eliminate some of the work involved in finding the roots. Let us try 2.

$$\begin{array}{l} 1 + 5 + 7 + 17 + 10 - 40 \,\lfloor 2 \\ \quad + 2 + 14 + 42 + 118 + 256 \\ \hline 1 + 7 + 21 + 59 + 128 + 216 \end{array}$$

Hence, 2 is not a root, but is an upper bound. This eliminates the possible roots 4, 5, 8, 10, 20, and 40.

Now we shall substitute $-x$ for x in the equation and determine the lower bound.

$$f(-x) = (-x)^5 + 5(-x)^4 +$$
$$7(-x)^3 + 17(-x)^2 + 10(-x) - 40 = 0.$$
$$-x^5 + 5x^4 - 7x^3 + 17x^2 - 10x - 40 = 0.$$

Multiply both members of the equation by -1 and we have

$$x^5 - 5x^4 + 7x^3 - 17x^2 + 10x + 40 = 0.$$

Then let us try 5 in this equation, and thereby test the value -5.

$$\begin{array}{l} 1 - 5 + 7 - 17 + 10 + 40 \,\lfloor 5 \\ \quad + 5 + 0 + 35 + 90 + 500 \\ \hline 1 + 0 + 7 + 18 + 100 + 540 \end{array}$$

Thus, -5 is not a root, but is a lower bound. This lower bound eliminates the possible integral roots -5, -8, -10, -20, and -40. These two operations have eliminated twelve of the possible sixteen choices for integral roots, leaving only -4, -2, -1, and 1 as those that may be integral roots of the equation.

Going back to the original equation, let us try 1.

$$\begin{array}{l} 1 + 5 + 7 + 17 + 10 - 40 \,\lfloor 1 \\ \quad + 1 + 6 + 13 + 30 + 40 \\ \hline 1 + 6 + 13 + 30 + 40 + 0 \end{array}$$

Thus, 1 *is* a root and our first depressed equation is $x^4 + 6x^3 + 13x^2 + 30x + 40 = 0$. With this equation let us try -4.

$$\begin{array}{l} 1 + 6 + 13 + 30 + 40 \,\lfloor -4 \\ \quad - 4 - 8 - 20 - 40 \\ \hline 1 + 2 + 5 + 10 + 0 \end{array}$$

Hence, -4 is also a root, and the second depressed equation is $x^3 + 2x^2 + 5x + 10 = 0$. With this equation let us try -2.

$$\begin{array}{l} 1 + 2 + 5 + 10 \,\lfloor -2 \\ \quad - 2 + 0 - 10 \\ \hline 1 + 0 + 5 + 0 \end{array}$$

And -2 is a root, leaving as our third depressed equation $x^2 + 5 = 0$. Solving this quadratic equation we get the roots $\pm i\sqrt{5}$, both of which are imaginary.

Thus, the only real roots are 1, -2 and -4.

9. Find the real roots of the equation

$$f(x) = 32x^4 + 24x^3 - 44x^2 - 6x + 9 = 0.$$

Let us look for an upper bound by using 1.

$$\begin{array}{l} 32 + 24 - 44 - 6 + 9 \,\lfloor 1 \\ \quad + 32 + 56 + 12 + 6 \\ \hline 32 + 56 + 12 + 6 + 15 \end{array}$$

Thus, 1 is not a root, nor are there any roots greater than 1.

Substitute $-x$ for x in the original equation and search for a lower bound.

$$f(-x) = 32(-x)^4 + 24(-x)^3 - 44(-x)^2 - 6(-x) + 9 = 0.$$
$$32x^4 - 24x^3 - 44x^2 + 6x + 9 = 0.$$

Now let us divide synthetically by 2 to test the possible root -2.

$$\begin{array}{l} 32 - 24 - 44 + 6 + 9 \,\lfloor 2 \\ \quad + 64 + 80 + 72 + 156 \\ \hline 32 + 40 + 36 + 78 + 165 \end{array}$$

Thus, -2 is not a root, but since all signs in the quotient are positive, -2 is a lower bound. Since -2 is a lower bound and 1 is an upper bound, the only possible integral roots are 0 and -1. Zero is not a root because both members of the original equation are not exactly divisible by x. Let us try -1 in the original equation.

$$\begin{array}{l} 32 + 24 - 44 - 6 + 9 \,\lfloor -1 \\ \quad - 32 + 8 + 36 - 30 \\ \hline 32 - 8 - 36 + 30 - 21 \end{array}$$

Thus, -1 is not a root, for we have a remainder. This eliminates all possible integral roots, and we must try those that can be represented by $\frac{a_n}{a_0}$ or $\frac{9}{32}$. Of these there are thirty-six possibilities. Let us try $\frac{1}{2}$.

$$\begin{array}{l} 32 + 24 - 44 - 6 + 9 \,\lfloor \tfrac{1}{2} \\ \quad + 16 + 20 - 12 - 9 \\ \hline 32 + 40 - 24 - 18 + 0 \end{array}$$

Thus, $\frac{1}{2}$ is a root, and our first depressed equation is $32x^3 + 40x^2 - 24x - 18 = 0$ which can be reduced to $16x^3 + 20x^2 - 12x - 9 = 0$ by dividing both members by 2.

By the rule of signs, it is possible that there is one more positive root. With the first depressed equation, let us try $\frac{3}{4}$.

$$\begin{array}{l} 16 + 20 - 12 - 9 \,\underline{|\, \frac{3}{4}} \\ \underline{\quad + 12 + 24 + 9} \\ 16 + 32 + 12 + 0 \end{array}$$

Thus, $\frac{3}{4}$ is a root, and the second depressed equation is $16x^2 + 32x + 12 = 0$ which reduces to $4x^2 + 8x + 3 = 0$. Then by factoring we obtain

$$(2x + 3)(2x + 1) = 0.$$

$$2x + 3 = 0. \qquad 2x + 1 = 0.$$

$$2x = -3. \qquad 2x = -1.$$

$$x = -\frac{3}{2}. \qquad x = -\frac{1}{2}.$$

The four roots, then, are $\frac{3}{4}$, $\frac{1}{2}$, $-\frac{1}{2}$, and $-\frac{3}{2}$.

10. Find the square root of 4,489 by the methods used in the preceding problems.

If the square of some number is 4,489, and we let x represent that number then, $x^2 = 4{,}489$. This equation can be written in the form $x^2 - 4{,}489 = 0$. Then finding the roots of that equation will give us the square root of the given number.

We know that $(70)^2 = 4{,}900$ and $(60)^2 = 3{,}600$. The number we are seeking lies somewhere between 60 and 70. Either 63 or 67, if squared, will produce the last digit in the given number. Since that given number is considerably closer to 4,900 than to 3,600 it would seem reasonable that 67 is the answer. Let us try 67.

$$\begin{array}{l} 1 + \ 0 - 4{,}489 \,\underline{|\, 67} \\ \underline{\quad + 67 + 4{,}489} \\ 1 + 67 + \quad 0 \end{array}$$

Thus, 67 is a root of the equation and our first depressed equation is $x + 67 = 0$. Solving that linear equation for x gives us $x = -67$.

There are two square roots of any number, one positive and one negative. Thus, we have two square roots of 4,489 namely, -67 and $+67$.

We could also use -67 as a choice in the original equation and have

$$\begin{array}{l} 1 + \ 0 - 4{,}489 \,\underline{|\, -67} \\ \underline{\quad - 67 + 4{,}489} \\ 1 - 67 + \quad 0 \end{array}$$

Then, again, we have shown that -67 is one of the square roots of the original number.

This may have seemed a somewhat trivial problem, but the same technique will be valuable in the next section as we seek to determine irrational roots of numbers.

IRRATIONAL ROOTS OF HIGHER-DEGREE EQUATIONS

One of the easier methods for approximating irrational roots is known as **Horner's Method,** and even this process is tedious. Horner's Method can be best explained in terms of the steps involved in a specific problem. In Problem 1 below each step is accompanied by an explanation.

1. For the equation $f(x) = x^3 + 4x^2 - 7x - 1 = 0$, find the positive root greater than 1 and less than 2.

Before we proceed with the solutions, let us sketch the graph of this equation in the coordinate plane. (See Fig. 66.) Graphically, the roots of the equation will be those points at which the graph crosses the X-axis. Since this problem calls for a root between $x = 1$ and $x = 2$, let us determine the value of the relation $y = f(x)$ for these two values of x.

$$y = f(x) = x^3 + 4x^2 - 7x - 1.$$

At the point where $x = 1$, $y = -3$. Thus $(1, -3)$ is an ordered pair satisfying the relation. Then, if $x = 2$, $y = 9$, and $(2, 9)$ is a second ordered pair that satisfies the equation. It can be observed that as the value of x increases from 1 to 2, the graph of the relation crosses the X-axis. At the point at which the graph crosses the X-axis is the root we are seeking because, at this point, y is equal to zero; and, therefore, at this point $f(x)$ is equal to zero. Thus,

$$y = f(x) = x^3 + 4x^2 - 7x - 1.$$

When $y = 0$,

$$0 = f(x) = x^3 + 4x^2 - 7x - 1.$$

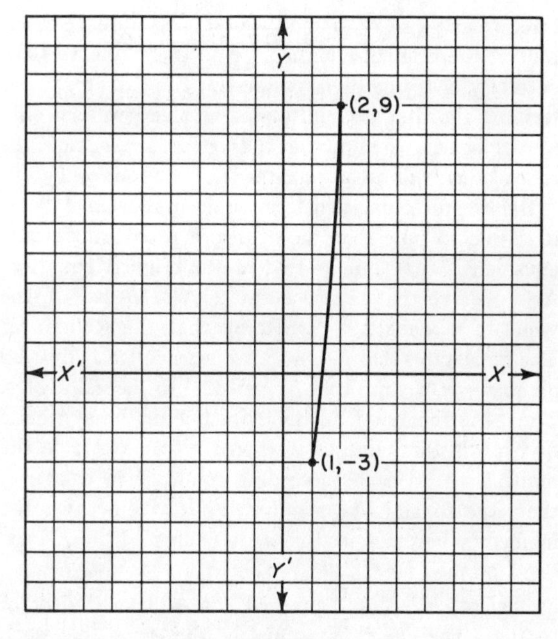

Fig. 66 (Prob. 1)

In this problem, it is important to remember that any value of y below the X-axis is a negative value; and that any value of Y above the X-axis is a positive value. The importance of this is as follows: In the ensuing discussion, we shall be finding values of y which become closer and closer to zero, and we do not want to pass zero. For it is at $y = 0$ that our root lies.

Since we know that $1 < x < 2$, we know that the root will be 1 plus some decimal. The procedure, then, will be one of finding successive decimal places of the root. Our first step in the solution is to use the 1 in the synthetic division process. Rather than stop at the place where we found the depressed equation in the preceding problems we continue this process, using that same 1 with those values obtained when we added. This process is continued, leaving the last one each time, until there are only two terms. This is shown as follows:

$$\begin{array}{l}
1+4-7-1 \;\lfloor 1 \\
\quad +1+5-2 \\ \hline
1+5-2-3 \\
\quad +1+6 \\ \hline
1+6+4 \\
\quad +1 \\ \hline
1+7
\end{array}$$

Observe the circled terms. These in succession, reading from left to right, are the coefficients of a new equation which we shall call our first **transformed equation** and which we shall designate by T_1. Thus, T_1 is

$$x^3+7x^2+4x-3=0.$$

The roots of this transformed equation are exactly one less than the roots of the original equation. And, therefore, we may use T_1 to find the first decimal place of the root of the original equation. And this eliminates the necessity of trying divisors like 1.1, 1.2, 1.3, etc. to determine the first decimal place. Instead we try only tenths, bearing in mind that the root we are seeking, the root of $f(x)$, will be one greater than the root of T_1.

Before proceeding to determine the first decimal place, it is well to note that the degree of each transformed equation is the same as that of the original equation. Also observe that the constant term of T_1 is -3, the value of y when $x=1$. Recall that the Remainder Theorem states that $f(r)=R$. But when we evaluate y given $x=r$ we find that $f(r)$ is equal to the value of y. Thus, the remainder in the third line of the synthetic division is the value of y when the value of x is the multiplier. In each succeeding transformed equation, then, the constant term must be negative because the moment y becomes positive we have crossed the X-axis.

To find the first decimal place, then, we must use the coefficients of T_1 and proceed with a trial divisor of tenths. The one chosen must be the largest one such that the constant term remains negative. Let us try 0.5.

$$\begin{array}{l}
1+7\;\;+4\;\;\;-3\;\;\;\lfloor .5 \\
\quad +0.5+3.75+3.875 \\ \hline
1+7.5+7.75+0.875
\end{array}$$

Thus 0.5 is too much since the constant (the value of y when $x=1.5$) is positive. Let us try 0.4.

$$\begin{array}{l}
1+7\;\;+4\;\;\;-3\;\;\;\lfloor .4 \\
\quad +0.4+2.96+2.784 \\ \hline
1+7.4+6.96-0.216
\end{array}$$

This leaves a negative value for y, so we will proceed to obtain T_2, our second transformed equation, as we did above to find T_1.

$$\begin{array}{l}
1+7\;\;+\;\;4\;\;\;-3\;\;\;\lfloor .4 \\
\quad +0.4+\;\;2.96+2.784 \\ \hline
1+7.4+\;\;6.96-0.216 \\
\quad +0.4+\;\;3.12 \\ \hline
1+7.8+10.08 \\
\quad +0.4 \\ \hline
1+8.2
\end{array}$$

Thus T_2 is $x^3+8.2x^2+10.08x-0.216=0$, and the value of the root of $f(x)$, thus far, is the sum, $1+0.4$ or 1.4. To find hundredths place, we proceed in the same manner using T_2 and a hundredths value as multiplier.

Using 0.03 gives us a positive number in the last place. Thus we cannot use a multiplier that large. Using 0.02 we get the following

$$\begin{array}{l}
1+8.2\;\;+10.08\;\;\;\;-0.216\;\;\;\lfloor .02 \\
\quad +0.02+\;\;0.1644+0.204888 \\ \hline
1+8.22+10.2444-0.011112 \\
\quad +0.02+\;\;0.1648 \\ \hline
1+8.24+10.4092 \\
\quad +0.02 \\ \hline
1+8.26
\end{array}$$

Then T_3 is $x^3+8.26x^2+10.4092x-0.011112=0$. Thus, the value of our root, at this stage, is $1+0.4+0.02$ or 1.42. If we wish to find the value of the root to the next decimal place we continue in a similar manner using T_3 and the largest thousandth that does not give us a positive value for the constant.

If we wish to find the value to the *nearest* hundredth we proceed as before using the coefficients of T_3 and a divisor of 0.005 to determine whether we leave the root at the hundredth value found above or raise it by 0.01. To do this we need only observe the sign of the constant at the end of the first line. (Proceed no further.) In this problem, we kept that constant negative. If it remains negative, the y-value corresponding to $x=1.425$ is not quite enough to bring the graph to the X-axis. If the constant term is positive when we divide T_3 by $x-.005$, the y-value is above the X-axis and we have too much. In this case, we leave the hundredth digit found above.

Thus

$$\begin{array}{l}
1+8.26\;\;+10.4092\;\;\;\;\;-0.011112\;\;\;\lfloor .005 \\
\quad +0.005+\;\;0.041325+0.052252625 \\ \hline
1+8.265+10.450525+0.041140625
\end{array}$$

Since the constant is positive 0.005 is too much and we retain the .02. Thus our root to the nearest hundredth is 1.42.

2. Find, to the nearest hundredth, the root between 1 and 2 in the equation $x^3 - 6x^2 + 3x + 4 = 0$.

The following are ordered pairs that satisfy this equation: $(-1,-6)$, $(0,4)$, $(1,2)$ and $(2,-6)$. A rough sketch is shown in Fig. 67.

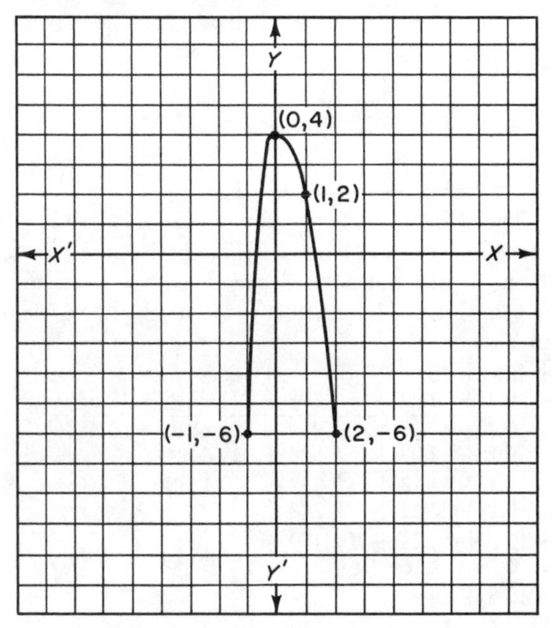

FIG. 67 (Prob. 2)

Observe that values of y, as we proceed from $x = 1$ on the left to the x-intercept on the right are positive. In this problem, then, the last term in the third line of the synthetic division (the constant of the transformed equation) must always be positive; for if it were negative, the graph would have crossed the X-axis.

We start by using 1 as multiplier, and continue until we have our first transformed equation.

$$\begin{array}{l} 1 - 6 + 3 + 4 \;\lfloor 1 \\ \quad + 1 - 5 - 2 \\ \hline 1 - 5 - 2 + 2 \\ \quad + 1 - 4 \\ \hline 1 - 4 - 6 \\ \quad + 1 \\ \hline 1 - 3 \end{array}$$

Hence, T_1 is $x^3 - 3x^2 - 6x + 2 = 0$. If, with T_1, we try 0.3 we have too much (y is negative), so we will use 0.2.

$$\begin{array}{l} 1 - 3 \quad - 6 \quad + 2 \qquad \lfloor .2 \\ \quad + 0.2 - 0.56 - 1.312 \\ \hline 1 - 2.8 - 6.56 + 0.688 \\ \quad + 0.2 - 0.52 \\ \hline 1 - 2.6 - 7.08 \\ \quad + 0.2 \\ \hline 1 - 2.4 \end{array}$$

Then, T_2 is $x^3 - 2.4x^2 - 7.08x + 0.688 = 0$, and our root, so far, is 1.2.

$$\begin{array}{l} 1 - 2.4 \quad - 7.08 \quad + 0.688 \qquad \lfloor .09 \\ \quad + 0.09 - 0.2079 - 0.655911 \\ \hline 1 - 2.31 - 7.2879 + 0.032089 \\ \quad + 0.09 - 0.1998 \\ \hline 1 - 2.22 - 7.4877 \\ \quad + 0.09 \\ \hline 1 - 2.13 \end{array}$$

Thus T_3 is $x^3 - 2.13x^2 - 7.4877x + 0.032089 = 0$. With T_3 we shall try 0.005.

$$\begin{array}{l} 1 - 2.13 \quad - 7.4877 \quad + 0.032089 \qquad \lfloor .005 \\ \quad + 0.005 - 0.010625 - 0.037491625 \\ \hline 1 - 2.125 - 7.498325 - 0.005402625 \end{array}$$

Thus 0.005 is too much and we leave our final answer at 1.29.

3. Find, to the nearest hundredth, the negative root between 0 and -1 for Problem 2.

Let us observe that part of the graph in Fig. 67 to the left of the X-axis. As we move from $x = 0$ on the *right* to $x = -1$ on the *left*, y is positive until we cross the X-axis. Thus, in our use of synthetic division we must keep the last term of the transformed equations positive. We start with the original equation in our quest for the proper digit for tenths place when the integral part of the root is zero.

If we use -0.6 as a divisor, that last term is negative so we shall use -0.5 and obtain

$$\begin{array}{l} 1 - 6 \quad + 3 \quad + 4 \qquad \lfloor -0.5 \\ \quad - 0.5 + 3.25 - 3.125 \\ \hline 1 - 6.5 + 6.25 + 0.875 \\ \quad - 0.5 + 3.5 \\ \hline 1 - 7.0 + 9.75 \\ \quad - 0.5 \\ \hline 1 - 7.5 \end{array}$$

Thus, T_1 is $x^3 - 7.5x^2 + 9.75x + 0.875 = 0$. If we use -0.09 the last term is negative so we shall use -0.08 with T_1.

$$\begin{array}{l} 1 - 7.5 \;+\; 9.75 \quad + 0.875 \qquad \lfloor -.08 \\ \quad - 0.08 + 0.6064 - 0.828512 \\ \hline 1 - 7.58 + 10.3564 + 0.046488 \\ \quad - 0.08 + 0.6128 \\ \hline 1 - 7.66 + 10.9692 \\ \quad - 0.08 \\ \hline 1 - 7.74 \end{array}$$

Then T_2 is $x^3 - 7.74x^2 + 10.9692x + 0.046488 = 0$. To find whether the hundredths digit is closer to 8 or 9 we use -0.005 in T_2.

$$\begin{array}{l} 1 - 7.74 \;+ 10.9692 \quad + 0.046488 \qquad \lfloor -0.005 \\ \quad - 0.005 + 0.038725 - 0.055039625 \\ \hline 1 - 7.745 + 11.007925 - 0.008551625 \end{array}$$

Since that last term is negative, we retain the -0.08 found previously and our root is -0.58.

4. By using Horner's Method find the cube root of 54 to the nearest hundredth.

$$\text{Let } x = \sqrt[3]{54}.$$

$$\text{Then } x^3 = 54$$

and

$$x^3 - 54 = 0.$$

The cube root of 64 is 4 and the cube root of 27 is 3. Thus, the cube root of 54 is greater than 3 and we shall start our solution with the integer 3.

$$\begin{array}{l}
1 + 0 + \ \ 0 - 54 \,\lfloor 3 \\
\quad\ + 3 + \ \ 9 + 27 \\
\hline
1 + 3 + \ \ 9 - 27 \\
\quad\ + 3 + 18 \\
\hline
1 + 6 + 27 \\
\quad\ + 3 \\
\hline
1 + 9
\end{array}$$

Hence, T_1 is $x^3 + 9x^2 + 27x - 27 = 0$. Since this last term is preceded by a negative sign, we must have a negative sign each time or the digit in the root will be too much. Using 0.8 with T_1 gives us too much, so we shall use 0.7.

$$\begin{array}{l}
1 + \ \ 9 \ \ + 27 \quad\ - 27 \qquad \lfloor 0.7 \\
\quad + \ 0.7 + \ \ 6.79 + 23.653 \\
\hline
1 + \ \ 9.7 + 33.79 - \ \ 3.347 \\
\quad + \ 0.7 + \ \ 7.28 \\
\hline
1 + 10.4 + 41.07 \\
\quad + \ 0.7 \\
\hline
1 + 11.1
\end{array}$$

Then T_2 is $x^3 + 11.1x^2 + 41.07x - 3.347 = 0$. Using 0.08 with this transformed equation gives us too much, so we shall use 0.07.

$$\begin{array}{l}
1 + 11.1 \ \ + 41.07 \quad\ - 3.347 \qquad \lfloor 0.07 \\
\quad + \ \ 0.07 + \ \ 0.7819 + 2.929633 \\
\hline
1 + 11.17 + 41.8519 - 0.417367 \\
\quad + \ \ 0.07 + \ \ 0.7868 \\
\hline
1 + 11.24 + 42.6387 \\
\quad + \ \ 0.07 \\
\hline
1 + 11.31
\end{array}$$

Hence, T_3 is $x^3 + 11.31x^2 + 42.6387x - 0.417367$. To determine whether the hundredths digit is closer to 7 or 8 we use 0.005 with T_3 and obtain

$$\begin{array}{l}
1 + 11.31 \ \ + 42.6387 \quad\ - 0.417367 \qquad \lfloor 0.005 \\
\quad + \ \ 0.005 + \ \ 0.056575 + 0.213476375 \\
\hline
1 + 11.315 + 42.695275 - 0.203890625
\end{array}$$

The negative sign preceding the last term tells us that 0.005 is not too much, so we raise the seven hundredths to eight hundredths. Our root is 3.78 to the nearest hundredth.

5. By the use of Horner's Method find the fifth root of 54 to the nearest hundredth.

$$x^5 = 54.$$

$$x^5 - 54 = 0.$$

The fifth root of 32 is 2 and of 243 is 3. Thus, we shall start our solution with the multiplier 2.

$$\begin{array}{l}
1 + \ \ 0 + \ \ 0 + \ \ 0 + \ \ 0 - 54 \,\lfloor 2 \\
\quad + \ \ 2 + \ \ 4 + \ \ 8 + 16 + 32 \\
\hline
1 + \ \ 2 + \ \ 4 + \ \ 8 + 16 - 22 \\
\quad + \ \ 2 + \ \ 8 + 24 + 64 \\
\hline
1 + \ \ 4 + 12 + 32 + 80 \\
\quad + \ \ 2 + 12 + 48 \\
\hline
1 + \ \ 6 + 24 + 80 \\
\quad + \ \ 2 + 16 \\
\hline
1 + \ \ 8 + 40 \\
\quad + \ \ 2 \\
\hline
1 + 10
\end{array}$$

Then T_1 is $x^5 + 10x^4 + 40x^3 + 80x^2 + 80x - 22 = 0$. Since the constant term of T_1 is negative, the constant term in each transformed equation must be negative. When we try 0.3 we find it is too much, so we use 0.2 in T_1.

$$\begin{array}{l}
1 + 10. \ \ + 40. \quad\ + \ \ 80. \quad\ + \ \ 80. \qquad - 22. \qquad \lfloor 0.2 \\
\quad + \ \ 0.2 + \ \ 2.04 + \ \ \ 8.408 + \ \ 17.6816 + 19.53632 \\
\hline
1 + 10.2 + 42.04 + \ \ 88.408 + \ \ 97.6816 - \ \ 2.46368 \\
\quad + \ \ 0.2 + \ \ 2.08 + \ \ \ 8.824 + \ \ 19.4464 \\
\hline
1 + 10.4 + 44.12 + \ \ 97.232 + 117.1280 \\
\quad + \ \ 0.2 + \ \ 2.12 + \ \ \ 9.248 \\
\hline
1 + 10.6 + 46.24 + 106.480 \\
\quad + \ \ 0.2 + \ \ 2.16 \\
\hline
1 + 10.8 + 48.40 \\
\quad + \ \ 0.2 \\
\hline
1 + 11.0
\end{array}$$

Thus, T_2 is $x^5 + 11.0x^4 + 48.40x^3 + 106.48x^2 + 117.128x - 2.46368 = 0$. Then, if we use 0.03 with T_2, we find it is too large, so we use 0.02.

$$\begin{array}{l}
1 + 11.0 \ + 48.40 \ \ + 106.48 \qquad + 117.128 - 2.46368 \,\lfloor 0.02 \\
\quad + 0.02 + \ 0.2204 + \quad 0.972408 + \quad 2.149 + 2.38554 \\
\hline
1 + 11.02 + 48.6204 + 107.452408 + 119.277 - 0.07814 \\
\quad + 0.02 + \ 0.2208 + \quad 0.976824 + \quad 2.169 \\
\hline
1 + 11.04 + 48.8412 + 108.429232 + 121.446 \\
\quad + 0.02 + \ 0.2212 + \quad 0.981248 \\
\hline
1 + 11.06 + 49.0624 + 109.410480 \\
\quad + 0.02 + \ 0.2216 \\
\hline
1 + 11.08 + 49.2840 \\
\quad + 0.02 \\
\hline
1 + 11.10
\end{array}$$

Then T_3 is $x^5 + 11.10x^4 + 49.2840x^3 + 109.4105x^2 + 121.446x - 0.07814 = 0$. To determine whether the nearest hundredth is 0.02 or 0.03, we use 0.005 with T_3.

(It will be observed that some of the decimals in the places beyond the third are being eliminated. Care must be taken in doing this especially if the sum or difference in the last term is extremely small.)

$$\begin{array}{l}
1 + 11.10 \ + 49.2840 + 109.4105 + 121.446 \ - 0.07814 \,\lfloor 0.005 \\
\quad + 0.005 + \ 0.0555 + \quad 0.2467 + \quad 0.5483 + 0.60997 \\
\hline
1 + 11.105 + 49.3395 + 109.6572 + 121.9943 + 0.53183
\end{array}$$

Since the last term is positive, we know that 0.005 is too much. Our root, to the nearest hundredth, then, is 2.22.

SUPPLEMENTARY PROBLEMS

Perform the following divisions by synthetic division:

1. $(x^4 + 4x^3 - 24x^2 + 13x - 12) \div (x - 3)$
2. $(3x^5 + 12x^4 - 2x^3 - 8x^2 + 3x + 12) \div (x + 4)$
3. $(x^3 - 1) \div (x - 1)$
4. $(x^4 + 2x^3 - 4x^2 - 4x - 3) \div (x + 3)$
5. $(x^5 - 3x^4 - 18x^3 - x + 6) \div (x - 6)$

By use of the remainder theorem, find the remainders in the following:

6. $(2x^3 - x^2 - 7x + 5) \div (x - 2)$
7. $(3x^4 - 7x^3 + 4x^2 - 2x - 71) \div (x + 3)$
8. Find the real roots of the equation
$$x^3 + x^2 - 14x - 24 = 0.$$
9. Find the real roots of the equation
$$6x^4 + 5x^3 - 60x^2 - 45x + 54 = 0.$$
10. Find the real roots of the equation
$$6x^3 - 43x^2 + 37x + 30 = 0.$$
11. Find the real roots of the equation $x^4 - 2x^2 + 1 = 0$.
12. Find all of the roots of the equation
$$x^4 + x^3 - x^2 + 5x - 30 = 0.$$
13. Find the cube root of 93, to the nearest hundredth, by use of Horner's Method.
14. Find, by use of Horner's Method, the square root of 307.863 to the nearest thousandth.
15. Use Horner's Method to find, to the nearest hundredth, the negative root between -3 and -4 of the equation
$$x^4 - x^3 - 6x^2 + 20x - 16 = 0.$$

Chapter 19

Classified Verbal Problems

Age Problems

1. Robert is four years older than Bruce. Twenty years ago Robert was twice as old as Bruce. How old is each at the present time?

Let x = Bruce's present age;
$x + 4$ = Robert's present age;
$x - 20$ = Bruce's age twenty years ago;
$x + 4 - 20$ = Robert's age twenty years ago.

$$2(x - 20) = x + 4 - 20.$$
$$2x - 40 = x - 16.$$
$$2x - x = 40 - 16.$$
$$x = 24.$$
$$x + 4 = 28.$$

Check:

$28 - 24 = 4.$ $\quad 4 = 4.$

$2(24 - 20) = 28 - 20.$ $\quad 2(4) = 8.$ $\quad 8 = 8.$

2. Mary is twice as old as Helen, and Sue is twice as old as Mary. In ten years their combined ages will total fifty-eight years. How old is each at the present time?

Let x = Helen's present age;
$2x$ = Mary's present age;
$4x$ = Sue's present age;
$x + 10$ = Helen's age in ten years;
$2x + 10$ = Mary's age in ten years;
$4x + 10$ = Sue's age in ten years;
$7x + 30$ = their combined ages in ten years.

$$7x + 30 = 58.$$
$$7x = 28.$$
$$x = 4.$$
$$2x = 8.$$
$$4x = 16.$$

Check:

$2(4) = 8.$ $\quad 8 = 8.$

$2(8) = 16.$ $\quad 16 = 16.$

$(4 + 10) + (8 + 10) + (16 + 10) = 58.$
$14 + 18 + 26 = 58.$
$58 = 58.$

3. Frank is as old as the combined ages of his two brothers, Harry and Edward. Edward is two years older than Harry. The combined ages of the three last year was three-fourths their combined ages at present. How old is each at the present time?

Let x = Harry's present age;
$x + 2$ = Edward's present age;
$2x + 2$ = Frank's present age;
$4x + 4$ = their combined ages at present;
$x - 1$ = Harry's age last year;
$x + 1$ = Edward's age last year;
$2x + 1$ = Frank's age last year;
$4x + 1$ = their combined ages last year.

$$4x + 1 = \tfrac{3}{4}(4x + 4).$$
$$16x + 4 = 12x + 12.$$
$$16x - 12x = 12 - 4.$$
$$4x = 8.$$
$$x = 2.$$
$$x + 2 = 4.$$
$$2x + 2 = 6.$$

Check:

$2 + 4 = 6.$ $\quad 6 = 6.$

$4 - 2 = 2.$ $\quad 2 = 2.$

$\tfrac{3}{4}(2 + 4 + 6) = 1 + 3 + 5.$
$\tfrac{3}{4}(12) = 9.$
$9 = 9.$

Number Problems

4. If the sum of the squares of two positive consecutive integers is equal to the square of the next positive integer, what are the three integers?

Let x = the first integer;
$x + 1$ = the second integer;
$x + 2$ = the third integer.

$$x^2 + (x + 1)^2 = (x + 2)^2.$$
$$x^2 + x^2 + 2x + 1 = x^2 + 4x + 4.$$
$$x^2 - 2x - 3 = 0.$$
$$(x - 3)(x + 1) = 0.$$
$$x - 3 = 0. \qquad x + 1 = 0.$$
$$x = 3. \qquad x = -1.$$

The -1 will be discarded as it does not meet the specification in the data which requires positive integers.

$$x = 3.$$
$$x + 1 = 4.$$
$$x + 2 = 5.$$

Check:

$3^2 + 4^2 = 5^2.$
$9 + 16 = 25.$
$25 = 25.$

Hence, the integers are 3, 4, and 5.

5. Find three consecutive odd integers such that the sum of the first two and three times the third equals the square of the first.

Let $2x + 1$ = the first integer;
$2x + 3$ = the second integer;
$2x + 5$ = the third integer.

$$(2x + 1)^2 = (2x + 1) + (2x + 3) + 3(2x + 5).$$
$$4x^2 + 4x + 1 = 2x + 1 + 2x + 3 + 6x + 15.$$
$$4x^2 - 6x - 18 = 0.$$
$$2x^2 - 3x - 9 = 0.$$
$$(2x + 3)(x - 3) = 0.$$
$$2x + 3 = 0. \qquad x - 3 = 0.$$
$$2x = -3. \qquad x = 3.$$
$$x = -\tfrac{3}{2}.$$

Discard the $-\frac{3}{2}$, as it will not satisfy the given conditions.

$$2x + 1 = 7.$$
$$2x + 3 = 9.$$
$$2x + 5 = 11.$$

Check:

$7 + 9 + 3(11) = 7^2.$
$7 + 9 + 33 = 49.$
$49 = 49.$

Hence, the integers are 7, 9, and 11.

6. The sum of the digits of a two-digit number is 9. If the digits are reversed the number is greater than the original number by 27. What was the original number?

Let u = units' digit;
t = tens' digit.

$$t + u = 9. \qquad (1)$$
$$10t + u = 10u + t - 27.$$
$$9t - 9u = -27.$$
$$t - u = -3. \qquad (2)$$

Add equations (1) and (2).

$$t + u = 9 \qquad (1)$$
$$t - u = -3 \qquad (2)$$
$$2t = 6$$
$$t = 3. \qquad (3)$$

Substitute the value of t from equation (3) in equation (1).

$$t + u = 9. \qquad (1)$$
$$3 + u = 9.$$
$$u = 6.$$

Check:

$36 = 63 - 27.$
$36 = 36.$

Hence, the original number was 36.

Mixture Problems

7. How many pounds each of 60-cent candy and 90-cent candy must be mixed together to make 50 pounds of candy that can sell for 72 cents per pound?

Let x = amount of 60-cent candy used;
$50 - x$ = amount of 90-cent candy used;
$0.60x$ = value of 60-cent candy used in mixture;
$0.90(50 - x)$ = value of 90-cent candy used in mixture $= 45 - 0.90x$;
$45 - 0.30x$ = total value of both kinds of candy used in mixture.
$(50)(0.72)$ = total value of mixture $= 36$.

$$45 - 0.30x = 36.$$
$$-0.30x = -9.$$
$$x = 30 \text{ pounds}.$$
$$50 - x = 20 \text{ pounds}.$$

Check:

$30(0.60) + 20(0.90) = 50(0.72).$
$18 + 18 = 36.$
$36 = 36.$

8. How many pounds of 75-cent coffee must be added to 50 pounds of 90-cent coffee to make a mixture that can sell for 80 cents per pound?

Let x = amount of 75-cent coffee to be added;
$0.75x$ = value of 75-cent coffee used;
$0.90(50)$ = value of 90-cent coffee used $= 45$;
$0.75x + 45$ = total value of mixture;
$50 + x$ = amount in new mixture;
$0.80(50 + x)$ = value of new mixture $= 40 + 0.80x$.

$$40 + 0.80x = 0.75x + 45.$$
$$0.05x = 5.$$
$$x = 100 \text{ pounds}.$$

Check:

$100(0.75) + 50(0.90) = 150(0.80).$
$75 + 45 = 120.$
$120 = 120.$

9. A laboratory technician has a 70% solution of alcohol and an 80% solution of alcohol. He wishes to make ten gallons of a 74% solution by mixing the two. How much of each solution will he use?

Let x = amount of 70% solution used;
$10 - x$ = amount of 80% solution used;
$0.70x$ = amount of pure alcohol in 70% solution used;

$(0.80)(10 - x) =$ amount of pure alcohol in 80% solution used
$= 8 - 0.80x;$
$0.70x + 8 - 0.80x =$ amount of pure alcohol in both parts used
$= 8 - 0.10x;$
$10(0.74) =$ amount of pure alcohol in new solution
$= 7.4.$
$8 - 0.10x = 7.4.$
$-0.10x = -0.6.$
$x = 6$ gallons.
$10 - x = 4$ gallons.

Check:
$6(0.70) + 4(0.80) = 10(0.74).$
$4.2 + 3.2 = 7.4.$
$7.4 = 7.4.$

10. A merchant wishes to mix a quantity of peanuts, walnuts, and pecans. The peanuts sell for 30 cents per pound; the walnuts, 45 cents per pound; and the pecans, 50 cents per pound. How many pounds of each will he use to make a mixture of 100 pounds that will sell for 40 cents a pound if he uses equal weights of peanuts and walnuts?

Let $x =$ weight of peanuts used;
$x =$ weight of walnuts used;
$100 - 2x =$ weight of pecans used;
$0.30x =$ value of peanuts used;
$0.45x =$ value of walnuts used;
$0.50(100 - 2x) =$ value of pecans used
$= 50 - x;$
$50 - 0.25x =$ total value before mixing;
$100(0.40) =$ total value after mixing
$= 40.$
$50 - 0.25x = 40.$
$-0.25x = -10.$
$x = 40$ pounds.
$x = 40$ pounds.
$100 - 2x = 20$ pounds.

Check:
$40(0.30) + 40(0.45) + 20(0.50) = 100(0.40).$
$12 + 18 + 10 = 40.$
$40 = 40.$

Business Problems

11. A man invests \$500 more at 5% than he already has invested at $4\frac{1}{2}\%$. The total annual income from the two investments is \$329. How much has he invested in each?

Let $x =$ amount invested at $4\frac{1}{2}\%$;
$x + \$500 =$ amount invested at 5%;
$0.045x =$ annual income from $4\frac{1}{2}\%$ investment;
$(0.05)(x + \$500) =$ annual income from 5% investment
$= 0.05x + \$25;$
$0.05x + \$25 + 0.045x =$ total annual income from both investments
$= 0.095x + \$25.$
$0.095x + \$25 = \$329.$
$0.095x = \$304.$
$x = \$3,200.$
$x + \$500 = \$3,700.$

Check:
$(0.045)(\$3,200) + (0.05)(\$3,700) = \$329.$
$\$144 + \$185 = \$329.$
$\$329 = \$329.$

12. A foreman gets 35 cents per hour more than his assistant. Their combined wage for a 40-hour week is \$198. What is the hourly wage of each?

Let $x =$ hourly wage of assistant;
$x + 0.35 =$ hourly wage of foreman;
$2x + 0.35 =$ combined hourly wage;
$(2x + 0.35)(40) =$ combined weekly wage
$= 80x + \$14.$
$80x + \$14 = \$198.$
$80x = \$184.$
$x = \$2.30.$
$x + 0.35 = \$2.65.$

Check:
$40(\$2.30) + 40(\$2.65) = \$198.$
$\$92 + \$106 = \$198.$
$\$198 = \$198.$

13. Two men invest in a partnership, one providing \$1,000 more capital than the other. At the end of the first year they find that they have a net profit of \$4,050, which represents a $22\frac{1}{2}\%$ profit on their investment. How much did each invest in the partnership?

Let $x =$ amount first man invested;
$x + \$1,000 =$ amount second man invested;
$2x + \$1,000 =$ total investment;
$0.225(2x + \$1,000) =$ profit.
$0.225(2x + \$1,000) = \$4,050.$
$0.45x + \$225 = \$4,050.$
$0.45x = \$3,825.$
$x = \$8,500.$
$x + 1,000 = \$9,500.$

Check:
$0.225(\$8,500) + 0.225(\$9,500) = \$4,050.$
$\$1912.50 + \$2,137.50 = \$4,050.$
$\$4,050 = \$4,050.$

14. A man paid \$220.80 state income tax, which was 4% of his adjusted net income after all deductions. The state allows him a ten-percent deduction on his gross income for incidentals and \$600 each for himself, his

wife, and two children for dependency deductions. What was his gross income?

$$\begin{aligned}
\text{Let } x &= \text{gross income;}\\
\$2{,}400 &= \text{family deductions;}\\
0.10x &= \text{deductions for incidentals;}\\
0.10x + \$2{,}400 &= \text{total deductions;}\\
x - (0.10x + \$2{,}400) &= \text{adjusted net income}\\
&= 0.90x - \$2{,}400;\\
0.04(0.90x - \$2{,}400) &= \text{amount of tax}\\
&= 0.036x - \$96.\\
0.036x - \$96 &= \$220.80.\\
0.036x &= \$316.80.\\
x &= \$8{,}800.
\end{aligned}$$

Check:

$$\begin{aligned}
\$880 &= \text{the deductions for incidentals;}\\
\$2{,}400 &= \text{family deductions;}\\
\$3{,}280 &= \text{total deductions.}\\
\$8{,}800 - \$3{,}280 &= \text{taxable income}\\
&= \$5{,}520.\\
0.04(\$5{,}520) &= \text{amount of tax.}\\
\$220.80 &= \$220.80.
\end{aligned}$$

Measurement Problems

15. A rectangle is 8 feet longer than its width, and its area is 308 square feet. What are the dimensions of the rectangle?

$$\begin{aligned}
\text{Let } x &= \text{width;}\\
x + 8 &= \text{length;}\\
x(x + 8) &= \text{area.}\\
x(x + 8) &= 308.\\
x^2 + 8x &= 308.\\
x^2 + 8x - 308 &= 0.\\
(x + 22)(x - 14) &= 0.
\end{aligned}$$

$$x + 22 = 0. \qquad x - 14 = 0.$$
$$x = -22. \qquad x = 14.$$

Discard the -22, as the width must be a positive number.

$$x + 8 = 22 \text{ feet.}$$

Check:

$$(14)(22) = 308.$$
$$308 = 308.$$

Hence, the dimensions of the rectangle are 14 feet by 22 feet.

16. A family has a back yard 30 feet long and 20 feet wide. They decide to construct a swimming pool in the yard, leaving a uniform strip of grass around the pool. If the pool is to have an area equal to that of the grassy plot around it, what will be the dimensions of the pool? (See Fig. 68.)

$$\begin{aligned}
\text{Let } x &= \text{length of pool;}\\
y &= \text{width of pool;}\\
d &= \text{width of grassy plot.}
\end{aligned}$$

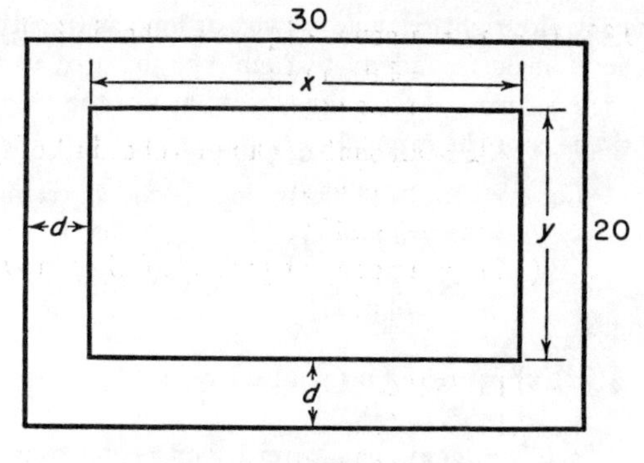

Fig. 68

$$xy = \tfrac{1}{2}(20)(30). \qquad (1)$$
$$xy = 300.$$
$$y = \frac{300}{x}. \qquad (1a)$$
$$x + 2d = 30. \qquad (2)$$
$$y + 2d = 20. \qquad (3)$$

Subtract (3) from (2).

$$\begin{aligned}
x + 2d &= 30\\
y + 2d &= 20\\
\hline
x - y \quad &= 10 \qquad (4)
\end{aligned}$$

Substitute the value of y from (1a) in (4).

$$x - \frac{300}{x} = 10.$$
$$x^2 - 300 = 10x.$$
$$x^2 - 10x - 300 = 0.$$

$$\begin{aligned}
x &= \frac{-(-10) \pm \sqrt{(-10)^2 - 4(1)(-300)}}{2(1)}\\
&= \frac{10 \pm \sqrt{100 + 1200}}{2}\\
&= \frac{10 \pm \sqrt{1300}}{2}\\
&= \frac{10 \pm 36.06}{2}\\
&= 5 \pm 18.03.
\end{aligned}$$

$$x = 23.03 \text{ or } 23 \text{ feet.} \quad (5) \qquad x = -13.03.$$

Discard the negative root, since the width of a pool must be positive.

Substitute the value of x from (5) in (4).

$$x - y = 10.$$
$$23 - y = 10.$$
$$y = 13 \text{ feet.}$$

Check:

$$(23)(13) = 300. \qquad 30 - 23 = 20 - 13.$$
$$299 = 300. \qquad 7 = 7.$$

To 2 figures,

$$300 = 300.$$

Hence, the pool will be 23 feet by 13 feet.

17. One leg of a right triangle is twice as long as the other. If eight inches is added to each leg, the area of the triangle is increased six times. What are the lengths of the legs of the original triangle.

Let x = length of shorter leg of original triangle;
$2x$ = length of longer leg of original triangle.

Consider the x as the base and $2x$ as the altitude of the triangle.

$$\tfrac{1}{2}(2x)(x) = \text{area of original triangle}$$
$$= x^2.$$
$$x + 8 = \text{length of shorter leg of second triangle;}$$
$$2x + 8 = \text{length of longer leg of second triangle;}$$
$$\frac{(x+8)(2x+8)}{2} = \text{area of second triangle}$$
$$= \frac{2x^2 + 24x + 64}{2}$$
$$= x^2 + 12x + 32.$$
$$6x^2 = x^2 + 12x + 32.$$
$$5x^2 - 12x - 32 = 0.$$
$$(5x + 8)(x - 4) = 0.$$
$$5x + 8 = 0. \qquad x - 4 = 0.$$
$$5x = -8. \qquad x = 4 \text{ inches.}$$
$$x = -\tfrac{8}{5}.$$

Discard the negative value, as the length of a leg must be a positive number.

$$2x = 8 \text{ inches.}$$

Check:
$$6(4)(8) = (4 + 8)(8 + 8).$$
$$192 = (12)(16).$$
$$192 = 192.$$

Hence, the legs of the original triangle are 4 inches and 8 inches.

18. The perimeter of a rectangle is 54 feet. If the area is 176 square feet, what are the dimensions of the rectangle?

Let x = length;
$27 - x$ = width;
$x(27 - x)$ = the area.

$$27x - x^2 = 176.$$
$$-x^2 + 27x - 176 = 0.$$
$$x^2 - 27x + 176 = 0.$$
$$(x - 16)(x - 11) = 0.$$
$$x - 16 = 0. \qquad x - 11 = 0.$$
$$x = 16. \qquad x = 11.$$
$$27 - x = 11. \qquad 27 - x = 16.$$

Check:
$$2(11) + 2(16) = 54. \qquad (11)(16) = 176.$$
$$22 + 32 = 54. \qquad 176 = 176.$$
$$54 = 54.$$

Hence, the dimensions of the rectangle are 16 feet by 11 feet.

Lever Problems

In the study of physics, problems involving the lever are frequently found. The support for the lever is called the **fulcrum.** The lever will be balanced when the sum of the products of the weights on one side and their distances from the fulcrum is equal to the sum of the product of the weights on the opposite side and their distances from the fulcrum.

19. Two weights, 140 and 160 pounds respectively, are placed on opposite sides of a 20-foot steel bar supported by a fulcrum midway between the ends of the bar. If the 140-pound weight is 8 feet from the fulcrum, what is the distance between the fulcrum and the 160-pound weight when the bar is balanced?

Let x = distance between 160-pound weight and fulcrum.

$$x(160) = 8(140).$$
$$160x = 1{,}120.$$
$$x = 7 \text{ feet.}$$

Check:
$$7(160) = 8(140).$$
$$1{,}120 = 1{,}120.$$

20. A fulcrum is placed midway between the ends of a 24-foot steel bar. At one end a 75-pound weight is suspended. How far from the other end of the bar must a 100-pound weight be suspended to balance the bar?

Let x = distance between end of bar and 100-pound weight;
$12 - x$ = distance between fulcrum and 100-pound weight.

$$(12 - x)(100) = 12(75).$$
$$1200 - 100x = 900.$$
$$-100x = -300.$$
$$x = 3 \text{ feet.}$$

Check:
$$9(100) = 12(75).$$
$$900 = 900.$$

21. A 16-foot lever, supported by a fulcrum at the center, has a 150-pound weight on its left end and a 100-pound weight on its right end. How far from the right end must a 75-pound weight be placed to balance the lever?

Let x = distance between right end of bar and 75-pound weight;
$8 - x$ = distance between fulcrum and 75-pound weight.

$$(8 - x)(75) + 8(100) = 8(150).$$
$$600 - 75x + 800 = 1{,}200.$$
$$-75x = -200.$$
$$x = 2\tfrac{2}{3} \text{ feet.}$$

Check:

$$5\tfrac{1}{3}(75) + 8(100) = 8(150).$$
$$400 + 800 = 1{,}200.$$
$$1{,}200 = 1{,}200.$$

22. A man wishes to move a large stone by using an iron bar 7 feet long. He places a block of wood under the iron bar to serve as the fulcrum. If the block of wood is 2 feet from the stone and he is just able to lift the stone when he exerts a force of 200 pounds, how heavy is the stone?

Let x = the weight of the stone.

$$(x)(2) = (200)(5).$$
$$2x = 1{,}000.$$
$$x = 500 \text{ pounds.}$$

Check:

$$(500)(2) = (200)(5).$$
$$1{,}000 = 1{,}000.$$

Uniform Motion Problems

23. Two automobiles start from the same place, the second one two hours later than the first. If the first averages 40 miles per hour and the second averages 50 miles per hour, how long will it take the second to overtake the first?

Let t = time first one travels;
$t - 2$ = time second one travels.

$$d = rt.$$
$$d_1 = 40t.$$
$$d_2 = 50(t - 2) = 50t - 100.$$

Since they both travel the same distance, let the distance the second car travels equal the distance the first car travels.

$$50t - 100 = 40t.$$
$$10t = 100.$$
$$t = 10 \text{ hours.}$$
$$t - 2 = 8 \text{ hours.}$$

Check:

$$(8)(50) = (10)(40).$$
$$400 = 400.$$

Hence, it will take the second car 8 hours to overtake the first car.

24. Two cars start traveling in opposite directions at the same time and from the same point. They are 600 miles apart after 6 hours. If one of them averaged 45 miles per hour, what was the average speed of the second car?

Let r_1 = average speed of first car;
r_2 = average speed of second car;
t = time they travel;
$r_1t + r_2t$ = distance between them.

$$r_1t + r_2t = 600.$$
$$(45)(6) + (r_2)(6) = 600.$$
$$270 + 6r_2 = 600.$$
$$6r_2 = 330.$$
$$r_2 = 55 \text{ miles per hour.}$$

Check:

$$(45)(6) + (55)(6) = 600.$$
$$270 + 330 = 600.$$
$$600 = 600.$$

25. An airplane flying with the wind made a 400-mile trip in 2 hours. It made the return trip flying against the same wind in $2\frac{1}{2}$ hours. What was the speed of the plane in still air? What was the speed of the wind?

Let x = speed of plane in still air;
y = speed of wind;
$x + y$ = speed of plane with wind;
$x - y$ = speed of plane against wind.

$$2(x + y) = 400. \quad (1)$$
$$\tfrac{5}{2}(x - y) = 400. \quad (2)$$
$$x + y = 200. \quad (1a)$$
$$x - y = 160. \quad (2a)$$

Add equations (1a) and (2a).

$$x + y = 200$$
$$x - y = 160$$
$$2x = 360$$
$$x = 180 \text{ miles per hour.}$$

Substitute this value of x in equation (1a).

$$x + y = 200.$$
$$180 + y = 200.$$
$$y = 20 \text{ miles per hour.}$$

Check:

If the plane's speed is 180 miles per hour in still air, it travels 200 miles per hour with the wind and 160 miles per hour against the wind.

$$2(200) = 400. \qquad \tfrac{5}{2}(160) = 400.$$
$$400 = 400. \qquad 400 = 400.$$

26. An airplane flies 2,000 miles with the wind in the same time that it flies 1,800 miles against the same wind. If the plane's speed is 285 miles per hour in still air, what is the speed of the wind?

Let x = speed of wind;
$285 + x$ = speed of plane with wind;
$285 - x$ = speed of plane against wind;
t = time of flight.

$$t(285 + x) = 2{,}000. \quad (1)$$
$$t(285 - x) = 1{,}800. \quad (2)$$

Divide equation (2) by equation (1).

$$\frac{t(285 - x)}{t(285 + x)} = \frac{1{,}800}{2{,}000}.$$
$$\frac{285 - x}{285 + x} = \frac{9}{10}.$$

$$2565 + 9x = 2850 - 10x.$$
$$19x = 285.$$
$$x = 15 \text{ miles per hour.}$$

Check:

If the time is the same for each flight, the ratios of distance to speed must be equal.

$$\frac{2{,}000}{285 + 15} = \frac{1{,}800}{285 - 15}.$$
$$\frac{2{,}000}{300} = \frac{1{,}800}{270}.$$
$$\tfrac{20}{3} = \tfrac{20}{3}.$$

Work Problems

27. A man and his son work together to mow the lawn of their large estate. The son uses a lawn mower that would take him 12 hours to complete the job alone. The father uses a power mower with which he could do the entire job in 4 hours. How long will it take them to complete the job working together?

Let x = time required if they work together;

$\frac{1}{x}$ = part they can do together in 1 hour;

$\frac{1}{12}$ = part son could do in 1 hour;

$\frac{1}{4}$ = part father could do in 1 hour;

$$\frac{1}{12} + \frac{1}{4} = \frac{1}{x}.$$
$$\frac{x}{12x} + \frac{3x}{12x} = \frac{12}{12x}.$$
$$4x = 12.$$
$$x = 3 \text{ hours.}$$

Check:

$$\frac{1}{12} + \frac{1}{4} = \frac{1}{x}.$$
$$\tfrac{1}{12} + \tfrac{1}{4} = \tfrac{1}{3}.$$
$$\tfrac{1}{12} + \tfrac{3}{12} = \tfrac{1}{3}.$$
$$\tfrac{4}{12} = \tfrac{1}{3}.$$
$$\tfrac{1}{3} = \tfrac{1}{3}.$$

28. If one man can do a piece of work in 5 working days and a second man can do the same job in 4 working days, how long will it take them to do it if they both work at the same time?

Let x = time required if they work together;

$\frac{1}{x}$ = part they can do together in 1 day;

$\frac{1}{5}$ = part first man could do in 1 day;

$\frac{1}{4}$ = part second man could do in 1 day.

$$\frac{1}{5} + \frac{1}{4} = \frac{1}{x}.$$
$$\frac{4x}{20x} + \frac{5x}{20x} = \frac{20}{20x}.$$
$$4x + 5x = 20.$$
$$9x = 20.$$
$$x = 2\tfrac{2}{9} \text{ working days.}$$

Check:

$$\frac{1}{5} + \frac{1}{4} = \frac{1}{x}.$$

Convert $2\frac{2}{9}$ days to $\frac{20}{9}$ days.

$$\frac{1}{5} + \frac{1}{4} = \frac{1}{\frac{20}{9}}.$$
$$\tfrac{1}{5} + \tfrac{1}{4} = \tfrac{9}{20}.$$
$$\tfrac{4}{20} + \tfrac{5}{20} = \tfrac{9}{20}.$$
$$\tfrac{9}{20} = \tfrac{9}{20}.$$

29. A man and his son were working together on a job which they knew would take them 2 hours. If the son became ill after working for 1 hour, and the father had to work $1\frac{1}{2}$ hours more to complete the job, how long would it have taken each to do the job alone?

Let x = time required if son works alone;

y = time required if father works alone;

$\frac{1}{x}$ = part son can do in 1 hour;

$\frac{1}{y}$ = part father can do in 1 hour;

$$\frac{1}{x} + \frac{1}{y} = \frac{1}{2}. \quad (1)$$
$$\frac{1}{x} + \frac{\frac{5}{2}}{y} = 1. \quad (2)$$
$$\frac{1}{x} + \frac{5}{2y} = 1. \quad (2a)$$

Solve equation (1) and (2a) by subtraction.

$$\frac{1}{x} + \frac{1}{y} = \frac{1}{2} \quad (1)$$
$$\frac{1}{x} + \frac{5}{2y} = 1 \quad (2a)$$
$$\frac{1}{y} - \frac{5}{2y} = -\frac{1}{2}$$
$$2 - 5 = -y.$$
$$y = 3 \text{ hours.} \quad (3)$$

Substitute the value of y from equation (3) in equation (1).

$$\frac{1}{x} + \frac{1}{y} = \frac{1}{2}. \quad (1)$$
$$\frac{1}{x} + \frac{1}{3} = \frac{1}{2}.$$
$$\frac{1}{x} = \frac{3}{6} - \frac{2}{6}$$
$$= \tfrac{1}{6}.$$
$$x = 6 \text{ hours.}$$

Check:

$$\frac{1}{x} + \frac{1}{y} = \frac{1}{2}. \quad (1)$$
$$\tfrac{1}{6} + \tfrac{1}{3} = \tfrac{1}{2}.$$
$$\tfrac{1}{6} + \tfrac{2}{6} = \tfrac{1}{2}.$$
$$\tfrac{3}{6} = \tfrac{1}{2}.$$
$$\tfrac{1}{2} = \tfrac{1}{2}.$$

$$\frac{1}{x} + \frac{5}{2y} = 1. \quad (2)$$
$$\tfrac{1}{6} + \tfrac{5}{6} = 1.$$
$$\tfrac{6}{6} = 1.$$
$$1 = 1.$$

30. A grain elevator at the railroad station has two pipes through which wheat can be transferred to the boxcar for shipment. If the smaller pipe can fill the boxcar in 12 hours and the larger one can do it in 10 hours, how long would it take to fill the boxcar if both pipes were used at the same time?

Let x = time required if both pipes are used;

$\frac{1}{x}$ = part filled in 1 hour by both pipes;

$\frac{1}{12}$ = part smaller pipe could fill in 1 hour;

$\frac{1}{10}$ = part larger pipe could fill in 1 hour.

$$\frac{1}{12} + \frac{1}{10} = \frac{1}{x}.$$

$$\frac{5x}{60x} + \frac{6x}{60x} = \frac{60}{60x}.$$

$$5x + 6x = 60.$$

$$11x = 60.$$

$$x = 5\tfrac{5}{11} \text{ hours or 5 hours, } \tfrac{300}{11} \text{ minutes}$$

$$= \text{approximately 5 hours, 27 minutes.}$$

Check:

Convert the $5\frac{5}{11}$ hours to $\frac{60}{11}$.

$$\frac{1}{12} + \frac{1}{10} = \frac{1}{\frac{60}{11}}.$$

$$\frac{5}{60} + \frac{6}{60} = \frac{11}{60}.$$

$$\frac{11}{60} = \frac{11}{60}.$$

SUPPLEMENTARY PROBLEMS

1. How much water must be added to 40 gallons of 40% alcohol to reduce it to a 30% solution?

2. The length of a rectangle is 2 feet more than its width. If the width is increased by 3 feet and the length by 5 feet the area is increased by 85 square feet. What are the dimensions of the new rectangle?

3. Two boys sit on opposite ends of a see-saw 18 feet long. One boy weighs 120 pounds and the other 110 pounds. How far from the heavier boy must the fulcrum be placed if they balance each other?

4. The sum of the ages of two boys is 24 years. The younger one is three-fifths the age of the older one. What are the ages of the boys?

5. The sum of the squares of two consecutive positive integers is 60 less than the square of their sum. What are the numbers?

6. The sum of the squares of three consecutive odd numbers is 130 more than the square of the smallest number. What are the numbers? (You should find two such triples—one consisting of positive numbers and one consisting of negative numbers.)

7. One leg of a right triangle is one half as long as the other and three less than the other. The area is equal to the square of the shorter leg. What are the lengths of the legs of the triangle if the measurements are in feet?

8. The sum of the digits of a two-digit number is 12 and their difference is 2. What is the number represented by the two digits? (You should find that there are two answers to this problem.)

9. A man invests an amount of money in one stock that pays him 4% annually and three times as much in one that pays $4\frac{1}{2}$%. The two investments pay him $700 annually. How much does he have invested at each rate?

10. Two automobiles travel towards each other, one at an average speed of 60 miles per hour and the other at an average of 50 miles per hour. If the faster one leaves one hour after the slower one starts and they have traveled an equal distance when they meet, how far apart were they before the first one started?

11. How much force must a person exert on one end of a 9 foot bar, placed on a fulcrum, to raise a weight of 300 pounds, if the fulcrum is 8 feet 4 inches from the end upon which the force is exerted?

12. A positive number and its square differ by 56. What is the number?

13. A rectangular piece of metal 19 inches long and 12 inches wide has small equal squares cut out of each corner after which the ends are bent upwards and the corners welded together to form a box that will hold 245 cubic inches of water. What was the length of each side of the squares that were removed?

14. A circular flower garden has a walk surrounding it such that the area of the walk is equal to the area of the garden. If the radius of the garden is r, by what would that radius have to be multiplied to obtain the width of the walk?

15. A dairyman sells two grades of milk, one with a butterfat content of 5% and the other 4%, and one grade of cream that contains 20% butterfat. How many gallons of cream and how many of 4% milk would he have to mix together to obtain 60 gallons of 5% milk?

Answers to Supplementary Problems

CHAPTER 1

1. The set of integers greater than -3 and less than 3.
2. The set whose elements consist of positive prime numbers less than 15.
3. $A = \{\frac{1}{2}, \frac{1}{3}, \frac{2}{3}\}$
4. $\{-1,0,1\}; \{-1\}; \{0\}; \{1\}; \{-1,0\}; \{-1,1\}; \{0,1\}; \phi.$
5. $\{a,b,c,d\}$; $\{a\}$; $\{b\}$; $\{c\}$; $\{d\}$; $\{a,b\}$; $\{a,c\}$; $\{a,d\}$; $\{b,c\}$; $\{b,d\}$; $\{c,d\}$; $\{a,b,c\}$; $\{a,b,d\}$; $\{a,c,d\}$; $\{b,c,d\}$; ϕ.
6. $8x^3y - 2x^2y^2 - 6xy^3 - 2$
7. $4m^2 - 3m + 9n^3 - 5$
8. $2a^3 - 3a^2b + 2ab^2 + b^3 + 6$
9. $-5ax - 5ac + ad - 2$
10. $8x^2 - xy - 2y^2 + 6$
11. $2x^5 - 11x^4 + 23x^3 - 31x^2 + 17x - 12$
12. $-4d^3x + 7d^2x^2 - 8d^2x - 3d - 3dx^3 + 3x + 6dx^2 - 6$
13. $x^3 - 3x^2 + 4x - 3$
14. $2a^4 - 7a^2 + 3a - 2$
15. $2x^2y^2 - 5xy + 6$, remainder $21xy - 15$
16. $4 + m - 2n$
17. 1
18. $4m - 24$
19. $-2 - 2d$
20. $-2x + 6y - 2x^2 + 2x^2y + 6xy - 4$

CHAPTER 2

1. $am - 2bm + cm$
2. $12c + 9ac - 15bc$
3. $d^2 - m^2$
4. $4b^2 - 9c^2$
5. $4x^2 + 4xy + y^2$
6. $25k^2 - 30k + 9$
7. $2x^2 + 7xz + 3z^2$
8. $12m^2 - 7mn - 10n^2$
9. $m^3 - 8n^3$
10. $27 + 64d^3$
11. $dm + dn - am - an$
12. $2a - 6y - az + 3yz$
13. $4a(x - y + 2z)$
14. $-3bc(m - 2n + 3p)$
15. $2a(5x + 7y)(5x - 7y)$
16. $3cd(2c + 5d)(2c - 5d)$
17. $(x - 4y)(x - 4y)$
18. $(2x + 5z)(2x + 5z)$
19. $3(m - 4p)(m - 4p)$
20. $(3p - y)(2p + 3y)$
21. $(4 - 3d)(3 + 2d)$
22. $(3m - 5a)(4m - 3a)$
23. $3(a - 3x)(2a + 7x)$
24. $a^2b(ab - 2c)(2ab + 3c)$
25. $(m - d)(n + c)$
26. $(3 - 2a)(b + c)$
27. $2(p + q)(r - x)$
28. $(d^2 + 2dc + 2c^2)(d^2 - 2dc + 2c^2)$
29. $(x^2 + 2x + 4)(x^2 - 2x + 4)$
30. $4adm(1 + 2d + 2d^2)(1 - 2d + 2d^2)$

CHAPTER 3

1. $\dfrac{4x + 3}{2x}$
2. $3m + 4$
3. $3r + 2s$
4. $9x - 4y$
5. $x^2 - 3x + 9$
6. $\dfrac{m + 2n}{2}$
7. $3(x + y)(x - y)(x - y)$
8. $x^2(x - 2)(x + 4)(x - 5)$
9. $(m - n)(m - n)(a - b)(a - b)$
10. $(s + 5)(s - 1)(s - 1)(s^2 - 5s + 25)$
11. 1
12. $\dfrac{p - 3}{6p + 5}$
13. 1
14. $\dfrac{4c - 3d}{2(c - 2d)}$
15. $-\dfrac{2b + 3}{4}$
16. $\dfrac{3x^2 + 3xy - 2y^2}{x^2 - y^2}$
17. $-\dfrac{r^2 + r + 2}{r + 1}$
18. $-\dfrac{c^2 + 4cd}{(c + d)(c - 2d)}$
19. $-\dfrac{3(m^2 + 4mn - n^2)}{m(m + n)(m - n)}$
20. $\dfrac{2b^2}{a(a - b)}$
21. $\dfrac{1}{x - y - 1}$
22. $\dfrac{a^2 - b}{b - 3a}$
23. 1
24. $\dfrac{a}{b}$
25. 1

CHAPTER 4

1. $\{6\}$
2. $\{2\}$
3. $\{-3\}$
4. $\{4\}$
5. $\{-5\}$
6. $\{-5\}$
7. $\{0\}$
8. $\{15\}$
9. $\{-\frac{1}{2}\}$
10. $\{15\}$
11. -5
12. 4
13. $-\frac{2}{7}$
14. $-\frac{7}{4}$
15. $\frac{1}{8}$
16. -5
17. $a + 4$
18. $\frac{5a}{4}$
19. $-\frac{2 + 3y}{2}$
20. $\frac{5 - 6m}{2}$
21. 55 miles per hour
22. \$7,000 at $3\frac{1}{2}\%$; \$8,000 at $4\frac{1}{2}\%$
23. 60 pounds of 80-cent coffee and 40 pounds of 95-cent coffee
24. 6 nickels, 19 dimes, and 11 quarters
25. 40

CHAPTER 5

1. 5
2. -2
3. 7
4. 9
5. 4
6. 3
7. -1
8. $\frac{5}{8}$
9. $-\frac{1}{3}$
10. -1
11. $\frac{5a + 2}{6}$
12. 4
13. $\frac{4 - 3y}{10}$
14. $\frac{12 + 5m}{3}$
15. $\frac{yz}{3y - 2z}$
16. 15 feet by 11 feet
17. 24 nickels, 16 dimes, and 9 quarters
18. 56 and 72
19. 9
20. \$3.60 for meat, \$2.40 for canned goods, and \$1.20 for fresh vegetables

CHAPTER 6

1. $x = 3$. $y = -1$.
2. $x = 2$. $y = 3$.
3. $x = -1$. $y = -2$.
4. $x = 3$. $y = 5$.
5. $x = 4$. $y = 3$.
6. $x = \frac{1}{2}$. $y = -\frac{1}{2}$.
7. $x = 2$. $y = 3$.
8. $x = \frac{1}{3}$. $y = -\frac{2}{3}$.
9. $x = 1$. $y = -3$. $z = 5$.
10. $x = -2$. $y = 1$. $z = -1$.
11. $x = 2$. $y = 3$. $z = 0$.
12. $x = 3$. $y = -1$. $z = -2$.
13. $x = \frac{2}{3}$. $y = \frac{1}{2}$. $z = \frac{3}{2}$.
14. $x = -\frac{1}{3}$. $y = \frac{2}{3}$. $z = \frac{1}{3}$.
15. $x = 3$. $y = -2$.
16. $x = -1$. $y = 3$.
17. $x = 1$. $y = -1$. $z = 0$.
18. $x = 2$. $y = -2$. $z = 3$.
19. $a = 2$. $b = -1$. $c = 1$. $d = 0$.
20. $w = 2$. $x = -1$. $y = -1$. $z = 3$.

CHAPTER 7

1. $t = \frac{I}{Pr}$.
2. $b = y - mx$.
3. $r^3 = \frac{3V}{4\pi h}$.
4. $b = \frac{2A - hb'}{h}$.
5. $b = \frac{ac}{a - c}$.
6. $w = \frac{V}{lh}$.
7. $r = \frac{C}{2\pi}$.
8. $h = \frac{V}{\pi r^2}$.
9. $c = \frac{ab}{a + b}$.
10. $r^2 = \frac{V}{\pi h}$.
11. $l = \frac{2S - an}{n}$.
12. $x = \frac{y - b}{m}$.
13. $l = \frac{P - 2w}{2}$.
14. $r = \sqrt[3]{\frac{3V}{4\pi}}$.
15. $t = \frac{A - P}{Pr}$.
16. $V = \sqrt{\frac{FR}{m}}$.
17. 2143.57 cubic inches (approximately)
18. 10^0
19. 14 inches
20. 7%
21. 42 feet by 56 feet

CHAPTER 8

1. $\frac{1}{x^3}$
2. x
3. $\frac{1}{x^{12}}$
4. $\frac{12}{x^4}$
5. $\frac{x}{y^2}$
6. $5a^2b^2$
7. $\frac{x^2}{2m}$
8. $\frac{1}{m^6n^4}$
9. $(a-1)^2$
10. $\frac{1+3b}{b^2}$
11. $\frac{d+c}{c}$
12. $\frac{3}{y}$
13. $m^{\frac{1}{3}}$
14. $\frac{x^{\frac{1}{4}}}{d^{\frac{1}{4}}}$
15. $\frac{x}{y}$
16. $8\sqrt{2}$
17. $-2\sqrt[3]{9}$
18. $8\sqrt{3}$
19. $5xy^2\sqrt{xy}$
20. $6x^{\frac{5}{6}}$
21. $7x^2$
22. $18\sqrt{2}$
23. $3\sqrt{5}$
24. 15
25. $\sqrt[6]{864}$
26. $2\sqrt[3]{30}$
27. $2\sqrt[6]{32}$
28. 4
29. $\frac{5\sqrt{21}}{6}$
30. $\frac{\sqrt{3}-\sqrt{15}}{3}$
31. $\frac{\sqrt{35}-\sqrt{21}}{2}$
32. $-\frac{11+\sqrt{21}}{4}$
33. $2ab\sqrt[6]{ab^2}$
34. $3\sqrt[6]{x^2y^5}$
35. $b\sqrt[3]{a^2}$

CHAPTER 9

1. i
2. $\frac{1}{i}$
3. i
4. 2
5. $3i\sqrt{3}$
6. $8i$
7. $ai\sqrt{3ab}$
8. $-2\sqrt{15}$
9. $-3ab$
10. $\sqrt{5}$
11. $\sqrt{2}$
12. $-3\sqrt{5}$
13. $x = 2.$
    $y = 1.$
14. $x = 1.$
    $y = 3.$
15. $x = 2.$
    $y = -\frac{1}{3}.$
16. $x = 2.$
    $y = -1.$
17. $3x + i$
18. $x + i$
19. $(2x-1) - (1+y)i$
20. $x^2 + 2 - ix$
21. $4x^2 + y^2$
22. $x^2 + 2xyi - y^2 - 1$
23. $\frac{6+\sqrt{3}}{12} + \frac{2\sqrt{3}-3}{12}i$
24. $\frac{1}{10} + \frac{13}{10}i$
25. $-\frac{17}{19} - \frac{6\sqrt{2}}{19}i$

CHAPTER 10

1. $3, -8$
2. $4, 9$
3. $-5, -11$
4. $-\frac{3}{2}, 6$
5. $\frac{2}{3}, -\frac{3}{4}$
6. $-\frac{1}{3}, -\frac{5}{6}$
7. $\frac{7-\sqrt{37}}{2}, \frac{7+\sqrt{37}}{2}$
8. $\frac{-2+\sqrt{22}}{2}, \frac{-2-\sqrt{22}}{2}$
9. $\frac{4+\sqrt{10}}{3}, \frac{4-\sqrt{10}}{3}$
10. $5, 4$
11. $\frac{3}{2}, 1$
12. $\frac{-2 \pm i\sqrt{11}}{3}$
13. $2, -\frac{7}{2}$
14. $\frac{3 \pm 2\sqrt{6}}{5}$
15. $\frac{4 \pm \sqrt{37}}{3}$
16. $\frac{-3 \pm 3\sqrt{6}}{10}$
17. -1
18. 2
19. 2
20. 3
21. $x^2 + x - 6 = 0.$
22. $10x^2 + x - 2 = 0.$
23. $6x^2 + 13x + 6 = 0.$
24. $x^2 + 3 = 0.$
25. $4x^2 - 12x + 11 = 0.$
26. ± 8
27. ± 2
28. 16
29. $\frac{9}{4}$
30. 4

CHAPTER 11

1. Relation: $\{(-2,-4), (-1,-1), (0,2), (1,5)\}$
   Domain: $\{-2,-1,0,1\}$
   Range: $\{-4,-1,2,5\}$
2. Relation: $\{(-4,0), (-3,1), (-2,2), (-1,3), (0,4), (1,5)\}$
   Domain: $\{-4,-3,-2,-1,0,1\}$
   Range: $\{0,1,2,3,4,5\}$
3. Relation: $\{(-3,-1), (-2,4), (2,4), (3,-1)\}$
   Domain: $\{-3,-2,2,3\}$
   Range: $\{-1,4\}$
4. Relation: $\{(-4,1), (-2,2), (0,3), (2,4), (4,5)\}$
   Domain: $\{-4,-2,0,2\}$
   Range: $\{1,2,3,4\}$
5. Relation: $\{(-4,-3), (-3,-1), (-2,1), (-1,3), (0,5)\}$
   Domain: $\{-4,-3,-2,-1,0\}$
   Range: $\{-3,-1,1,3,5\}$
6. $m = -\frac{9}{5}.$
7. $m = \frac{1}{3}.$
8. $m = \frac{9}{10}.$
9. $y = -\frac{1}{3}x + \frac{4}{3}.$
10. $y = -\frac{3}{20}x + \frac{1}{5}.$

11. $m = \frac{3}{2}$; $b = -\frac{9}{2}$.
12. $m = -\frac{4}{3}$; $b = \frac{10}{3}$.
13. $x + 3y - 8 = 0$.
14. $30x + 6y + 17 = 0$.
15. $-\frac{6}{7}$.
16. $10x + 11y + 17 = 0$.
17. $3x - 3y - 1 = 0$.
18. $\frac{x}{\frac{3}{7}} + \frac{y}{-3} = 1$.
19. $\frac{x}{\frac{27}{2}} + \frac{y}{-9} = 1$.
20. x-intercept is $-4\frac{1}{2}$, y-intercept is $2\frac{1}{8}$.

21. FIG. 69

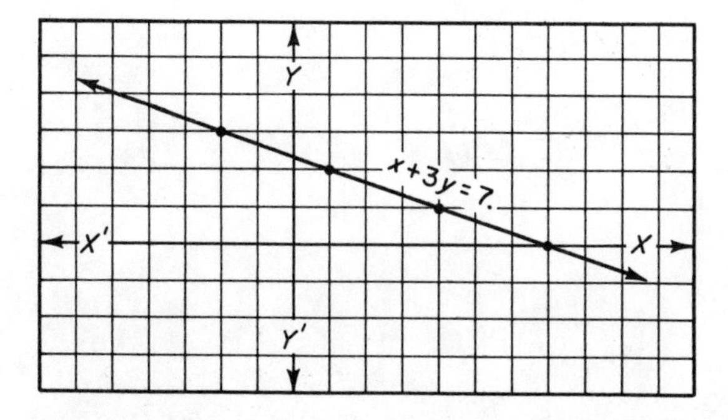

22. FIG. 70

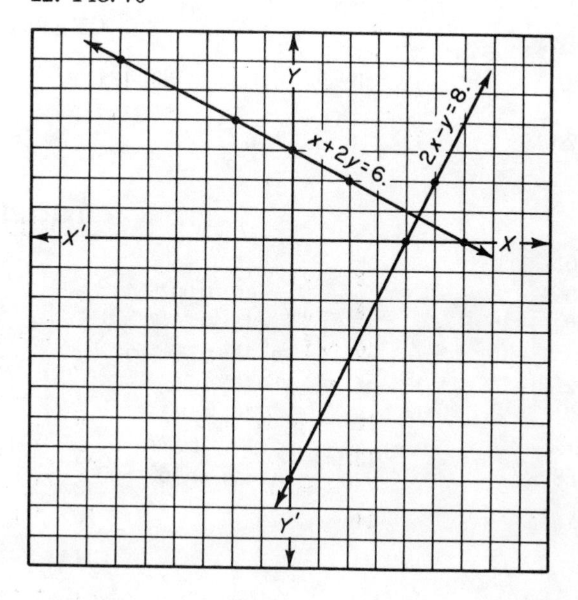

23. FIG. 71

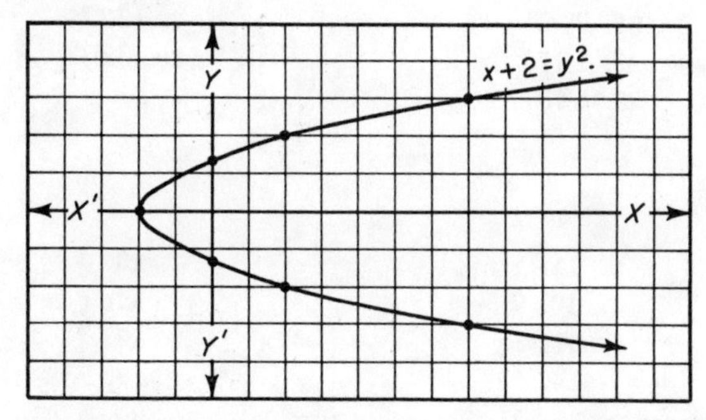

24. FIG. 72

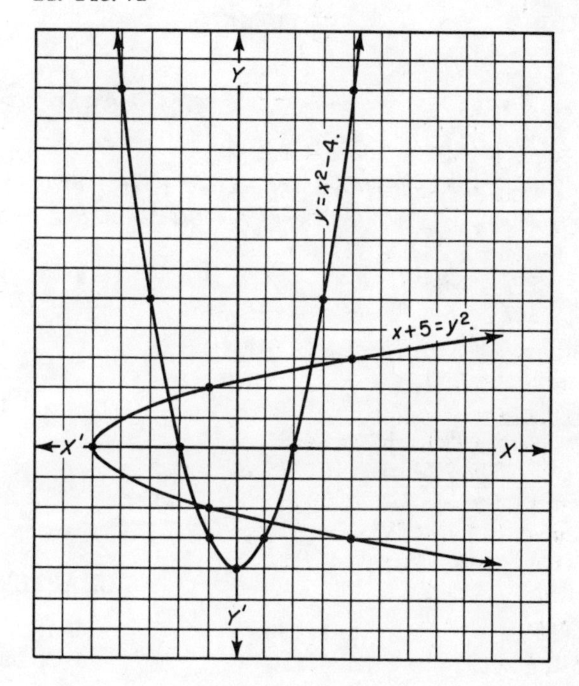

25. FIG. 73

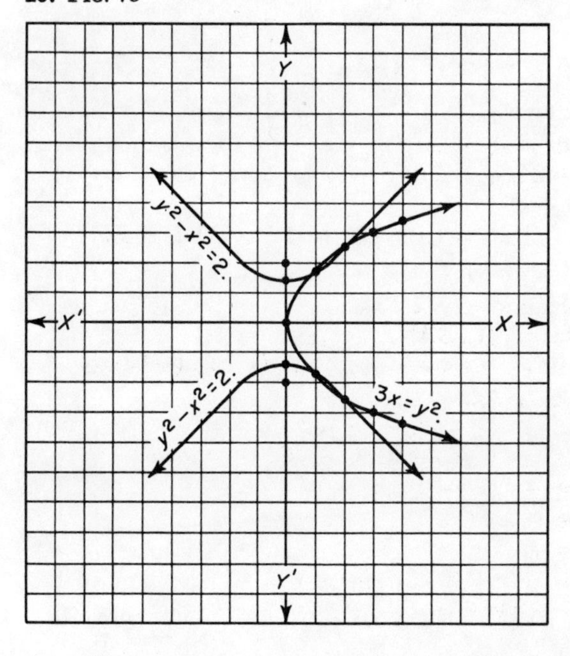

CHAPTER 12

1. $x > 2$.
2. $x < 4$.
3. $x > \frac{3}{2}$.
4. $x < 12$.
5. $x < -\frac{1}{6}$.
6. $\{6,5,4,3,2,1\}$
7. ϕ
8. $\{7,6,5,4,3,2,1\}$
9. $\{1,2,3,4,5,6,7,8,9\}$
10. U
11. All x in the interval $(0,\frac{2}{3})$
12. All x not in the interval $[1,\frac{5}{2}]$
13. All x not in the interval $(-5,-1)$
14. All x not in the interval $(-\frac{5}{3}, -\frac{3}{5})$
15. All x in the interval $(-\frac{2}{5},2)$

CHAPTER 13

1. $\frac{21}{4}$
2. $\frac{3}{4}$
3. $\pm 2\sqrt{3}$
4. 6
5. $\frac{4}{9}$
6. $\pm 2\sqrt{6}$
7. $6x$
8. $\pm 2\sqrt{x+1}$
9. $(x+1)^3$
10. $\frac{14}{3}$
11. 5
12. $\frac{8}{5}$
13. 16
14. $\frac{3}{2}$
15. $\frac{9}{16}$

CHAPTER 14

1. 1.5012
2. 2.6040
3. 9.8947 −10
4. 7.9855 −10
5. 5.9215 −10
6. 0.9947
7. 1,115
8. 47.68
9. 2.433
10. 0.002554
11. 0.8618
12. 0.001166
13. 1,369
14. 4,911
15. 160,778 or 160, 800
16. 25.91
17. 2.160
18. 3.428
19. 76.33
20. 124.7
21. 29.87
22. 1.178
23. 0.7077
24. 8,638
25. 15.91
26. 265.6
27. 100.4
28. 2.624
29. 26.74
30. 3.638

CHAPTER 15

1. 35
2. −32
3. −486
4. $\frac{128}{729}$
5. 12, 6, 3
6. $32, 21\frac{1}{3}, 14\frac{2}{9}, 9\frac{13}{27}$
7. 8, 11, 14, 17, 20, 23
8. $\frac{8}{7}, \frac{9}{7}, \frac{10}{7}, \frac{11}{7}, \frac{12}{7}, \frac{13}{7}$
9. 900
10. 316
11. $156\frac{15}{16}$
12. 9
13. $3\frac{1}{2}$
14. 13
15. 6
16. $\frac{4}{3}$
17. $\frac{2}{3}$
18. $\frac{3}{4}$
19. 23
20. 5

CHAPTER 16

1. $a^5 + 5a^4x + 10a^3x^2 + 10a^2x^3 + 5ax^4 + x^5$
2. $m^7 - 7m^6t + 21m^5t^2 - 35m^4t^3 + 35m^3t^4 - 21m^2t^5 + 7mt^6 - t^7$
3. $32m^5 + 240m^4p + 720m^3p^2 + 1080m^2p^3 + 810mp^4 + 243p^5$
4. $81d^4 - 108d^3 + 54d^2 - 12d + 1$
5. $\frac{d^5}{243} + \frac{5d^4c}{162} + \frac{5d^3c^2}{54} + \frac{5d^2c^3}{36} + \frac{5dc^4}{48} + \frac{c^5}{32}$
6. $\frac{16m^4}{81} - \frac{16m^3p}{9} + 6m^2p^2 - 9mp^3 + \frac{81p^4}{16}$.
7. $\frac{1}{a^4 - 4a^3b + 6a^2b^2 - 4ab^3 + b^4}$
8. $1024 - \frac{1280x}{3} + \frac{640x^2}{9} - \frac{160x^3}{27} + \frac{20x^4}{81} - \frac{x^5}{243}$
9. $10{,}206a^5b^4$
10. $-20x^3y^3$
11. $x^3 - 3x^2y + 3xy^2 + y^3 + 3x^2z - 6xyz + 3y^2z + 3xz^2 - 3yz^2 + z^3$
12. $8a^3 - 12a^2b + 6ab^2 - b^3 + 12a^2c - 12abc + 3b^2c + 6ac^2 - 3bc^2 + c^3$
13. 1.159
14. 0.85
15. 10,816

CHAPTER 17

1. 362,880
2. 40,320
3. 17,280
4. 325
5. 17,250
6. 30,240
7. 4,320
8. 3,003
9. 240
10. 3,628,800

CHAPTER 18

1. $x^3 + 7x^2 - 3x + 4$
2. $3x^4 - 2x^2 + 3$
3. $x^2 + x + 1$
4. $x^3 - x^2 - x - 1$
5. $x^4 + 3x^3 - 1$
6. 3
7. 403
8. $4, -2, -3$
9. $3, -3, \frac{2}{3}, -\frac{3}{2}$
10. $6, \frac{5}{3}, -\frac{1}{2}$
11. $1, 1, -1, -1$
12. $2, -3, i\sqrt{5}, -i\sqrt{5}$
13. 4.53
14. 17.546
15. -3.24

CHAPTER 19

1. $13\frac{1}{3}$ gallons
2. 11 feet by 15 feet
3. $8\frac{14}{23}$ feet
4. 9 years and 15 years
5. 5 and 6
6. 5, 7, and 9; or -11, -9, and -7
7. 3 feet and 6 feet
8. 75 or 57
9. \$4,000 at 4% and \$12,000 at $4\frac{1}{2}$%
10. 600 miles
11. 24 pounds
12. 8
13. $2\frac{1}{2}$ inches
14. $\sqrt{2} - 1$
15. $3\frac{3}{4}$ gallons cream, $56\frac{1}{4}$ gallons 4% milk

Appendix

Table I. Powers and Roots

No.	Square	Cube	Square Root	Cube Root	No.	Square	Cube	Square Root	Cube Root
1	1	1	1.000	1.000	51	2601	132651	7.141	3.708
2	4	8	1.414	1.260	52	2704	140608	7.211	3.733
3	9	27	1.732	1.442	53	2809	148877	7.280	3.756
4	16	64	2.000	1.587	54	2916	157464	7.348	3.780
5	25	125	2.236	1.710	55	3025	166375	7.416	3.803
6	36	216	2.449	1.817	56	3136	175616	7.483	3.826
7	49	343	2.646	1.913	57	3249	185193	7.550	3.849
8	64	512	2.828	2.000	58	3364	195112	7.616	3.871
9	81	729	3.000	2.080	59	3481	205379	7.681	3.893
10	100	1000	3.162	2.154	60	3600	216000	7.746	3.915
11	121	1331	3.317	2.224	61	3721	226981	7.810	3.936
12	144	1728	3.464	2.289	62	3844	238328	7.874	3.958
13	169	2197	3.606	2.351	63	3969	250047	7.937	3.979
14	196	2744	3.742	2.410	64	4096	262144	8.000	4.000
15	225	3375	3.873	2.466	65	4225	274625	8.062	4.021
16	256	4096	4.000	2.520	66	4356	287496	8.124	4.041
17	289	4913	4.123	2.571	67	4489	300763	8.185	4.062
18	324	5832	4.243	2.621	68	4624	314432	8.246	4.082
19	361	6859	4.359	2.668	69	4761	328509	8.307	4.102
20	400	8000	4.472	2.714	70	4900	343000	8.367	4.121
21	441	9261	4.583	2.759	71	5041	357911	8.426	4.141
22	484	10648	4.690	2.802	72	5184	373248	8.485	4.160
23	529	12167	4.796	2.844	73	5329	389017	8.544	4.179
24	576	13824	4.899	2.884	74	5476	405224	8.602	4.198
25	625	15625	5.000	2.924	75	5625	421875	8.660	4.217
26	676	17576	5.099	2.962	76	5776	438976	8.718	4.236
27	729	19683	5.196	3.000	77	5929	456533	8.775	4.254
28	784	21952	5.292	3.037	78	6084	474552	8.832	4.273
29	841	24389	5.385	3.072	79	6241	493039	8.888	4.291
30	900	27000	5.477	3.107	80	6400	512000	8.944	4.309
31	961	29791	5.568	3.141	81	6561	531441	9.000	4.327
32	1024	32768	5.657	3.175	82	6724	551368	9.055	4.344
33	1089	35937	5.745	3.208	83	6889	571787	9.110	4.362
34	1156	39304	5.831	3.240	84	7056	592704	9.165	4.380
35	1225	42875	5.916	3.271	85	7225	614125	9.220	4.397
36	1296	46656	6.000	3.302	86	7396	636056	9.274	4.414
37	1369	50653	6.083	3.332	87	7569	658503	9.327	4.431
38	1444	54872	6.164	3.362	88	7744	681472	9.381	4.448
39	1521	59319	6.245	3.391	89	7921	704969	9.434	4.465
40	1600	64000	6.325	3.420	90	8100	729000	9.487	4.481
41	1681	68921	6.403	3.448	91	8281	753571	9.539	4.498
42	1764	74088	6.481	3.476	92	8464	778688	9.592	4.514
43	1849	79507	6.557	3.503	93	8649	804357	9.644	4.531
44	1936	85184	6.633	3.530	94	8836	830584	9.695	4.547
45	2025	91125	6.708	3.557	95	9025	857375	9.747	4.563
46	2116	97336	6.782	3.583	96	9216	884736	9.798	4.579
47	2209	103823	6.856	3.609	97	9409	912673	9.849	4.595
48	2304	110592	6.928	3.634	98	9604	941192	9.899	4.610
49	2401	117649	7.000	3.659	99	9801	970299	9.950	4.626
50	2500	125000	7.071	3.684	100	10000	1000000	10.000	4.642

APPENDIX

Table II. Four-Place Common Logarithms

N	0	1	2	3	4	5	6	7	8	9
10	0000	0043	0086	0128	0170	0212	0253	0294	0334	0374
11	0414	0453	0492	0531	0569	0607	0645	0682	0719	0755
12	0792	0828	0864	0899	0934	0969	1004	1038	1072	1106
13	1139	1173	1206	1239	1271	1303	1335	1367	1399	1430
14	1461	1492	1523	1553	1584	1614	1644	1673	1703	1732
15	1761	1790	1818	1847	1875	1903	1931	1959	1987	2014
16	2041	2068	2095	2122	2148	2175	2201	2227	2253	2279
17	2304	2330	2355	2380	2045	2430	2455	2480	2504	2529
18	2553	2577	2601	2625	2648	2672	2695	2718	2742	2765
19	2788	2810	2833	2856	2878	2900	2923	2945	2967	2989
20	3010	3032	3054	3075	3096	3118	3139	3160	3181	3201
21	3222	3243	3263	3284	3304	3324	3345	3365	3385	3404
22	3424	3444	3464	3483	3502	3522	3541	3560	3579	3598
23	3617	3636	3655	3674	3692	3711	3729	3747	3766	3784
24	3802	3820	3838	3856	3874	3892	3909	3927	3945	3962
25	3979	3997	4014	4031	4048	4065	4082	4099	4116	4133
26	4150	4166	4183	4200	4216	4232	4249	4265	4281	4298
27	4314	4330	4346	4362	4378	4393	4409	4425	4440	4456
28	4472	4487	4502	4518	4533	4548	4564	4579	4594	4609
29	4624	4639	4654	4669	4683	4698	4713	4728	4742	4757
30	4771	4786	4800	4814	4829	4843	4857	4871	4886	4900
31	4914	4928	4942	4955	4969	4983	4997	5011	5024	5038
32	5051	5065	5079	5092	5105	5119	5132	5145	5159	5172
33	5185	5198	5211	5224	5237	5250	5263	5276	5289	5302
34	5315	5328	5340	5353	5366	5378	5391	5403	5416	5428
35	5441	5453	5465	5478	5490	5502	5514	5527	5539	5551
36	5563	5575	5587	5599	5611	5623	5635	5647	5658	5670
37	5682	5694	5705	5717	5729	5740	5752	5763	5775	5786
38	5798	5809	5821	5832	5843	5855	5866	5877	5888	5899
39	5911	5922	5933	5944	5955	5966	5977	5988	5999	6010
40	6021	6031	6042	6053	6064	6075	6085	6096	6107	6117
41	6128	6138	6149	6160	6170	6180	6191	6201	6212	6222
42	6232	6243	6253	6263	6274	6284	6294	6304	6314	6325
43	6335	6345	6355	6365	6375	6385	6395	6405	6415	6425
44	6435	6444	6454	6464	6474	6484	6493	6503	6513	6522
45	6532	6542	6551	6561	6571	6580	6590	6599	6609	6618
46	6628	6637	6646	6656	6665	6675	6684	6693	6702	6712
47	6721	6730	6739	6749	6758	6767	6776	6785	6794	6803
48	6812	6821	6830	6839	6848	6857	6866	6875	6884	6893
49	6902	6911	6920	6928	6937	6946	6955	6964	6972	6981
50	6990	6998	7007	7016	7024	7033	7042	7050	7059	7067
51	7076	7084	7093	7101	7110	7118	7126	7135	7143	7152
52	7160	7168	7177	7185	7193	7202	7210	7218	7226	7235
53	7243	7251	7259	7267	7275	7284	7292	7300	7308	7316
54	7324	7332	7340	7348	7356	7364	7372	7380	7388	7396

APPENDIX

Table II. (Continued)

N	0	1	2	3	4	5	6	7	8	9
55	7404	7412	7419	7427	7435	7443	7451	7459	7466	7474
56	7482	7490	7497	7505	7513	7520	7528	7536	7543	7551
57	7559	7566	7574	7582	7589	7597	7604	7612	7619	7627
58	7634	7642	7649	7657	7664	7672	7679	7686	7694	7701
59	7709	7716	7723	7731	7738	7745	7752	7760	7767	7774
60	7782	7789	7796	7803	7810	7818	7825	7832	7839	7846
61	7853	7860	7868	7875	7882	7889	7896	7903	7910	7917
62	7924	7931	7938	7945	7952	7959	7966	7973	7980	7987
63	7993	8000	8007	8014	8021	8028	8035	8041	8048	8055
64	8062	8069	8075	8082	8089	8096	8102	8109	8116	8122
65	8129	8136	8142	8149	8156	8162	8169	8176	8182	8189
66	8195	8202	8209	8215	8222	8228	8235	8241	8248	8254
67	8261	8267	8274	8280	8287	8293	8299	8306	8312	8319
68	8325	8331	8338	8344	8351	8357	8363	8370	8376	8382
69	8388	8395	8401	8407	8414	8420	8426	8432	8439	8445
70	8451	8457	8463	8470	8476	8482	8488	8494	8500	8506
71	8513	8519	8525	8531	8537	8543	8549	8555	8561	8567
72	8573	8579	8585	8591	8597	8603	8609	8615	8621	8627
73	8633	8639	8645	8651	8657	8663	8669	8675	8681	8686
74	8692	8698	8704	8710	8716	8722	8727	8733	8739	8745
75	8751	8756	8762	8768	8774	8779	8785	8791	8797	8802
76	8808	8814	8820	8825	8831	8837	8842	8848	8854	8859
77	8865	8871	8876	8882	8887	8893	8899	8904	8910	8915
78	8921	8927	8932	8938	8943	8949	8954	8960	8965	8971
79	8976	8982	8987	8993	8998	9004	9009	9015	9020	9025
80	9031	9036	9042	9047	9053	9058	9063	9069	9074	9079
81	9085	9090	9096	9101	9106	9112	9117	9122	9128	9133
82	9138	9143	9149	9154	9159	9165	9170	9175	9180	9186
83	9191	9196	9201	9206	9212	9217	9222	9227	9232	9238
84	9243	9248	9253	9258	9263	9269	9274	9279	9284	9289
85	9294	9299	9304	9309	9315	9320	9325	9330	9335	9340
86	9345	9350	9355	9360	9365	9370	9375	9380	9385	9390
87	9395	9400	9405	9410	9415	9420	9425	9430	9435	9440
88	9445	9450	9455	9460	9465	9469	9474	9479	9484	9489
89	9494	9499	9504	9509	9513	9518	9523	9528	9533	9538
90	9542	9547	9552	9557	9562	9566	9571	9576	9581	9586
91	9590	9595	9600	9605	9609	9614	9619	9624	9628	9633
92	9638	9643	9647	9652	9657	9661	9666	9671	9675	9680
93	9685	9689	9694	9699	9703	9708	9713	9717	9722	9727
94	9731	9736	9741	9745	9750	9754	9759	9763	9768	9773
95	9777	9782	9786	9791	9795	9800	9805	9809	9814	9818
96	9823	9827	9832	9836	9841	9845	9850	9854	9859	9863
97	9868	9872	9877	9881	9886	9890	9894	9899	9903	9908
98	9912	9917	9921	9926	9930	9934	9939	9943	9948	9952
99	9956	9961	9965	9969	9974	9978	9983	9987	9991	9996

Index